T0362167

Domino Effects in the Process Industries

Domino Effects in the Process Industries
Modeling, Prevention and Managing

Edited by

Genserik Reniers
Centre for Economics and Corporate Sustainability (CEDON), HUB,
KULeuven, Brussels, Belgium

Valerio Cozzani
LISES, Dipartimento di Ingegneria Civile, Chimica, Ambientale e dei
Materiali, Alma Mater Studiorum – Università di Bologna, Bologna,
Italy

AMSTERDAM • BOSTON • HEIDELBERG • LONDON • NEW YORK
OXFORD • PARIS • SAN DIEGO • SAN FRANCISCO • SINGAPORE
SYDNEY • TOKYO

ELSEVIER

Elsevier
225 Wyman Street, Waltham, MA, 02451, USA
The Boulevard, Langford Lane, Kidlington, Oxford OX5 1GB, UK
Radarweg 29, PO Box 211, 1000 AE Amsterdam, The Netherlands

Notice
No responsibility is assumed by the publisher for any injury and/or damage to persons or property as a matter of products liability, negligence or otherwise, or from any use or operation of any methods, products, instructions or ideas contained in the material herein. Because of rapid advances in the medical sciences, in particular, independent verification of diagnoses and drug dosages should be made

British Library Cataloguing-in-Publication Data
A catalogue record for this book is available from the British Library

Library of Congress Cataloging-in-Publication Data
A catalog record for this book is available from the Library of Congres

ISBN: 978-0-444-54323-3

For information on all Elsevier publications visit
our website at www.store.elsevier.com

Printed and bound by CPI Group (UK) Ltd, Croydon, CR0 4YY

Working together
to grow libraries in
developing countries

www.elsevier.com • www.bookaid.org

Contents

Preface

The chemical industry provides most of the necessities for our modern day lives. Owing to an ever-increasing population and to the associated need for more goods and materials, and due to an increasing requirement for innovation in products and processes, the production of chemicals has known a steep increase during the past decades. Hence, the use, storage, processing, and transportation of hazardous substances are also characterized with an increasing trend. As a result, ever more chemical plants are being built around the world, mostly settled together in integrated industrial parks or in so-called chemical clusters. Thus, a combined increase in activities involving hazardous chemicals and in population densities can be observed on a global scale.

This worldwide situation leads to more wealth and easier lives, but also to an increase of potential hazards. Even if the safety scores of the chemical industry are generally high, major accidents, if they would happen, are likely to cause severe consequences in this industrial sector. The concentration of activities within chemical clusters may result in accidents having a simultaneous impact on several plant units, resulting in loss of life, environmental contamination, huge asset damage as well as in important financial consequences and in the disruption of community life by the interruption of lifelines and fuel supplies. Among the most destructive major accidents are those where a "domino effect" takes place, causing the escalation of a primary accident and the propagation of the primary event possibly involving multiple equipment and plant units.

The hazard due to "domino effect" is well known and addressed in safety standards and legislation. Catastrophic accidents, as that in Mexico City in 1984 where an entire plant was almost entirely destroyed and more than 500 persons died, led to a high perception of the hazard due to this specific category of accidents.

However, models for the assessment of domino effects are demanding because of the complexity of the accident scenario and evolution (simulation of the source term and of the primary scenario, damage of secondary units, consequence assessment of simultaneous primary and secondary scenarios, role of safety barriers, etc.) and due to the high level of detail of the input data required. This is the reason—in combination with the extremely low probabilities of such accidents—that leads often to leave out from the safety assessment of chemical activities, the quantitative assessment and the management of risks due to domino scenarios.

Nonetheless, recent events as those related to the 2011 Tōhoku Tsunami in Japan require the safety practitioners to an even more important necessity to explicitly prevent, model and manage the risks due to *high-impact low-probability* events as domino scenarios. Thus, both the assessment and management of risk due to domino scenarios, and the academic and industrial research on domino effects, are ever more priority topics. A book collecting the available approaches to the modeling, prevention and management of such possible devastating events represented, therefore, a much needed challenge for the industrial world as well as for the research community worldwide.

This volume presents the most state-of-the-art and advanced insights, models, theories, concepts, frameworks, technologies, and methodologies to deal with domino effects and to tackle their prevention, modeling and management in the chemical and process industry. It is intended to become a standard support tool for every professional who has to handle escalating events in process and plant safe design and operation, as well as a reference point stating the state-of-the-art for further research on the topic.

Prof. Genserik Reniers and Prof. Valerio Cozzani

List of Contributors

Bahman Abdolhamidzadeh   Center for Process Design, Safety and Loss Prevention (CPSL), Sharif University of Technology, Tehran, Iran

Paul Amyotte   Department of Process Engineering and Applied Science, Dalhousie University, Halifax, Nova Scotia, Canada

Giacomo Antonioni   LISES, Dipartimento di Ingegneria Civile, Chimica, Ambientale e dei Materiali, Alma Mater Studiorum - Università di Bologna, Bologna, Italy

Michael Birk   Department of Mechanical and Materials Engineering, Queen's University, Kingston, Ontario, Canada

Sarah Bonvicini   LISES, Dipartimento di Ingegneria Civile, Chimica, Ambientale e dei Materiali, Alma Mater Studiorum - Università di Bologna, Bologna, Italy

Joaquim Casal   Department of Chemical Engineering, Centre for Studies on Technological Risk (CERTEC), Universitat Politècnica de Catalunya, Barcelona, Catalonia, Spain

Valerio Cozzani   LISES, Dipartimento di Ingegneria Civile, Chimica, Ambientale e dei Materiali, Alma Mater Studiorum – Università di Bologna, Bologna, Italy

Rosa-Mari Darbra   Department of Chemical Engineering, Centre for Studies on Technological Risk (CERTEC), Universitat Politècnica de Catalunya, Barcelona, Catalonia, Spain

Robby Faes   3M Environmental Health & Safety, Zwijndrecht, Belgium

Pol Hoorelbeke   Total Refining & Chemicals, Vice President Safety Division Brussels, Belgium, Visiting Professor South Chine University of Technology, China

Nima Khakzad   Safety and Risk Engineering Research Group, Faculty of Engineering and Applied Science, Memorial University of Newfoundland, St. John's, Newfoundland, Canada

Faisal Khan   Safety and Risk Engineering Research Group, Faculty of Engineering and Applied Science, Memorial University of Newfoundland, St. John's, Newfoundland, Canada

Elisabeth Krausmann   European Commission, Joint Research Centre, Institute for the Protection and Security of the Citizen, Ispra, Varese, Italy

Gabriele Landucci   Dipartimento di Ingegneria Civile e Industriale, Università di Pisa, Pisa, Italy

Genserik Reniers   Centre for Economics and Corporate Sustainability (CEDON), HUB, KULeuven, Brussels, Belgium

Ernesto Salzano   Istituto di Ricerche sulla Combustione, Consiglio Nazionale delle Ricerche (CNR), Napoli, Italy

Gigliola Spadoni   LISES, Dipartimento di Ingegneria Civile, Chimica, Ambientale e dei Materiali, Alma Mater Studiorum – Università di Bologna, Bologna, Italy

Jerome Taveau   Scientific advisor at Fike Corporation, Industrial Explosion Protection Group, Blue Springs, USA

Alessandro Tugnoli   LISES, Dipartimento di Ingegneria Civile, Chimica, Ambientale e dei Materiali, Alma Mater Studiorum - Università di Bologna, Bologna, Italy

1 Historical Background and State of the Art on Domino Effect Assessment

Valerio Cozzani*, Genserik Reniers†

* LISES, Dipartimento di Ingegneria Civile, Chimica, Ambientale e dei Materiali, Alma Mater Studiorum – Università di Bologna, Bologna, Italy
† Centre for Economics and Corporate Sustainability (CEDON), HUB, KULeuven, Brussels, Belgium

Chapter Outline

1.1 Historical Background and Importance of Research on Domino Effects

Since the early times of process safety, the potential hazard coming from accidents involving a "domino effect" has been recognized. The concentration of plants active within the chemical and process industries into huge integrated sites (the so-called "chemical clusters"), started in the 1960s, and the increase in the production potentialities of single plants required by the reduction of fixed and operating costs, led to a progressive increase in complexity and extension of industrial sites. This corresponded to a relevant growth of the inventory of hazardous substances present in wide industrial sites, thus increasing the potential for severe accidents involving the so-called "domino effects" (indicating the escalation and propagation of an accident where—due to knock-on effects—possibly extended portions or the entire industrial site can be involved).

In general, it is regretfully obvious that major accidents[1] in the chemical industry have occurred worldwide. Among them, a relevant number of severe "domino"

---

[1] The definition of *major accident* within the European Directive 96/82/EC is "an occurrence such as a major emission, fire, or explosion resulting from uncontrolled developments in the course of the operation of any establishment covered by the Directive, and leading to serious danger to human health and/or the environment, immediate or delayed, inside or outside the establishment, and involving one or more dangerous substances".

Domino Effects in the Process Industries. http://dx.doi.org/10.1016/B978-0-444-54323-3.00001-4

accidents are documented for chemical and process plants (Lees, 1996; Rasmussen, 1996; Delvosalle, 1996; Kourniotis et al., 2000; Darbra et al., 2010; Abdolha-midzadeh et al., 2011). Major accidents, in combination with the (rather self-complacent) attitude of the industry, at a certain point in time (during the 1960s and 1970s, depending on the world region) led to a change in safety perception of the chemical process industry. The Flixborough accident, that took place in 1974, and the Seveso accident, that took place in 1976, (e.g. see Lees, 1996) represent such a milestone in Europe. The society had come to perceive the risks of chemical industrial activities as very high and it was only a matter of time before politicians decided that a more active prevention policy was necessary within the industry. Nonetheless, the perceived risk regarding chemical industrial activities should be put into perspective. Although it is straightforward that preventing major accidents within the chemical industry is absolutely necessary if the industrial sector wants to gain and maintain its license to operate, the average statistical probability of dying as a result of a major accident in the chemical industry can be situated as 80 times lower than dying from being struck by lightning (CBS, 2003).

Nevertheless, the high risk perception due to the fatalities and huge destruction experienced in some chemical accidents, led in Europe to the adoption of an advanced legislation for the control of major accident hazards. The first "Seveso" Directive (Directive 82/501/EEC) asked for a comprehensive assessment of safety in chemical and process sites. The need to assess the possibility of "domino" accidents was already cited in this Directive, although no technical approach or specific tool existed at that time to allow the identification and assessment of such scenarios.

The huge destruction caused by the Mexico City disaster in 1984 (Pietersen, 1988), possibly one of the better known domino accidents, raised specific concern on escalation hazard. Demonstration of the potential severity of accidents involving domino effects led to important efforts for the prevention of domino accident scenarios since then. The technical standards and legislation concerned with the control of major accident hazard include measures to assess, control and prevent domino effects. Several technical standards introduce preventive measures, such as safety distances, thermal insulation or emergency water deluges, in order to control and reduce the probability of domino events (Mecklenburgh, 1985).

As stated above, in Europe, the hazard of domino effect is recognized in the legislation since the first Seveso Directive (Directive 82/501/EEC), which required the assessment of domino hazards in all plants falling under the obligation to issue a "safety report". Article 8 of Seveso-II Directive (Directive 96/82/EC) and presently Article 9 of the Seveso-III European Directive (Directive, 2012/18/EU) are specifi-cally dedicated to domino effects, requiring to assess domino scenarios that may propagate a primary accident to nearby plants. Table 1.1 summarizes the requirements of Article 9 of the latest Seveso Directive.

Domino effects are thus officially recognized in the European Directive in relation to major accident hazards. The article, being preventive legislation, clearly relates to cross-company exchangeable information concerning measures to prevent major accidents. The Directive does not provide an unambiguous definition of what constitutes a domino effect, although Article 24 foresees the possibility that guidance

Table 1.1 Requirements of Article 9 of Directive 2012/18/EU Concerning Domino Effect

1. Member States shall ensure that the competent authority, using the information received from the operators in accordance with Articles 7 and 10, or following a request for additional information from the competent authority, or through inspections pursuant to Article 20, identifies all lower-tier and upper-tier establishments or groups of establishments where the risk or consequences of a major accident may be increased because of the geographical position and the proximity of such establishments, and their inventories of dangerous substances.

2. Where the competent authority has additional information to that provided by the operator pursuant to point (g) of Article 7(1), it shall make this information available to that operator, if it is necessary for the application of this Article.

3. Member States shall ensure that operators of the establishments identified in accordance with paragraph 1:

   a. exchange suitable information to enable those establishments to take account of the nature and extent of the overall hazard of a major accident in their MAPP, safety management systems, safety reports and internal emergency plans, as appropriate;

   b. cooperate in informing the public and neighboring sites that fall outside the scope of this Directive, and in supplying information to the authority responsible for the preparation of external emergency plans.

is provided by the Commission on this specific point and, more generally, on the issue of domino effect.

In the practical implementation of the Seveso Directives, the Commission requires periodic reports to member states concerning the application of these Directives, based on questionnaires (document 2002/605/EC—European Commission Decision of July 17, 2002 concerning the questionnaire relating to Council Directive 96/82/EC). Table 1.2 gives an overview of the requirements concerning domino effect in the questionnaire prepared for the first reporting period (2003–2005).

Land use planning (LUP) is also highly concerned with domino accidents. On the one hand, safety distances among different plants and critical units are crucial to prevent escalation of accident scenarios. On the other hand, the consequences of domino accidents are potentially among the most severe that may be considered credible. Thus, accounting for domino scenarios is a key point for the protection of population from the consequences of major accidents, as required e.g. by Article 13 of the European Seveso-III Directive on LUP related to major accident hazards.

Despite the increasing interest in defining escalation threshold criteria and safety distances for domino accidents in the legislation of European member states in the past decades, some stakeholders are not familiar with domino effects and consider the occurrence probability of a domino accident as being too low to take preventive measures. Moreover, in complex industrial areas, protection from domino effect can be obtained as a result of a coordinated effort of all the operating companies in the area. Actually, if only a single plant takes precautions against domino effects, it causes inconvenience to the company without having much chance of benefiting the cluster as a whole. Nevertheless, if many companies were to decide to proactively deal with domino risks, industrial areas would be substantially safer, the reason being

Table 1.2 EU Member States Report Requirements Concerning Domino Effects (EC, 2002)

Required Information	Year 1	Year 2	Year 3
a. General background information	Nonnumerical		
b. How many groups of establishments?	✔	✔	✔
c. Average number of establishments per group?*	✔	✔	✔
d. Number of establishments in the smallest group?*	✔	✔	✔
e. Number of establishments in the biggest group?*	✔	✔	✔
f. Strategy for ensuring that suitable information is exchanged?	Nonnumerical		

✔ = numerical.
* Optional.

that many companies acting together (through collaboration and intelligent precaution taking) would prove significant. Therefore, to trigger this process, it is important that state-of-the-art information and guidance are available, to offer risk decision-makers technology and management recommendations about how to adequately protect against domino effects. In this framework, research efforts to fill knowledge gaps are still of fundamental importance.

The main conceptual issue posed by domino effect accident scenarios to process safety science is complexity. Actually a domino scenario is beyond the threshold where the necessity to consider the functioning of an entire complex system is required in order to prevent its failure, and not only the functioning of its parts (Leveson, 2004). A number of pioneering studies recognized this issue and addressed the analysis of domino effects. Possibly, the first systematic study on the subject present in the literature is that of Bagster and Pitblabo, which dates back to 1991 and proposed a first approach to escalation assessment (Bagster and Pitblado, 1991). Several other studies followed after 1995, mainly concerned with qualitative methodologies for domino assessment (Contini et al., 1996; Gledhill and Lines, 1998; Delvosalle, 1998). Important contributions were dedicated to some specific aspects of the problem, e.g. domino frequency estimation (Bagster and Pitblado, 1991; Gledhill and Lines, 1998; Pettitt et al., 1993), deterministic estimation of domino effect due to radiation (Latha et al., 1992; Morris et al., 1994), and models for accident propagation (Bagster and Pitblado, 1991; Khan and Abbasi, 1998a; Cozzani et al., 2005). However, two main issues hindered the quantitative assessment of domino effects in the 1990s: (1) the insufficient development of integrated software tools, able to cope with geographical information in direct relation to consequence assessment and (2) the lack of knowledge on structural damage leading to equipment failure, which is a key point for the assessment of escalation possibility and/or probability. Only in the past 10 years, several research breakthroughs paved the way toward the quantitative assessment of domino scenarios aimed at the prevention of escalation and at the management of risk associated with domino accidents. Started by the relevant work of Khan and Abbasi (Khan and Abbasi, 1998b, 2000, 2001), tools for the quantitative assessment of risk due to domino scenarios supported on Geographical Information Systems were developed (Cozzani et al., 2005, 2006; Antonioni et al., 2009). A new approach to the identification and analysis of domino clusters was proposed (Reniers et al., 2005; Reniers and Dullaert, 2007, 2008). Game theory was introduced as a mathematical tool to help decision-makers decide whether or not to take precautions against domino effects—in collaboration with others or not—in chemical clusters (Reniers, 2010). Moreover, improved models were developed for the assessment of equipment damage (Cozzani and Salzano, 2004a, 2004b; Zhang and Jiang, 2008; Landucci et al., 2009). The progress achieved allowed the consolidation of methodologies for quantitative risk assessment of domino accidents as well as for the identification and management of risk related to domino scenarios. In this perspective, further results are expected from the application of approaches based on inherent safety (Cozzani et al., 2007) and on the integration of safety and security for the assessment of domino effects (Reniers, 2010), as well as from the continuous improvement of concepts, theories, methods and tools.

Actually, relevant work carried out in the field allowed the achievement of ready-to-use and up-to-date tools for domino risk assessment, able to provide the support for a further step toward the design and operation of safer and more sustainable chemical industrial facilities, infrastructures, and industrial parks.

1.2 Safety and Security: Both Important for the Prevention of Domino Effects

Risk assessments and subsequent preventive measures for safety and security are largely similar, but there are some differences. These differences can be understood by a more thorough explanation of the difference between safety risks and security risks. The term "risk" implies uncertainty, for example expressed as likelihood, either probability or frequency, rather than certainty.

In CCPS (2000), safety risk is defined as "*a measure of human injury, environmental damage, or economic loss in terms of both the incident likelihood and the magnitude of the loss or injury*". The definition of a safety risk thus bears the suggestion of being accidental. The incident likelihood may be either a frequency (the number of specified incidents occurring per unit of time) or a probability (the probability of a specified incident following a prior incident), depending on the circumstances. The magnitude of the loss or injury determines how the risk is described on a continuous scale from diminutive to catastrophic. As pointed out by Kirchsteiger (1998), risk is not simply a product-type function between likelihood and consequence values, but an extremely complex multiparametric function of "all" circumstantial factors surrounding the event source of occurrence.

Somebody who explicitly intends to cause damage to chemical facilities or perpetrate a theft of chemicals makes for a very different security risk analysis than is typically conducted to assess accidental safety risks, since a security risk (suggesting intentionality) is an expression of "*the likelihood that a defined threat will exploit a specific vulnerability of a particular attractive target or combination of targets to cause a given set of consequences*" (CCPS, 2003). According to the Department of Homeland Security (DHS, 2010), security risks are composed of consequences, vulnerabilities, and threats.

Safety and security are thus different in the nature of incidents: safety is non-intentional, whereas security is intentional (Holtrop and Kretz, 2008; Hessami, 2004). This implies that in the case of security an aggressor is present who is influenced by the physical environment and personal factors (Randall, 2008). These parameters should thus be taken into account during security assessments. The aggressor may act from within the organization (internal) and from outside the organization (external). Probabilities in terms of security are very hard to determine (Johnston, 2004). Hence, the identification of threats and the development of measures in terms of security is a challenging task, which is largely qualitative.

Safety and security may also differ in their proactive approach. In case of safety risk assessments, risks are detected and analyzed by using consequences and probabilities (or frequencies). In case of security risk assessments, threats are detected and analyzed by using consequences, vulnerabilities and target attractiveness (Holtrop and Kretz, 2008). The different proactive approach sometimes leads to the need for different and complementary protection measures in case of safety and security. Table 1.3 provides an overview of different characteristics related to safety and security.

Despite the differences in the required proactive approaches to such issues, safety and security have a lot of important similarities. First, the approach by which the consequences of a certain action, whether it is accidental or intentional, are treated is the same for safety and security. This can be explained by the fact that the effects of accidental or intentional actions are often comparable: e.g. a fire which was caused by an industrial accident will be extinguished in the same way as a fire which is the result of a criminal action. However, it is important to take into account that criminals deliberately search for the best manner to execute their plans and that their goal is to cause as much damage as possible. Especially in the case of domino effects this may be a very important fact. Second, according to the study of Holtrop and Kretz (2008), effective preventive safety measures may decrease security risks as well.

Table 1.3 Nonexhaustive List of Differences between Safety and Security

Safety	Security
– The nature of an incident is an inherent risk	– The incident is caused by a human act
– Nonintentional	– Intentional
– No human aggressor	– Human aggressor
– Quantitative probabilities and frequencies of safety-related risks are available	– In case of less-common security risks (e.g. terrorism), only qualitative (expert opinion based) likelihood of security-related risks may be available
– Risks are of rational nature	– Threats may be of symbolic nature

An important exception (or counter example) is the availability of information. It is clear that safety measures such as risk maps and labeling of dangerous substances have a negative impact on security (Holtrop and Kretz, 2008; CCPS, 2003).

Prior to the New York World Trade Center attacks in 2001, a successful intentional attack (e.g. by terrorists) on a chemical facility was believed to be extremely unlikely (although terrorist threats had been recognized for a long time prior to this date). The ramifications of the post-9/11 era include heightened security risk with regard to physical and economic damage. For the chemical industry in particular, being subject to risks which could potentially lead to major accidents (such as domino effects), the security implications for everyday operations might prove to be very significant in the prevention of incidents intentionally designed to cause damage. Security with respect to domino effects is therefore also further discussed further in this book, in Chapter 7.

1.3 Domino Effects and Chemical Industrial Areas

As already indicated in the beginning of this chapter, in the (petro)chemical industry, economies of scope, environmental factors, social motives and legal requirements often force companies to "cluster". Therefore, chemical plants are most often physically located in groups and are rarely located separately. Clearly, such industrial areas are characterized by reciprocal danger between equipment and infrastructures being part of the area. Since the first study of the Canvey area (HSE, 1978) and in several further studies in Europe (HSE, 1981; Rijnmond Public Authority, 1982; Egidi et al., 1995), risk in such chemical clusters was analyzed considering entire industrial areas and not the single chemical plants.

Within chemical clusters, intangible interdependencies between equipment and infrastructures may exist from a safety (and security) point of view. Every chemical installation represents a hazard depending on the amount of substances present, the physical and toxic properties of the substances and the specific process conditions. Hence, such installations present—to a greater or lesser extent—a danger to the other installations in the neighborhood.

Although many chemical companies are grouped into industrial parks, safety and security efforts are currently concentrated on individual chemical facilities. At present, very limited approaches or concepts are available for enhancing cross-company collaboration concerning safety and security topics (see also Reniers, 2010). However, dealing with cross-company-related threats might prove very important in reducing domino risks. More attention should therefore be devoted to the need of cross-company disaster management (dealing with preventing and mitigating domino effects potentially involving more than one plant), from a safety as well as a security viewpoint. Management strategies for countering cross-company major hazards should be developed in chemical industrial parks through systematic and guided collaboration between the companies composing the cluster.

A picture is slowly emerging of chemical industrial clusters that will set their own safety and security standards through intensive collaboration. The growing complexity of chemical processes, organizations and chemical logistics, global companies with independent business units, corporate goal-setting policies with local implementation, intracluster outsourcing and increased involvement by the public are all trends that chemical industrial parks have to take into account when dealing with potential domino scenarios.

Summarizing, all the theories, concepts, definitions, tools, and methodologies discussed in this book can be applied to chemical industrial areas. An industrial area can (most obviously) be a chemical plant, but it can also be several chemical plants at once or an entire chemical cluster. When reading this book, the reader should keep this in mind.

1.4 Contents of the Book

This book is intended to give the state-of-the-art information on the identification and assessment of domino scenarios, both in the safety and the security context. The reader is introduced to both recent but well-established methodologies and open issues in academic research. Part 1 (Chapters 2–7) reports the basic definitions needed to define a framework for the assessment of domino scenarios, based also on the analysis of past accidents, and introduces to methodologies for the analysis of escalation events, providing methods and models for equipment damage assessment. Part 2 (Chapters 8–11) provides a specific framework for the quantitative assessment of domino scenarios. Part 3 (Chapters 12–14) is concerned with the management of risk due to domino hazards, both in design and operations. Finally, Annexes 1 and 2 provide tutorials demonstrating the application of the methodologies described in the book.

References

Abdolhamidzadeh, B., Abbasi, T., Rashtchian, D., Abbasi, S.A., 2011. Domino effect in process-industry – an inventory of past events and identification of some patterns. Journal of Loss Prevention in the Process Industries 24, 575–593.

Antonioni, G., Spadoni, G., Cozzani, V., 2009. Application of domino effect quantitative risk assessment to an extended industrial area. Journal of Loss Prevention in the Process Industries 22, 614–624.

Bagster, D.F., Pitblado, R.M., 1991. The estimation of domino incident frequencies: an approach. Proceedings of Safety and Environment 69, 196.

CBS, Centraal Bureau voor de Statistiek, 2003. Vademecum gezondheidsstatistiek en arbeidsinspectie. Ministerie van Volksgezondheid, Welzijn en Sport, Voorburg, The Netherlands.

CCPS, Center for Chemical Process Safety, 2000. Evaluating Process Safety in the Chemical Industry: A User's Guide to Quantitative Risk Analysis. American Institute of Chemical Engineers, New York (USA).

CCPS, Center for Chemical Process Safety, 2003. Guidelines for Analyzing and Managing the Security Vulnerabilities of Fixed Chemical Sites. American Institute of Chemical Engineers, New York (USA).

Contini, S., Boy, S., Atkinson, M., Labath, N., Banca, M., Nordvik, J.P., 1996. Domino effect evaluation of major industrial installations: a computer aided methodological approach. In: Proceedings of European Seminar on Domino Effects, Leuven (B), pp. 21–34.

Cozzani, V., Salzano, E., 2004a. The quantitative assessment of domino effects caused by overpressure. Part I: probit models. Journal of Hazardous Materials 107, 67–80.

Cozzani, V., Salzano, E., 2004b. The quantitative assessment of domino effect caused by overpressure. Part II: case-studies. Journal of Hazardous Materials 107, 81–94.

Cozzani, V., Gubinelli, G., Antonioni, G., Spadoni, G., Zanelli, S., 2005. The assessment of risk caused by domino effect in quantitative area risk analysis. Journal of Hazardous Materials 127, 14–30.

Cozzani, V., Antonioni, G., Spadoni, G., 2006. Quantitative assessment of domino scenarios by a GIS-based software tool. Journal of Loss Prevention in the Process Industries 19, 463.

Cozzani, V., Tugnoli, A., Salzano, E., 2007. Prevention of domino effect: from active and passive strategies to inherently safe design. Journal of Hazardous Materials 139, 209–219.

Darbra, R.M., Palacios, A., Casal, J., 2010. Domino effect in chemical accidents: main features and accident sequences. Journal of Hazardous Materials 183, 565–573.

Delvosalle, C., 1996. Domino effect phenomena: definition, overview and classification. In: Proceedings of European Seminar on Domino Effects, Leuven (B), pp. 5–10.

Delvosalle, C., 1998. A Methodology for the Identification and Evaluation of Domino Effects, Rep. CRC/MT/003, Belgian Ministry of Employment and Labour, Bruxelles (B).

DHS. URL: http://www.dhs.gov/xlibrary/assets/dhs-risk-lexicon-2010.pdf (last accessed in 2010).

Directive 82/501/EEC. Council Directive 82/501/EEC of 24 June 1982 on the Major Accident Hazards of Certain Industrial Activities. Official Journal of the European Communities L 230/25, Brussels, 5.8.82.

Directive 96/82/EC. Council Directive 96/82/EC of 9 December 1996 on the Control of Major-Accident Hazards Involving Dangerous Substances. Official Journal of the European Communities, L 10/13, Brussels, 14.1.97.

Directive 2012/18/EU. European Parliament and Council Directive 2012/18/EU of 4 July 2012 on Control of Major-Accident Hazards Involving Dangerous Substances, Amending and Subsequently Repealing Council Directive 96/82/EC. Official Journal of the European Communities, L 197/1, Brussels, 24.7.2012.

EC, 2002. Document 2002/605/EC, Official Journal of the European Communities, L 195/79, Brussels, 14.1.2002.

Egidi, D., Foraboschi, F.P., Spadoni, G., Amendola, A., 1995. The ARIPAR project: an analysis of the major accident risks connected with industrial and transportation activities in the Ravenna area. Reliability Engineering and System Safety 49, 75.

Gledhill, J., Lines, I., 1998. Development of Methods to Assess the Significance of Domino Effects from Major Hazard Sites, CR Report 183, Health and Safety Executive, London (UK).

Hessami, A.G., 2004. A system framework for safety and security: the holistic paradigm. Systems Engineering 7 (2), 99–112.

Holtrop D., Kretz D., 2008. Research Security & Safety: An Inventory of Policy, Legislation and Regulations. Research Report 141223/EA8/043/000603/sfo. Arcadis, The Netherlands (in Dutch).

HSE, Health and Safety Executive, 1978. Canvey: An Investigation of Potential Hazards from Operations in the Canvey Island/Thurrock Area. HM Stationary Office, London, (UK).

HSE, Health and Safety Executive, 1981. Canvey: A Second Report. A Review of the Potential Hazards from Operations in the Canvey Island/Thurrock Area Three Years after Publication of the Canvey Report. HM Stationery Office, London (UK).

Johnston, R.G., 2004. Adversarial safety analysis: borrowing the methods of security vulnerability assessments. Journal of Safety Research 35, 245–248.

Khan, F.I., Abbasi, S.A., 1998a. Models for domino effect analysis in chemical process industries. Process Safety Progress 17, 107.

Khan, F.I., Abbasi, S.A., 1998b. DOMIFFECT (DOMIno eFFECT): user-friendly software for domino effect analysis. Environmental Modelling and Software 13, 163–177.

Khan, F.I., Abbasi, S.A., 2000. Studies on the probability and likely impacts of chains of accident (domino effect) in a fertilizer industry. Process Safety Progress 19, 40–56.

Khan, F.I., Abbasi, S.A., 2001. An assessment of the likelihood of occurrence, and the damage potential of domino effect (chain of accidents) in a typical cluster of industries. Journal of Loss Prevention in the Process Industries 14, 283–306.

Kirchsteiger, C., 1998. Absolute and relative ranking approaches for comparing and communicating industrial accidents. Journal of Hazardous Materials 59, 31–54.

Kourniotis, S.P., Kiranoudis, C.T., Markatos, N.C., 2000. Statistical analysis of domino chemical accidents. Journal of Hazardous Materials 71, 239–252.

Landucci, G., Gubinelli, G., Antonioni, G., Cozzani, V., 2009. The assessment of the damage probability of storage tanks in domino events. Accident Analysis and Prevention 41, 1206–1215.

Latha, P., Gautam, G., Raghavan, K.V., 1992. Strategies for the quantification of thermally initiated cascade effects. Journal of Loss Prevention in the Process Industries 5, 18.

Lees, F.P., 1996. Loss Prevention in the Process Industries, second ed. Butterworth-Heinemann, Oxford (UK).

Leveson, N.G., 2004. A new accident model for engineering safer systems. Safety Science 42, 237–270.

Mecklenburgh, J.C., 1985. Process Plant Layout. George Goodwin, London (UK).

Morris, M., Miles, A., Copper, J., 1994. Quantification of escalation effects in offshore quantitative risk assessment. Journal of Loss Prevention in the Process Industries 7, 337.

Pettitt, G.N., Schumacher, R.R., Seeley, L.A., 1993. Evaluating the probability of major hazardous incidents as a result of escalation events. Journal of Loss Prevention in the Process Industries 6, 37.

Pietersen, C.M., 1988. Analysis of the LPG-disaster in Mexico city. Journal of Hazardous Materials 20, 85–107.

Randall, L.A., 2008. 21st Century Security and CPTED. CRC Press, Boca Raton (USA).

Rasmussen, K., 1996. The Experience with the Major Accident Reporting System from 1984 to 1993. EUR 16341 EN. Commission of the European Communities, Luxembourg (L).

Reniers, G.L.L., Dullaert, W., Soudan, K., Ale, B.J.M., 2005. Developing an external domino accident prevention framework: Hazwim. Journal of Loss Prevention in the Process Industries 18, 127–138.

Reniers, G.L.L., Dullaert, W., 2007. DomPrevPlanning©: user-friendly software for planning domino effects prevention. Safety Science 45, 1060–1081.

Reniers, G.L.L., Dullaert, W., 2008. Knock-on accident prevention in a chemical cluster. Expert Systems with Applications 34 (1), 42–49.

Reniers, G.L.L., 2010. An external domino effects investment approach to improve cross-plant safety within chemical clusters. Journal of Hazardous Materials 177, 167–174.

Rijnmond Public Authority, 1982. Risk Analysis of Six Potentially Hazardous Industrial Objects in the Rijnmond Area, A Pilot Study. D. Reidel Publishing Company, The Hague.

Zhang, M., Jiang, J., 2008. An improved probit method for assessment of domino effect to chemical process equipment caused by overpressure. Journal of Hazardous Materials 158, 280–286.

Part I

Causes of Domino Effects

2 Analysis of Past Accidents and Relevant Case-Histories

Joaquim Casal, Rosa-Mari Darbra

Department of Chemical Engineering, Centre for Studies on Technological Risk (CERTEC), Universitat Politècnica de Catalunya, Barcelona, Catalonia, Spain

2.1   Introduction

Failure is often a better teacher than success. Thus, past accidents always contain a lesson to learn from them. The survey of accidents is a way to search why and how accidents occur and to determine which are the most common sequences they follow. The study of specific accidents shows what went wrong and which measures should be applied to avoid it in the future. Both approaches are helpful in improving the safety and efficiency of industrial plants and activities.

This chapter analyzes the diverse historical surveys performed on domino accidents and the main characteristics of this phenomenon, commenting as well a few examples of such accidents.

Domino Effects in the Process Industries. http://dx.doi.org/10.1016/B978-0-444-54323-3.00002-6

2.2 The Analysis of Past Accidents

The study of accidents that have occurred in chemical process industries or in the transportation of hazardous materials is very interesting due to several reasons, first, because real accidents are a source of "experimental data" obtained from full-scale major accidents, a field in which experimental work is very difficult and expensive and, in many cases, practically impossible. Furthermore, such experimental data have been obtained as a consequence of severe accidents—in terms of both economical and human loses—and not using them would be rather absurd. And, finally, historical analysis of accidents can lead to the identification of a number of features, which can be very helpful for performing risk assessment and developing accident prevention strategies: initiating events, most frequent type of installation in which accidents have occurred, materials most frequently involved, consequences, etc. Thus, the results of such an analysis can be used to identify the main sources of risk, to establish which accidents are more prone to happen and to define safer operating procedures. In fact, no one doubts nowadays that the lessons learned from past accidents are a very valuable resource.

In the case of domino effect, this type of survey serves to identify the more frequent origins of this effect and the frequency of the diverse event sequences that can occur. In fact, from the point of view of risk analysis, historical surveys could constitute in certain cases a useful approach to the quantitative assessment of domino effect.

However, although an increasing interest on the domino effect can be clearly observed in the literature in the past decade, historical analysis has been treated only in a rather reduced number of communications.

2.3 Domino Effect Surveys

Analysis of the domino effect and its importance in accidents involving hazardous materials is a complex task. Accident databases often contain incomplete information, and, in some cases, it is difficult to determine whether an accident involved a domino effect and, if so, whether it was a first- or second-level domino effect. Given these difficulties, a conservative approach should be adopted and only those accidents with clear evidence of a domino effect should be considered when performing a historical survey.

A pioneering historical survey was the one published by Delvosalle et al. (1998), who analyzed a relatively reduced number of domino accidents (41) that occurred between 1944 and 1995 (66% of them during the past 20 years of this period). These authors classified the type of domino effect according to (1) the type of primary and secondary installations involved and (2) the nature of primary and secondary physical effects produced.

Abdolhamidzadeh et al. (2009) studied a set of 73 domino accidents that occurred between 1917 and 2008 in industrial plants and transportation (21% of the accidents analyzed had occurred in transportation). They analyzed the type of activity, the substances involved, the level of domino effect and the impact on the affected population. The study focused on a relatively small number of accidents (including accidents involving conventional explosives), distributed over a wide period during which the process industry structure and safety changed significantly. The ratio between first-level and second-level domino sequences obtained by these authors was 2.2.

Darbra et al. (2010) analyzed 225 accidents involving domino effect that occurred in process/storage plants and in the transportation of hazardous materials; the selected accidents had occurred between 1961 and 2007. The aspects analyzed included the accident location, the type of accident, the materials involved, the causes and consequences and the most common accident sequences. The analysis showed that the most frequent causes were external events (31%) and mechanical failures (29%). Storage areas (35%) and process plants (28%) were by far the most common setting for domino accidents. About 89% of the accidents involved flammable materials, the most common of which was liquefied petroleum gas. The domino effect sequences were analyzed using relative probability event trees. The more frequent ones were explosion → fire (27.6%), fire → explosion (27.5%) and fire → fire (17.8%).

Finally, Abdolhamidzadeh et al. (2011) published an inventory of 224 major process industry accidents involving domino effect that occurred between 1917 and 2009 (175 of them occurred in the period 1970–2009). Although most of the accidents had occurred in process plants, some of them had occurred in transportation and a few of them involved ammunition or conventional explosives. The inventory included the following information: year of occurrence, location, and type of plant. It also reported, if available, the type of unit and chemicals involved, sequence of the accident, consequences (deaths, injuries, and other impacts), and the source of the record. The authors analyzed, among other aspects, the type of materials involved, the primary (initiating) events, the contribution of transportation and fixed plants, the frequencies of the accident sequence lengths and the number of fatalities, and they reached the corresponding conclusions. A certain number of the accidents analyzed had occurred in countries not economically fully developed; obviously, this introduces a clear heterogeneity in the inventory, as the safety standards of the plants and activities are different in these countries and this fact should increase both the frequency and the consequences of domino accidents.

Other authors have analyzed sets of accidents with and without domino effect. Kourniotis et al. (2000) examined 207 major chemical accidents that occurred between 1960 and 1998, all of them in economically developed countries and 114 of which involved a domino effect. These authors analyzed the likelihood or relative probability of accidents with first- or second-level domino sequences with respect to the total number of accidents involving substances in four categories: liquid fuels, vapor hydrocarbons, toxic substances and miscellaneous substances. They also studied the severity of the consequences on the affected population using p–N curves and a modified version of the Pareto distribution. According to the data published by Kourniotis et al. (2000), the ratio between the number of accidents involving one domino effect (80) and the number of accidents involving a sequence of at least two domino effects (34) was 2.3, a value very similar to that obtained by Abdolhamidzadeh et al. (2009).

Domino effect sequences were studied by Ronza et al. (2003) in the frame of a survey on major accidents in port areas. These authors applied for the first time the *relative probability event tree* to the analysis of the diverse sequences found when a sequence of accidents had occurred. They applied this new methodology to 675 accidents that occurred in ports, including accidents that originated from a ship-to-ship collision and accidents that originated in ship-loading operations. In 390 of these cases, only a release occurred. As for the other 285 cases, in 38 cases there was domino effect (first level). By using the relative probability event tree, once the frequency of the initial event had been ascertained (such as the release from a ship-loading pipeline), it was possible to ascribe a frequency value to the sequence scenarios, by multiplying the probability of occurrence by the frequency of the root event. The reliability of this procedure was proved by a wide range of historically documented accidents. This methodology was also used by Oggero et al. (2006) when analyzing the accidental sequences—with and without domino effect—that occurred in the transport of hazardous substances by road and rail.

Besides general surveys on all types of accidents, another useful approach is the analysis of specific events which are prone to originate domino effect. A significant case is that of jet fires. Fires were involved in a large proportion of the accidents that occur in industrial installations or in the transportation of hazardous materials. And, among the diverse types of fires, jet fires play a significant role. Although they are usually smaller than pool or flash fires, and per se often they are not severe accidents, they can be very intense locally and can imply very high heat fluxes if there is flame impingement; if a jet fire occurs in a relatively compact installation, it will probably affect equipment (e.g. a pipe or a tank) that may subsequently fail and ultimately amplify the scale of the accident. Gómez-Mares et al. (2008) performed a specific study of 84 accidental scenarios involving jet fires and found that in 50% of them jet fires had been the primary event of a domino sequence. The most common sequence was jet fire → explosion. In transportation, a typical sequence began after a derailment or a road accident, leading to a leak in the pipework, a jet fire and tank explosion, often followed by a fireball.

As a whole, there exists a reduced number of domino accident surveys and, furthermore, as always happens with this type of study, the information available is sometimes incomplete. Nevertheless, the data collected in these studies have given interesting results.

2.4 Significance of Domino Effect in the Frame of all Accidents

Knowing the percentage of accidental scenarios in which domino effect occurs is a rather difficult question. In order to get an answer, a large number of accidents should be analyzed. However, when doing this, often the information found is incomplete and it is not possible to ascertain if this effect really occurred. Nevertheless, experience shows that there is an important relative number of accidents in which domino effect has been involved. Kourniotis et al. (2000) observed that the likelihood of an accident to cause domino effects depends on the substance involved in it. For the whole set of accidents analyzed by these authors, domino effect existed in 39% of the cases, whereas when analyzing only accidents involving liquid fuels, this percentage increased to 49% and with vapor hydrocarbons, it reached 58%. For accidents involving toxic substances, this value was only 16%.

Concerning the accidents that occurred in the transportation of hazardous materials, the incidence of domino effect is significantly smaller, as usually no additional equipment (compact plants and fixed storage tanks) installations are affected. For accidents that occurred in sea ports (a scenario clearly associated with transportation), Ronza et al. (2003) obtained a very low percentage, 6%, but their analysis involved mainly low-volatility liquids. For road and rail transport, Oggero et al. (2006) also obtained a rather low value, 8%, in their analysis of 1932 accidents.

2.5   Characteristics of Accidents Involving Domino Effect

2.5.1   Origin and Causes of Domino Effect

In this and the following subsections, the main characteristics of the accidents in which the domino effect has taken place are explained: e.g. the origin and causes, the materials involved, the consequences, and the event sequences.

Domino effect can be found both in fixed plants and in transportation. Few authors have given data on the distribution between these two possibilities. Here a certain lack of accuracy is found again as, according to the authors' criterion, transportation may or may not include pipelines.

Abdolhamidzadeh et al. (2009), in a rather reduced sample of data (73 accidents), found that 21% of accidents involving domino effect had occurred in transportation. Gómez-Mares et al. (2008), in their specific analysis of jet fires, found that 44% of accidents had occurred in transportation, 11% in loading/unloading operations, 36% in process plants and 9% in storage. Thus, when the process plant and storage categories were considered together as fixed plants, they accounted for 45% of the total accidents, whereas the transport and loading/unloading categories combined accounted for 55%. In the survey performed by Darbra et al. (2010) on 225 accidents (Table 2.1), transport accounted for 18.7% of cases, whereas the percentage for "transfer" (essentially loading/unloading) was 13.3. Thus, transport plus transfer accounted for 32%. Fixed plants accounted for 68%: 35% storage, 28% process and 5% "other". Finally, Abdolhamidzadeh et al. (2011) (224 accidents) found that 20% of accidents involving domino effect had occurred in transportation and 80% in fixed installations. These authors analyzed the distribution among the diverse transport modes: 40% road, 39% rail, 13% shipping and 8% pipeline.

Table 2.1 Origin of the Domino Accidents

Source	Transport* (%)	Fixed Installations (%)
Abdolhamidzadeh et al. (2009)	21	79
Darbra et al. (2010)	32	68
Abdolhamidzadeh et al. (2011)	20	80

*Transport includes loading/unloading.

Table 2.2 General Causes of the Primary Accidents (Darbra et al., 2010)

Cause	No. of Events	%
External events	69	30.7
Mechanical failure	65	28.9
Human factor	47	20.9
Impact failure	40	17.8
Violent reaction	21	9.3
Instrument failure	8	3.6
Upset process conditions	5	2.2
Services failure	3	1.3

The causes of the primary accident were studied in detail in the survey made by Darbra et al. (2010). These authors classified the causes according to the categories established in the Major Hazardous Incident Data Service (MHIDAS) database: external events, mechanical failure, human error, impact failure, violent reaction (runaway reaction), instrument failure, upset process conditions and services failure. Among them, the category *human error* can lead to misunderstanding, because other causes such as mechanical failure or impact failure could themselves be caused by human error. The generic causes that initiated a domino effect are summarized in Table 2.2. The percentages sum up to more than 100 because some accidents were triggered by more than one generic cause.

External events and mechanical failure were the main causes, followed by human error and impact failure. The other causes had a smaller significance.

2.5.2   Materials Most Frequently Involved

Domino accidents generally involve more than one substance; for example, Darbra et al. (2010) identified 335 substances in the 225 accidents analyzed. However, the real number of substances was higher, as often only the substance involved in the primary accident is mentioned in the information source and sentences such as "the fire spread to storage tanks containing chemicals" are sometimes used in the databases to describe the secondary accident. A relatively small number of accidents involved only one substance as, for example, propane, when it was involved in a series of fires and explosions in a propane gas farm. Table 2.3 contains a summary of the results obtained in three surveys.

Table 2.3 Materials Most Frequently Involved in Domino Accidents (percentages)

Source	Flammable	Toxic	Miscellaneous
Kourniotis et al. (2000)	45	22	33
Darbra et al. (2010)	89*	22*	16*
Abdolhamidzadeh et al. (2011)	89	4	7

*Percentages sum up to more than 100 because some substances were simultaneously flammable and toxic.

Flammable substances were associated with most of the accidents. Darbra et al. (2010) and Abdolhamidzadeh et al. (2011) obtained exactly the same figure: in 89% of domino accidents flammable substances were involved. Kourniotis et al. (2000) obtained a lower value (45%). As shown in the table, toxic and "miscellaneous" materials have a significant lower frequency.

2.5.3   Consequences of the Accidents

Concerning the population affected by the accidents, only the number of fatalities will be commented in this chapter. The best way to represent the lethality of accidents is the p–N curve, in which the abscissa represents the severity of the accident (the number of fatalities, N) and the values of the ordinate represent the probability (p) that an accident with casualties will cause a number of fatalities equal to or greater than N (for $N = 1, p = 1$) (see Figure 2.1). For the accidents analyzed by Darbra et al. (2010), the best fit for a curve $p = N^b$ gave $b = -0.74$, which means that the probability of an accident involving a domino effect that causes 10 or more deaths is 5.5 ($=10^{-b}$), higher than the probability of an accident that causes 100 or more deaths.

Kourniotis et al. (2000) found slightly different results; the trend of their data followed a certain curvature, with values of b around -0.5. They compared domino accidents with all accidents, obtaining a higher severity for accidents involving domino effect.

Abdolhamidzadeh et al. (2011) analyzed the variation of the number of fatalities as a function of time, obtaining an increasing trend, which was significantly influenced, however, by the exceptional value corresponding to the accidents that occurred in the decade 1980–1990 (with the accidents of Tacoa, Venezuela (150 deaths) and Mexico City (503 deaths)).

2.5.4   Domino Sequences

An interesting aspect of domino effect accidents is the length of the accidental sequence and the events that occurred during it. Abdolhamidzadeh et al. (2011) found in their survey that 53% of accidents involved primary plus secondary events, whereas 47% included a tertiary or even higher level escalation events. Darbra et al. (2010) analyzed the domino sequences by using the relative probability event tree. To do this, the accidents were classified into four categories: release, fire, explosion and gas cloud. However, the "release" category is misleading, as this event is not registered in many databases, although most accidents are in fact initiated by a loss of containment. Therefore, in order to not underestimate this category, the authors considered only the other three categories and built the relative probability event tree shown in Figure 2.2.

In this tree, the branches indicate the different accident scenarios. The application of a simple statistical procedure gave the relative probability of occurrence of each sequence. The number of accidents and the relative probability of occurrence (in square brackets) are shown for each branch. The figures in square brackets represent the probability of occurrence with respect to the level immediately above. The values at the end of each branch represent the overall probability of occurrence of the specific accident sequence relative to all possible events. The primary events were fire

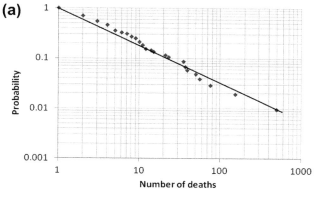

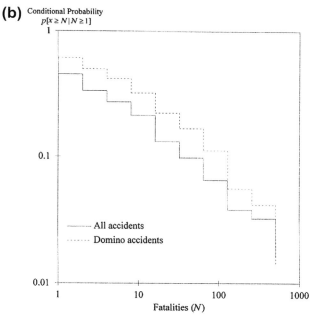

Figure 2.1 Accumulated probability vs. the number of deaths. (a) Darbra et al. (2010) and (b) Kourniotis et al. (2000).

(52.4%) and explosion (47.6%). A fourth-level accident was reported in a single case (with the sequence fire → explosion → fire → explosion).

The most frequent final domino sequences were explosion → fire (27.6%), fire → explosion (27.5%) and fire → fire (17.8%). Of the 225 accidents considered, 193 involved one domino effect (i.e. primary plus secondary accidents), whereas only 32 involved at least two domino effects (a sequence of primary plus secondary plus tertiary accidents). This gives a ratio between first-level and second-level domino

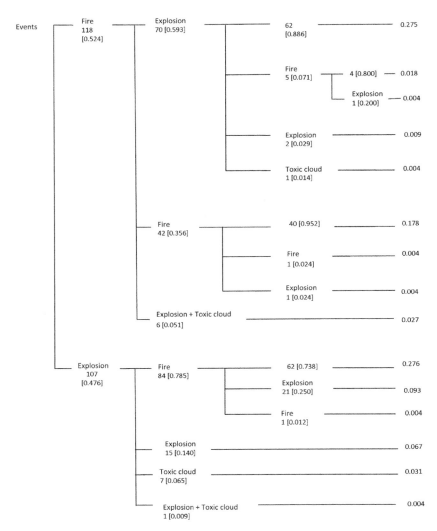

Figure 2.2 Relative probability tree showing the diverse domino effect sequences; in square brackets: relative probability (Darbra et al., 2010).

effect sequences of 6. These results are significantly higher than those reported by Kourniotis et al. (2000) and Abdolhamidzadeh et al. (2011). The difference could be attributed to the lack of accuracy in the description of the accidents in the databases, which often can lead to different interpretations.

Delvosalle et al. (1998) analyzed the relative occurrence of the different installation types in primary or secondary domino accidents. Their results (based on a reduced sample, 41 cases) have been summarized in Table 2.4 (loading/unloading areas were not included because too few data were available in the sample).

Table 2.4 Relative Occurrences of Installation Types in Primary or Secondary Accidents (Delvosalle et al., 1998)

Type of Installation	Primary Events	Secondary Events
Pressure storage tanks	30%	33%
Atmospheric or cryogenic storage tanks	28%	46%
Process equipment	30%	12%
Pipe networks	12%	–
Small conditionings	–	9%

2.5.5 The Frequency of Domino Accidents over the Past Decades

Concerning the evolution as a function of time, Darbra et al. (2010) observed an increasing trend between 1961 and 1980–1990 (Table 2.5), which decreased afterward until 2007. The first increase must be attributed to the continuous growth of chemical industry since the early 1960s, as well as to the improvement in the access to information on industrial accidents. A considerable number of accidents that occurred during the 1960s were not recorded and the information on them was lost.

Most historical surveys report a significant and continuous rise in the frequency of major accidents in the past 50 years. However, in this survey, the peak accident rate was reached in the late 1970s and the 1980s, after which there was a continuous decrease until 2007. This decreasing trend is in good agreement with the data published by Gómez-Mares et al. (2008) and Niemitz (2010). Niemitz analyzed the major accidents registered in the European Union's Major Accident Reporting System (MARS, 2009) between 1996 and 2004 and also found a decreasing trend. Several reasons would justify this trend: general improvements in the safety culture of the chemical industry brought about by strict new regulations, more effective operator training, or increasing automation of industrial facilities. Nevertheless, this trend is new and the results must be considered with caution, monitoring this trend over the coming years.

2.6 Relevant Case Histories

In this section, four cases are briefly commented. The first and the second concern a type of event often associated with large and severe accidents: the spill of fuel in a storage area,

Table 2.5 Frequency of Occurrence of Domino Effects (Darbra et al., 2010)

Period	No. of Accidents	%
1961–1970	49	22
1971–1980	70	31
1981–1990	63	28
1991–2000	24	11
2001–2007	19	8

with the formation of a large flammable cloud. The third one is an example of how a relatively small jet fire—implying high heat fluxes in the case of flame impingement—can start a chain of domino accidents. Finally, the fourth one is related to the series of explosions that often occur when dust is involved in the accidents.

2.6.1 Buncefield, 2005

The Buncefield depot is a large tank farm located at 3 miles from the center of Hemel Hempstead (United Kingdom). The site is surrounded by residential areas. It was the fifth largest oil storage site in the United Kingdom. The depot distributed fuels to diverse sites through three pipelines. Jet aviation fuel was distributed to Heathrow and Gatwick airports via two other pipelines. Furthermore, fuel was also distributed by road tankers. The accident sequence can be divided into the following steps:

Initial loss of containment. At approximately 19.00 on Saturday 10 December 2005, a delivery of unleaded gasoline (incorporating about 10% of butane) started to arrive at Tank 912 (BMIIB, 2008). At about 05.30 on 11 December, the tank was full but the safety systems failed to shut off the supply and petrol cascaded from the top of the tank, collecting in the bund. A vapor cloud—a mixture of volatile fractions of the fuel and air—was dispersed, flowing to west toward another industrial site, as well as to north and south. It has been estimated that approximately 300 t of petrol escaped from the tank, 10% of which got vaporized and fed the cloud.

First explosion. The area covered by the cloud ranged between 80,000 m² and 100,000 m², with a height of approximately 2 m (Figure 2.3). At 06.01:32 on Sunday 11 December, a massive explosion took place.

Fire and additional explosions. The explosion was followed by a huge fire, which engulfed 20 fuel storage tanks. There were also other smaller explosions, independent of the first one (probably, inside tanks affected by fire).

Domino sequence. The main steps in the event chain were

* Spillage of gasoline (overfilling of a tank)
* Development of a vapor cloud
* Ignition
* Vapor cloud explosion
* Fires engulfing tanks.

And, from the point of view of domino accident, the sequence can be simplified to: vapor cloud explosion → fires → smaller explosions.

The origin of the accident was the failure of the pumping shutdown system. The records for Tank 912 showed that just after 03.00 on 11 December the automatic tank gauging system indicated that the level remained static at about two-thirds full (BMIIB, 2008). However, the tank was receiving a flow rate of 550 m³/h. In fact, 7 min before the accident, this flow rate was increased up to 890 m³/h when another line was closed. It has been calculated that Tank 912 was completely full at approximately 05.20, overflowing thereafter. This means that at the moment of the explosion more than 300 t of gasoline had been spilled from the eight vents at the top of the tank. The falling liquid was partly fragmented into the atmosphere by the effect of the top deflector plate and of a wind girder located on the tank wall, causing fuel

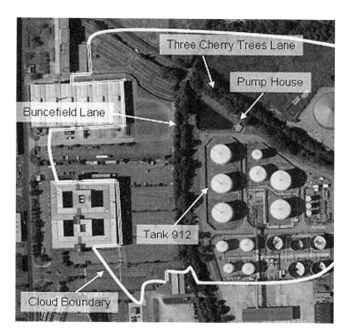

Figure 2.3 Buncefield site prior to explosion showing the approximate cloud boundary (Johnson, 2010). (For color version of this figure, the reader is referred to the online version of this book.)

droplets. This phenomenon improved the evaporation rate of the volatile fractions (butanes, pentanes, and hexanes).

Several potential ignition sources were identified by the research team investigating the accident: the fire pump house, the emergency generator cabin and car engines. The overpressures estimated from the explosion effects were extraordinarily high, significantly higher than those predicted by the methods usually applied (Figure 2.4).

The secondary explosions were significantly smaller than the main one. Their number cannot be known, as they were not detected seismically. They occurred between 7 min and half an hour after the main one (one or two smaller explosions could also have occurred just before the first one, e.g. inside the pump house). These secondary explosions were not further explosions of parts of the vapor cloud, but probably were internal tank explosions caused by the fire effects. The blast and thermal effects resulted in an important destruction of a significant part of the site, as well as in significant effects on neighboring sites and car parks (Figure 2.4).

2.6.2 Naples, 1985

This severe accident was similar to that of Buncefield. It occurred in the Naples fuel storage area in 1985: overfilling of a tank gave rise to a large vapor cloud, explosion and fire. The sequence of events is as follows.

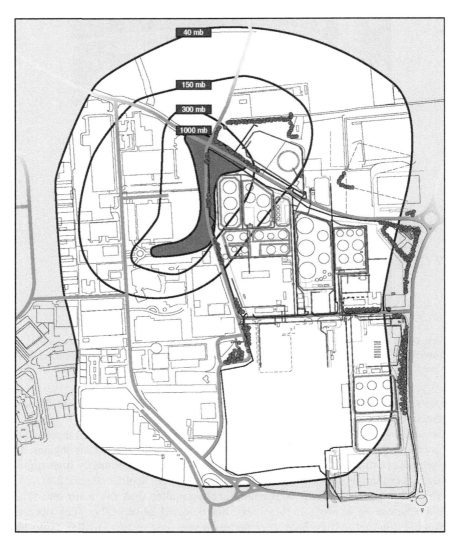

Figure 2.4 Map of overpressure isobars around the Buncefield site; in the grey area, over-pressures greater than 1000 mbar were reached (BMIIB, 2008).

Initial loss of containment. In the late afternoon of 20 December an oil tanker berthed in the harbor started to pump 780 m³/h of gasoline into the storage area (Russo et al., 1998). At 01.20 on 21 December the fuel flow was diverted to tanks no. 17 and 18 simultaneously; this was not the adequate procedure: the operators should have filled tank 17 first and then tank 18, but they arbitrarily decided to change the procedure. However, although the operators opened the first valve to tank 18, they were not aware that a second valve, usually (but not correctly) left open, was closed.

Thus, all flow reached tank 17. Filling of both tanks should have been completed at 06.30, but at about 04.00, tank 17 was completely filled and gasoline overflowed through the floating roof for more than 1 h: about 700 t of gasoline were spilled.

Vapor cloud explosion. A light breeze (2m/s), together with the significant ignition delay (more than 1 h) favored the formation of a large vapor cloud covering approximately 20,000 m^2. A total evaporated mass of about 135 t was estimated, giving rise to a flammable cloud with a volume of 45,000 m^3 containing 4 t of flammable vapor (Maremonti et al., 1999). The ignition was probably due to a small fire (seen by a train driver) outside the plant, near to the pumping area.

Fire. Following the explosion, there was an extensive fire which lasted over 1 week. The storage area was almost completely destroyed. The blast wave caused five casualties within the area, and minor effects were observed up to 5 km away.

Domino sequence. The steps in the event chain were

- Spillage of gasoline
- Development of a vapor cloud
- Ignition
- Vapor cloud explosion
- Fire engulfing tanks

These steps, from the point of view of domino accident, can be simplified to: vapor cloud explosion → fire.

2.6.3   Valero (Texas), 2007

On February 16 at 02.09, at a refinery in Texas, a jet of liquid propane started to flow from a high-pressure pipe, near an extractor tower (where liquid propane was used as a solvent to separate gas oil from asphalt). When the operators determined that the material released was propane, the fire alarm was activated (12.10) and the area was evacuated. A vapor cloud was formed which was soon ignited. The fire flashed back to the leak source. The flames impinged on diverse pipes near the extractor, releasing additional propane jets, which were also ignited. The fire impingement weakened a steel support column—not protected by fireproofing insulation—which buckled, causing the collapse of a rack that resulted in several pipe failures (USCSHIB, 2008). A butane sphere (1600 m^3) was affected by the thermal radiation. Three vessels containing chlorine were affected by the thermal radiation—and possibly by flame impingement—and released most of their content to the atmosphere. The refinery had been evacuated 15 min after the start of the accident. The accident sequence may be divided into the following steps.

Initial loss of containment. Release of liquid propane from a freeze-related failure of a high-pressure (35.5 bar) pipe at a control station that had not been in service for 15 years. Water contained in the propane had accumulated in a low point (elbow), freezing during cold weather and cracking the pipe (Figure 2.5). The propane initial leak rate was estimated to be 34 kg/s (USCSHIB, 2008).

Fire and jet fires. After the initial flash fire, flames flashed back and a highly turbulent jet fire was formed, with flames impinging on other pipes. Due to the very high heat fluxes, the pipes failed. New large propane jet fires were formed, with

Figure 2.5 A crack in an elbow, due to water freezing, the initiating step of a three-step domino accident (USCSHIB, 2008). (For color version of this figure, the reader is referred to the online version of this book.)

important thermal effects on surrounding equipment. Fire fighting was hampered by strong winds and rapid fire growth.

Toxic cloud. Three 1-ton chlorine containers were subjected to radiant heating and all three vented when their fusible plugs melted. One of them vented completely and another developed a leak that was repaired by emergency responders using self-contained breathing equipment for protection against the toxic gas. The third one ruptured. More than 2.5 t of chlorine were released into the atmosphere. There were also sulfuric acid leaks.

Domino sequence. The steps of the event were

- Release of pressurized liquefied propane
- Flash fire/jet fire
- Multiple jet fires
- Vessel venting and rupture
- Toxic cloud of chlorine.

And, as a domino accident this can be represented as jet fire → secondary jet fires → toxic cloud.

The fire was finally extinguished on February 18, 54 h after it spread. Most of the equipment in the area was destroyed, and the refinery remained completely shutdown for nearly 2 months.

2.6.4 Corbin (Kentucky), 2003

On February 20, 2003, at 7.30, a fire originated a series of phenolic resin dust explosions at CTA Acoustics, Inc. facility (USA), a plant manufacturing industrial and automotive insulation products (USCSHIB, 2005). The accident killed seven operators and injured 37 more. The explosions were followed by a fire. This severe accident is an important example of domino dust explosions.

The plant had already had problems with the ventilation and with the release of dust from the production lines. In line 405, ventilation air flow rates had decreased due to the poor condition of the fans, overloading of the baghouse and possible plugging of the ductwork. Large quantities of dust were released from this line, and phenolic resin dust migrated through door and wall openings and settled onto flat surfaces throughout the facility. The sequence of the event is explained below.

Initiating event. The accident originated at line 405. The curing oven of the line was operated with the doors open due to a malfunction of the temperature control equipment. This controller had been giving problems since 20 December 2002, causing overheating. Maintenance personnel had attempted to fix it on five occasions. Finally, 4 days prior to the accident, the control was switched to manual by the line operators, and the oven temperature was controlled by opening and closing the access doors on the oven sides. Small fires, extinguished by the operators, started from time to time in the oven when accumulated phenolic resin ignited due to sparks from the oven flight chain.

Due to a poor functioning of the baghouse there was an excessive release of dust from the production line. The baghouse was turned off for cleaning. The cleaning operation (with a compressed air lance) created a cloud of combustible dust around line 405. Just after the flow was started again across the baghouse at 7.30, flames shot up through the inlet duct. Operators located 24 m away from line 405 observed smoke coming from the line.

Dust explosions. Immediately after, an initial explosion took place in line 405 (Figure 2.6), which knocked down portions of the firewall on each side of the line. Two operators were knocked down, one with second-degree burns. A third one suffered third-degree burns and died several days later. A fourth employee suffered first-degree burns. The pressure wave blew dust around, and a second explosion took place in the confined area above the line 405 blending room. This explosion damaged significantly the building and the blast wave (as well as a fireball) propagated toward line 403. As a consequence, six employees were burned, and four of them died later. The baghouse of this line caught fire. Three operators in line 402 were burned; two of them died later. Finally, an additional explosion occurred in the southeast corner of line 401. Three operators were severely burned.

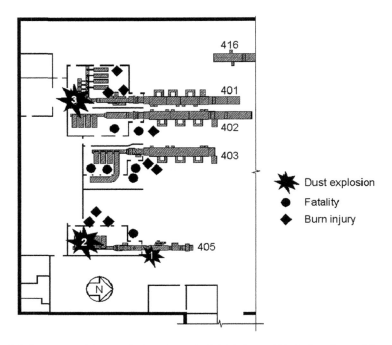

Figure 2.6 Sequence of dust explosions from the initial fire and likely locations of those killed and severely burned (USCSHIB, 2005).

Domino sequence. The sequence of the domino accident can be summarized as follows fire → explosion → explosion → explosion + fire.

The consequences of the accident were death of 7 employees and injury to 37, some of them very seriously. The plant was severely damaged.

2.7   The Analysis of Past Accidents, a Useful Resource

Only with the four cases presented in this chapter, a variety of causes can be identified for domino scenarios: failure in equipment or instrumentation, human error, lack of application of the established procedures, danger of nonroutine operations, bad management, etc. All of them had been identified also in the aforementioned historical surveys. In some occasions, the cause is related to very small failures or errors, but this may trigger a sequence of accidents through the domino effect, increasing significantly the overall severity and damage.

It is therefore obvious that this phenomenon should be considered in any risk analysis, which should take into account the potential increase in the magnitude of the consequences that domino effect can give rise to. The application of adequate safety measures can reduce substantially the frequency of this effect. Any information that can be learned from the past accidents can be helpful to implement such measures. In the following chapters, such information will be at the base of the analysis of domino scenarios.

References

Abdolhamidzadeh, B., Rashtchian, D., Morshedi, M., 2009. Statistical survey of domino past accidents. In: Proc. 8th World Congress of Chemical Engineering.

Abdolhamidzadeh, B., Abbasi, T., Rashtchian, D., Abbasi, S.A., 2011. Domino effect in process-industry – an inventory of past events and identification of some patterns. Journal of Loss Prevention in the Process Industries 24, 575–593.

BMIIB, 2008. The Buncefield Incident. The final report of the Major Incident Investigation Board. Crown copyright, London (UK).

Darbra, R.M., Palacios, A., Casal, J., 2010. Domino effect in chemical accidents: main features and accident sequences. Journal of Hazardous Materials 183, 565–573.

Delvosalle, C., Fievez, C., Benjelloun, F., 1998. Development of a methodology for the identification of potential domino effects in "Seveso" industries. In: Proc. 9th International Symposium on Loss Prevention and Safety Promotion in the Process Industries, vol. 3. AEIC, Barcelona (E), pp. 1252–1261.

Gómez-Mares, M., Zárate, L., Casal, J., 2008. Jet fires and the domino effect. Fire Safety Journal 43, 583–588.

Johnson, D.M., 2010. The potential for vapour cloud explosions. Lessons from the Buncefield accident. Journal of Loss Prevention in the Process Industries 23, 921–927.

Kourniotis, S.P., Kiranoudis, C.T., Markatos, N.C., 2000. Statistical analysis of domino chemical accidents. Journal of Hazardous Materials 71, 239–252.

Major Accident Reporting System, MARS, 2009. Major Accident Hazards Bureau. European Commission Joint Research Centre.

Maremonti, M., Russo, G., Salzano, E., Tufano, V., 1999. Post-accident analysis of vapour cloud explosions in fuel storage areas. Trans IChemE, Vol 77, Part B. 360–365.

Niemitz, K.J., 2010. Process safety culture or what are the performance determining steps? In: Workshop on Safety Performance Indicators, Ispra, March 17–March 19, 2010.

Oggero, A., Darbra, R.M., Muñoz, M., Planas, E., Casal, J., 2006. A survey of accidents occurring during the transport of hazardous substances by road and rail. Journal of Hazardous Materials 133, 1–7.

Ronza, A., Félez, S., Darbra, R.M., Carol, S., Vílchez, J.A., Casal, J., 2003. Predicting the frequency of accidents in port areas by developing event trees from historical analysis. Journal of Loss Prevention in the Process Industries 16, 551–560.

Russo, G., Maremonti, M., Salzano, E., Tufano, V., Ditali, S., 1998. Vapour cloud explosion in a fuel storage area: a case study. In: Proc. 9th International Symposium on Loss Prevention and Safety Promotion in the Process Industries, vol. 3. AEIC, Barcelona, pp. 1121–1130.

USCSHIB, 2005. Combustible Dust Fire and Explosions at CTA Acoustics, Inc. Investigation Report No. 2003-09-I-KY. U.S. Chemical Safety and Hazard Investigation Board.

USCSHIB, 2008. LPG Fire at Valero-Mckee Refinery. Investigation Report No. 2007-05-I-TX. 2008. U.S. Chemical Safety and Hazard Investigation Board.

3 Features of Escalation Scenarios

Genserik Reniers*, Valerio Cozzani [†]

* Centre for Economics and Corporate Sustainability (CEDON), HUB, KULeuven, Brussels, Belgium, [†] LISES, Dipartimento di Ingegneria Civile, Chimica, Ambientale e dei Materiali, Alma Mater Studiorum – Università di Bologna, Bologna, Italy

3.1   Elements of a Domino Accident

The case histories reported in the previous chapter evidenced the complexity that may affect domino scenarios. Actually, the complexity of these scenarios also affects the definition of "domino effect" and the identification of the features of domino accident scenarios.

The main element that identifies scenarios where a "domino effect" takes place is the "propagation" effect. It is universally recognized that in a "domino" accident the propagation (in space and/or in time) of a primary accident scenario should take place to start one or more than one secondary scenario. Thus, two further elements of a domino scenario may be identified in relation to the "propagation" element: the presence of a primary accident scenario and of one or more than one secondary scenarios.

Figure 3.1 shows alternative propagation patterns that may be assumed in the analysis of domino scenarios. A "simple" propagation may be assumed, defining a "one-to-one" correspondence, that is, a single primary scenario triggering a single secondary scenario. Alternatively, second-, third- and more in general multilevel propagation may be assumed, defining a so-called multilevel "domino chain": a first accident scenario triggers a second accident scenario, the second accident scenario triggers a third accident scenario, and so on.

Actually, in severe domino accidents, the propagation of the primary accident resulted in several simultaneous secondary scenarios triggered by the first primary

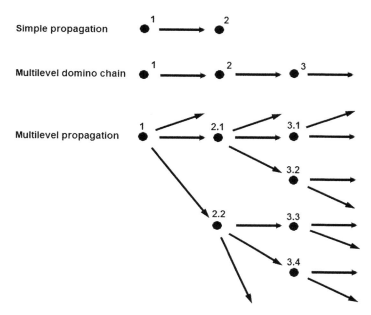

Figure 3.1 Examples of simple propagation, multiple-level domino chain and multilevel parallel propagation patterns.

accident. Secondary scenarios also triggered more than one scenario, defining a complex parallel multilevel propagation (multilevel propagation pattern in Figure 3.1). This is for example the case of the 1984 Mexico City accident (Pietersen, 1988) or of the 2005 Buncefield accident and the 2007 Valero Texas accident case histories presented in the previous chapter.

Simplification, usually required by computational limitation, led several previous studies and even some technical standards to consider only simple domino effects (Bagster and Pitblado, 1991; Pettitt et al., 1993; Khan and Abbasi, 1998a; Zhang and Chen, 2011) or multilevel domino chains (Reniers and Dullaert, 2007; Abdolhamidzadeh, 2010). More recently, few approaches were developed to provide the possibility to deal with multiple-level simultaneous secondary scenarios (Cozzani et al., 2005; Khakzad et al., 2013).

A further element that needs to be recognized in order to describe a domino accident is the cause of the propagation of the primary scenario. If multiple-level domino chains are considered, the propagation cause also needs to be identified for the further scenarios that cause propagation effects.

An accident scenario usually propagates due to the direct damage of other storage or process units, caused by the physical effects of the primary scenario or, in general, of the initiating scenario (e.g. fire radiation due to fire or overpressure caused by explosions). Seldom, indirect effects are responsible for the propagation (e.g. loss of control of the installation due to control room damage or evacuation in the case of explosions or toxic releases). This issue will be addressed in detail in Chapter 7. However, what needs to be recognized is that a further element required to have a domino effect is the presence of an "escalation vector", which leads to the propagation of the primary scenario.

Table 3.1 Elements Needed for a Domino Accident

Element	Definition
Primary scenario	An accident scenario that starts a domino effect propagating and escalating to other process or storage units, triggering one or several secondary accident scenarios.
Secondary scenario	An accident scenario caused by the impact of an escalation vector generated by a primary accident scenario.
Propagation	In case of a spatial domino effect, the propagation indicates the involvement of other units or equipment items, present at different positions with respect to that of the primary accident. In case of a temporally domino effect, there is propagation within the same unit or equipment item.
Escalation	The intensification of the overall consequences of an undesired event.
Escalation vector	A vector of physical effects generated by the primary accident scenario.

Table 3.1 summarizes the necessary elements of a domino accident and their definition. In the following discussion, for the sake of simplicity, the terms used will be those related to a simple propagation domino effect. Thus, only primary and secondary scenarios will be mentioned. However, all the concepts also apply to any further propagation step of a multilevel domino chain.

3.2   Escalation as a Fundamental Feature of Relevant Domino Accidents

A further specific element of domino accidents that needs to be discussed in detail is the issue of "escalation". Actually, propagation alone may not justify considering a scenario as a "domino accident". It is rather frequent that a severe primary event such as an explosion or a fire involves other units besides the ones where the accident was initiated. However, such secondary effects may be not relevant with respect to the damage caused by the primary event itself.

Escalation is implicitly required in order to consider an accident scenario as a domino accident. In a domino accident, it should be expected that the overall consequences of the domino event are more severe than those of the primary scenario taken alone. If this is not the case, the propagation and the secondary events may have only a forensic or a scientific relevance.

Thus, in domino accidents, propagation is associated with escalation, and secondary scenarios contribute to the overall consequences of the domino event.

3.3   Sources of Domino Accidents and Primary Scenarios

As shown in Chapter 2, past accidents evidence that any equipment item that may cause a loss of containment of hazardous substances, or that may directly release energy due to

Table 3.2 Physical Effects Responsible for Escalation in 100 Domino Accidents (Cozzani et al., 2005)

Primary Scenario	Events	Escalation Vector		
		Heat Radiation	Overpressure	Fragments
VCE	17	0	16	1
Mechanical explosion	17	0	10	7
BLEVE	13	0	0	13
Fireball	1	1	0	0
Jet fire	8	8	0	0
Pool fire	44	44	0	0
Flash fire	0	0	0	0

an internal explosion (confined explosion, boiling liquid expanding vapor explosion (BLEVE), etc.) may potentially be considered as a likely source of a domino accident. The true point concerning the identification of domino "triggers" is in the assessment of the escalation potential of primary accident scenarios. Gathering detailed data on this issue is not simple, since data stored in industrial accident databases have usually scarce details. Table 3.2 shows some data available on the categories of primary scenarios that triggered domino accidents. The data were obtained from the analysis of 100 domino accidents reported in the Major Hazard Incident Data Service (MHIDAS) database (MHIDAS, 2001). As shown in the table, some primary scenarios played a major role in triggering the domino accidents analyzed (pool fires and explosions), while others seem to have a minor escalation potential even if they are, unfortunately, rather frequent accident scenarios (e.g. jet fires, fireballs, etc.). Thus, the escalation potential is related to both the expected frequency of the primary accident scenarios and their characteristics. The limited duration of flash fires and fireballs limits the possibility that these scenarios cause structural damage (Cozzani et al., 2006). The limited severity of many jet fires limits the escalation due to structural damage from these events (Cozzani et al., 2006; Di Padova et al., 2011).

However, turning these observations in to quantitative criteria requires the identification and analysis of all the relevant primary scenarios that may be triggered from the primary unit of interest.

A ranking approach of domino potential sources based on inherent safety thresholds for escalation and on a simplified hazard and consequence analysis was introduced by Cozzani et al. (2009), proposing the calculation and the use of a domino chain potential index for a preliminary ranking of the domino hazard caused by a primary unit independent of the layout and a unit domino hazard index expressing the hazard due to escalation scenarios triggered by a given primary unit.

3.4 Identification and Relative Ranking of Domino Targets and Secondary Scenarios

Secondary scenarios in domino sequences are caused by the damage of one or more than one target unit due to the physical effects (escalation vector) generated by the primary accident. Thus, domino targets are plant items that have the potential, if damaged, to trigger a secondary scenario. However, as stated above, the secondary scenario originated by the damage of the target unit should be sufficiently severe to give an escalation. This leads to identify as relevant potential target units only equipment items that may cause, if damaged, a relevant release of hazardous substances. Coherently, Table 2.4 of Chapter 2 evidences that storage vessels are the plant items more frequently involved as domino targets.

A relative ranking aimed at the identification of the domino targets more exposed to damage by escalation may be based on indexes accounting for the number and distance of units triggering hazardous primary scenarios (Cozzani et al., 2009). A relative ranking of domino targets based on the potential hazard of secondary scenarios may be based on inherent safety indexes, as those proposed, e.g. by Heikkila (1999), Khan and Amyotte (2004), or Tugnoli et al. (2007). Furthermore, Reniers et al. (2012) provide a method to determine the relative position of chemical industrial areas with respect to their systemic risk behavior, thereby calculating a so-called safety and security index (taking domino effects into consideration) and a supply chain index.

3.4.1 Escalation Vectors

Starting from the pioneering work of Khan and Abbasi (1998b), the features of the propagation and escalation of primary accidents to cause a domino effect were systematically approached. Two main patterns were identified for propagation and escalation:

- Direct escalation
- Indirect escalation

Direct escalation is caused by the direct damage of target units due to the effect of radiation, blast waves and fragment projection. Table 3.3 shows the escalation vectors generated by different categories of primary scenarios. As shown in the table, three escalation vectors — often contemporary — have to be considered: heat radiation

Table 3.3 Direct Escalation: Escalation Vectors Generated by Different Categories of Primary Scenarios

Primary Scenario	Escalation Vector
Pool fire	Radiation, fire impingement
Jet fire	Radiation, fire impingement
Fireball	Radiation, fire impingement
Flash fire	Fire impingement
Mechanical explosion	Fragment projection, overpressure
Confined explosion	Fragment projection, overpressure
BLEVE	Fragment projection, overpressure
VCE	Overpressure, fire impingement
Toxic release	–

BLEVE: boiling liquid expanding vapor explosion; VCE: vapor cloud explosion.

and/or fire impingement, overpressure, and fragment projection. As discussed in the following sections, standard or advanced consequence analysis models may be used to assess the intensity of the escalation vector.

Indirect escalation scenarios may be triggered by the loss of control of units or plant sections due to the effect of the primary scenario. For example, the damage of a control room by a blast wave or the flee of untrained operators due to a toxic dispersion or a fire may lead to the loss of control of a process. Khan and Abbasi (2001) documented some examples of potential domino accidents caused by indirect escalation. Such accidents are more likely to take place if the primary event impacts on a nearby plant, run by a different company, where different types of accident scenarios are expected (e.g. impact of a toxic cloud on a plant where only flammable substances are present or impact of a blast wave on a plant where only a toxic hazard is present). Actually, in the absence of cluster safety management, or at least of an effective exchange of information among the site safety managers, the operators of the target plant may not be prepared to face the consequences of a primary scenario originated outside the premises of their own plant. However, even in this case, it should be remarked that the automatic or manual activation of widely adopted mitigation barriers, such as emergency shutdown systems, should be able to prevent loss of control leading to indirect propagation. Thus, the likelihood of indirect escalation should be assessed taking into account, among others, the plant's emergency management systems and the levels of information exchange between the companies present in the industrial area under consideration (Reniers and Dullaert, 2008; Reniers et al., 2009).

Direct escalation is by far the more likely and documented mechanism leading to domino accidents, as evidenced from past accident analyzes, discussed in Chapter 2. The credibility of indirect escalation is limited to very specific situations, mostly related to industrial clusters where several companies are operating (Reniers et al., 2009).

3.5 Domino Accident Definition

Due to the complex features that a domino incident may have, it is difficult to unambiguously define a "domino accident". It should be remarked that defining what should be considered as a domino accident is not a mere academic exercise, since several technical standards and the legislation require specifically to assess "domino effects"; e.g. Article 9 of the European Directive 2012/18/EU requires to identify establishments that may be affected by domino scenarios and to include such scenarios in safety reports and accident prevention policies (see also Chapter 1).

Table 3.4 reports several proposed definitions for a "domino effect" that may be applied. Although all the definitions in the table are valid and have a scientific and technical basis, in the present approach the following definition will be applied:

> *An accident in which a primary unwanted event propagates within an equipment ("temporally"), or/and to nearby equipment ("spatially"), sequentially or simultaneously, triggering one or more secondary unwanted events, in turn possibly triggering further (higher order) unwanted events, resulting in overall consequences more severe than those of the primary event.*

Table 3.4 Definitions Given for a "Domino Effect" or a "Domino Accident"

Author(s)	Domino Effect/Accident Definition
Third report of the Advisory Committee on Major Hazards (HSE, 1984)	The effects of major accidents on other plants on the site or nearby sites.
Bagster and Pitblado (1991)	A loss of containment of a plant item which results from a major incident on a nearby plant unit.
Lees (1996)	An event at one unit that causes a further event at another unit.
Khan and Abbasi (1998b)	A chain of accidents or situations when a fire/explosion/missile/toxic load generated by an accident in one unit in an industry causes secondary and higher order accidents in other units.
Delvosalle (1998)	A cascade of accidents (domino events) in which the consequences of a previous accident are increased by the following one(s), spatially as well as temporally, leading to a major accident.
Uijt de Haag and Ale (1999)	The effect that loss of containment of one installation leads to loss of containment of other installations.
CCPS (2000)	An accident which starts in one item and may affect nearby items by thermal, blast or fragment impact.
Vallee et al. (2002)	An accidental phenomenon affecting one or more installations in an establishment which can cause an accidental phenomenon in an adjacent establishment, leading to a general increase in consequences.
Council Directive 2003/105/EC (2003)	A loss of containment in a Seveso installation which is the result (directly and indirectly) of a loss of containment at a nearby Seveso installation. The two events should happen simultaneously or in very fast subsequent order, and the domino hazards should be larger than those of the initial event.
Post et al. (2003)	A major accident in a so-called "exposed company" as a result of a major accident in a so-called "causing company". A domino effect is a subsequent event happening as a consequence of a domino accident.
Lees (2005)	A factor to take account of the hazard that can occur if leakage of a hazardous material can lead to the escalation of the incident, e.g. a small leak which catches fire and damages by flame impingement a larger pipe or vessel with subsequent spillage of a large inventory of hazardous material.
Cozzani et al. (2006)	Accidental sequences having at least three common features: (1) a primary accidental scenario, which initiates the domino accidental sequence; (2) the propagation of the primary event, due to "an escalation vector" generated

Table 3.4 Definitions Given for a "Domino Effect" or a "Domino Accident"—*cont'd*

Author(s)	Domino Effect/Accident Definition
	by the physical effects of the primary scenario, that results in the damage of at least one secondary equipment item; and (3) one or more secondary events (i.e. fire, explosion and toxic dispersion), involving the damaged equipment items (the number of secondary events is usually the same as the damaged plant items).
Bozzolan and Messias de Oliveira Neto (2007)	An accident in which a primary event occurring in primary equipment propagates to nearby equipment, triggering one or more secondary events with severe consequences for industrial plants.
Gorrens et al. (2009)	A major accident in a so-called secondary installation which is caused by failure of a so-called external hazards source.
Antonioni et al. (2009)	The propagation of a primary accidental event to nearby units, causing their damage and further "secondary" accidental events, resulting in an overall scenario more severe than the primary event that triggered the escalation.

Source: Adapted from Reniers (2010) and Abdolhamidzadeh et al. (2011).

The definitions in Table 3.1 apply to the key terms in the above sentence, and the discussion in Sections 3.1–3.5 should be well in mind when identifying relevant domino scenarios. In particular, the key element of escalation is crucial to understand the relevance of a specific domino scenario.

3.6 Categorization of Domino Accidents

Several different types of domino accidents comply to the definition given in Section 3.5. Thus, it may be useful to introduce specific categories of domino accidents, in order to achieve a more detailed identification of possible domino scenarios.

If the recorded domino accidents are read in the light of the required elements of domino scenarios listed in Table 3.1, two quite different types of escalation may be identified:

1. Escalation of low-severity initiating events
2. Interaction of different "critical events"

These two categories of escalation may be specific to a single scenario, but may also take place at different time steps of a single accident. An example is given by an accident that took place in an Italian plant for ethylene and propylene production in 1985. The accident was initiated by the rupture of a small-diameter (2 in.) ethylene pipe caused by vibrations due to the toggling of a safety valve. The minor jet fire that followed impinged a 600-mm pipe. A full-bore rupture of the pipe took place, starting a major jet fire fed by the high-pressure release of C_2–C_3 hydrocarbons in the pipe. The jet impinged a vertical pressurized propane storage tank. A BLEVE took place after a few minutes, followed by the damage and release from three other pressurized storage vessels present in the tank park. The plant was almost completely destroyed.

In this accident, both types of escalation took place. A low-severity initiating event (a "minor" jet fire) was responsible for the first escalation, triggering the major jet fire. The second escalation (the "major" jet fire causing a BLEVE) and the other simultaneous escalations (BLEVE causing the damage of the other three pressurized tanks) clearly represent an interaction of different critical events (more properly, a first critical event that triggered the other scenarios).

Identifying these two different modes of escalation is of fundamental importance for the identification of possible domino effects.

In the first type of escalation, the low severity of the initiating event has two possible effects. One possible effect is that the potential for escalation may be overlooked, and some critical scenarios may thus not be considered in the risk assessment. In the above example, it should be considered that a 2-in. pipe is not likely to be classified as a "major accident" if only primary consequences are considered. Moreover, the full-bore rupture of a 600-mm pipe is usually not considered a credible event. The other effect is that the low severity of the event limits its potential for propagation, thus the potential for escalation only affects the unit where the primary event may take place.

In the second type of escalation, the severity of the primary event is high, thus the propagation in space is the main factor causing the escalation. Only for this type of escalation the propagation of the primary event to nearby units or outside the plant boundaries seems credible.

A classification of domino events into the various types that may occur can be further elaborated. By doing so, it is possible to unambiguously identify the character of the domino event under consideration. The various types of domino event are explained in Table 3.5 using four different parameters.

From the definitions of a direct or indirect domino event, it is not possible to deduce how many domino events have happened before the event under consideration. For this purpose, the concept of *domino cardinality* was introduced by Reniers (2010). This is a term used to indicate the domino event link number in a sequence of domino events, starting from the initiating event, with domino cardinality "0".

These categorizing definitions may be illustrated considering the hypothetical domino effects illustrated in Figure 3.2.

In case of simple propagation, the domino event can be categorized as internal, spatial, serial, and with cardinality 0. The different exemplary domino events of the multipropagation case can then be categorized as in Table 3.6.

A different representation of multilevel domino effect is provided in Figure 3.3, which reports a "domino event tree".

Domino events characterized with cardinality 0 are the initiating domino events or the so-called "primary domino events", whereas cardinality 1 refers to secondary domino events, cardinality 2 to tertiary domino events, etc. A multiplant domino effect represents an escalating accident involving more than one chemical plant.

Table 3.5 Categorization of Domino Events (Reniers, 2010)

Categorization of Domino Events

Type Number	Instances of Type	Definition of Type
Type 1	Internal	Begin and end of the escalation vector characterizing the domino event are situated inside the boundaries of the same chemical plant.
	External	Begin and end of the escalation vector characterizing the domino event are *not* situated inside the boundaries of the same chemical plant.
Type 2	Direct	The domino event happens as a direct consequence of the previous domino event.
	Indirect	The domino event happens as an indirect consequence of a preceding domino event, not being the previous one.
Type 3	Temporal	The domino event happens within the same area as the preceding event, but with a delay.
	Spatial	The domino event happens outside the area where the preceding event took place.
Type 4	Serial	The domino event happens as a consequent link of the only accident chain caused by the preceding event.
	Parallel	The domino event happens as one of several simultaneous consequent links of accident chains caused by the preceding event.

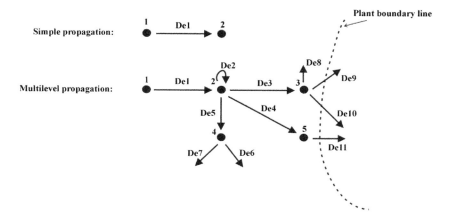

Figure 3.2 Explaining domino events' ("De") categorization.

Table 3.6 Type of the Multilevel Exemplary Domino Effect Depicted in Figure 3.2

	Type	De1	De2	De3	De4	De5	De6	De7	De8	De9	De10	De11
1	Internal	✔	✔	✔	✔	✔	✔	✔	✔			
	External									✔	✔	✔
2	Cardinality	0	1	1	1	1	2	2	2	2	2	2
3	Temporal		✔									
	Spatial	✔		✔	✔	✔	✔	✔	✔	✔	✔	✔
4	Serial	✔										✔
	Parallel		P1	P1	P1	P1	P2	P2	P3	P3	P3	

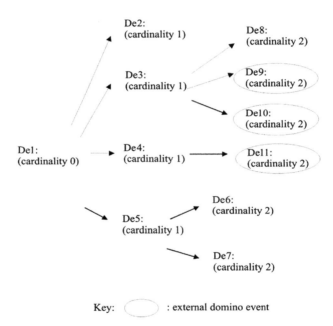

Figure 3.3 Domino event tree of the multilevel domino effect of Figure 3.2.

Hence, according to the categorization of domino events (Table 3.5), a multiplant accident can be considered to be a domino effect typified by external and spatial characteristics.

3.7 Conclusions

This chapter describes the essential elements constituting a domino accident and explains these elements. It is argued that any domino accident results from an initiating unwanted event which "propagates" and escalates into one or more larger scale event(s), with more disastrous consequences than the original ("primary") event. A generic and unambiguous definition for "domino effect" is provided, taking into consideration a myriad of definitions for the term that have been suggested since 1984. An approach is elaborated and discussed to allow drafting a comprehensive and full-scale categorization of events that may constitute a domino accident.

References

Abdolhamidzadeh, B., Abbasi, T., Rashtchian, D., Abbasi, S.A., 2010. A new method for assessing domino effect in chemical process industry. Journal of Hazardous Materials 182, 416–426.

Abdolhamidzadeh, B., Abbasi, T., Rashtchian, D., Abbasi, S.A., 2011. Domino effect in process-industry accidents – an inventory of past events and identification of some patterns. Journal of Loss Prevention in the Process Industries 24, 575–593.

Antonioni, G., Spadoni, G., Cozzani, V., 2009. Application of domino effect quantitative risk assessment to an extended industrial area. Journal of Loss Prevention in the Process Industries 22, 614–624.

Bagster, D.F., Pitblado, R.M., 1991. The estimation of domino incident frequencies: an approach. Process Safety Environment 69, 196.

Bozzolan, J.-C., Messias de Oliveira Neto, J., 2007. A Study on Domino Effects in Nuclear Fuel Cycle Facilities, International Nuclear Atlantic Conference – INAC 2007, Santos, Brazil.

CCPS, Center for Chemical Process Safety, 2000. Evaluating Process Safety in the Chemical Industry: A User's Guide to Quantitative Risk Analysis. American Institute of Chemical Engineers, New York (USA).

Council Directive 2003/105/EC, 2003. Seveso II Directive on the control of major-accident hazards involving dangerous substances with amendments. Official Journal of the European Union L345, 97–105.

Cozzani, V., Gubinelli, G., Antonioni, G., Spadoni, G., Zanelli, S., 2005. The assessment of risk caused by domino effect in quantitative area risk analysis. Journal of Hazardous Materials 127, 14–30.

Cozzani, V., Gubinelli, G., Salzano, E., 2006. Escalation thresholds in the assessment of domino accidental events. Journal of Hazardous Materials 129, 1–21.

Cozzani, V., Tugnoli, A., Salzano, E., 2009. The development of an inherent safety approach to the prevention of domino accidents. Accident Analysis and Prevention 41, 1216–1227.

Delvosalle, C., 1998. A Methodology for the Identification and Evaluation of Domino Effects. Rep. CRC/MT/003. Belgian Ministry of Employment and Labour, Bruxelles (B).

Di Padova, A., Tugnoli, A., Cozzani, V., Barbaresi, T., Tallone, F., 2011. Identification of fireproofing zones in oil & gas facilities by a risk-based procedure. Journal of Hazardous Materials 191, 83–93.

Gorrens, B., De Clerck, W., De Jongh, K., Aerts, M., 2009. Domino effecten van en naar Seveso-inrichtingen. Rep. 07.0007. Flemish Ministry of Environment, Nature and Energy, Brussels (Belgium).

Heikkilä, A., 1999. Inherent Safety in Process Plant Design. PhD thesis. VTT Publications n384, Espoo (SF).

HSE (Health and Safety Executive), 1984. The Control of Major Hazards, Third Report of the HSC Advisory Committee on Major Hazards. HMSO, London (UK).

Khakzad, N., Khan, F.I., Amyotte, P., Cozzani, V., 2013. Domino effect analysis using Bayesian networks. *Risk Analysis* 33, 292–306.

Khan, F.I., Abbasi, S.A., 1998a. DOMIFFECT (DOMIno eFFECT): user-friendly software for domino effect analysis. Environmental Modelling and Software 13, 163–177.

Khan, F.I., Abbasi, S.A., 1998b. Models for domino effect analysis in chemical process industries. Process Safety Progress 17, 107.

Khan, F.I., Abbasi, S.A., 2001. An assessment of the likelihood of occurrence, and the damage potential of domino effect (chain of accidents) in a typical cluster of industries. Journal of Loss Prevention in the Process Industries 14, 283–306.

Khan, F.I., Amyotte, P., 2004. Integrated inherent safety index (I2SI): a tool for inherent safety evaluation. Process Safety Progress 23, 136–148.

Lees, F.P., 1996. Loss Prevention in the Process Industries, second ed. Butterworth-Heinemann, Oxford (UK).

Lees, F.P., 2005. In: Mannan, S. (Ed.), Loss Prevention in the Process Industries, third ed. Butterworth-Heinemann, Oxford (UK).

MHIDAS, Major Hazard Incident Data Service, 2001. AEA Technology, Major Hazards Assessment Unit. Health and Safety Executive, London (UK).

Pettitt, G.N., Schumacher, R.R., Seeley, L.A., 1993. Evaluating the probability of major hazardous incidents as a result of escalation events. Journal of Loss Prevention in the Process Industries 6, 37.

Pietersen, C.M., 1988. Analysis of the LPG-disaster in Mexico city. Journal of Hazardous Materials 20, 85–107.

Post, J.G., Bottelberghs, P.H., Vijgen, L.J., Matthijsen, A.J.C.M., 2003. Instrument Domino Effecten. RIVM, Bilthoven (The Netherlands).

Reniers, G.L.L., 2010. An external domino effects investment approach to improve cross-plant safety within chemical clusters. Journal of Hazardous Materials 177, 167–174.

Reniers, G.L.L., Dullaert, W., 2007. DomPrevPlanning©: user-friendly software for planning domino effects prevention. Safety Science 45, 1060–1081.

Reniers, G.L.L., Dullaert, W., 2008. Knock-on accident prevention in a chemical cluster. Expert Systems with Applications 34 (1), 42–49.

Reniers, G.L.L., Dullaert, W., Karel, S., 2009. Domino effects within a chemical cluster: a game-theoretical modeling approach by using Nash-equilibrium. Journal of Hazardous Materials 167, 289–293.

Reniers, G., Sörensen, K., Dullaert, W., 2012. A multi-attribute systemic risk index for comparing and prioritizing chemical industrial areas. Journal of Reliability Engineering and System Safety 98 (1), 35–42.

Tugnoli, A., Cozzani, V., Landucci, G., 2007. A consequence based approach to the quantitative assessment of inherent safety. AIChE Journal 53, 3171–3182.

Uijt de Haag, P.A.M., Ale, B.J.M., 1999. Guidelines for Quantitative Risk Assessment (Purple Book). Committee for the Prevention of Disasters, The Hague (NL).

Vallee, A., Bernuchon, E., Hourtolou, D., 2002. MICADO: Méthode pour l'identification et la caractérisation des effets dominos, Rep. INERIS-DRA-2002–25472, Direction des Risques Accidentels, Paris (France).

Zhang, X.M., Chen, G.H., 2011. Modeling and algorithm of domino effect in chemical industrial parks using discrete isolated island method. Safety Science 49, 463–467.

4 Overpressure Effects

Ernesto Salzano*, Pol Hoorelbeke[†], Faisal Khan[‡],
Paul Amyotte[§]

[*] Istituto di Ricerche sulla Combustione, Consiglio Nazionale delle Ricerche
(CNR), Napoli, Italy, [†] Total Refining & Chemicals, Vice President Safety
Division, Visiting Professor South Chine University of Technology, China,
[‡] Safety and Risk Engineering Research Group, Faculty of Engineering and
Applied Science, Memorial University of Newfoundland, St. John's,
Newfoundland, Canada, [§] Department of Process Engineering and Applied
Science, Dalhousie University, Halifax, Nova Scotia, Canada

4.1 Introduction

An explosion may be defined as the rapid release of energy in the atmosphere with the generation of pressure waves causing damage to people and properties in the physical domain. Typical major industrial accidents in the process industries are vapor cloud explosions (e.g. Flixborough, 1974; Beek, 1975; BP Texas City, 2005; Buncefield, 2005), Boiling Liquid Expanding Vapor Explosions (BLEVEs) (e.g. Feyzin, 1966) and solid explosions (e.g. the ammonium nitrate explosion in Toulouse, 2001, or the explosion of 2,4,6-trinitrotoluene (TNT) charge).

A common characteristic of these accidents is the generation of an incidental pressure wave which moves radially away from the center of the explosion and impacts the structure and equipment that it encounters. The initial wave (pressure and impulse) is then reflected and can result in failure of the structure or equipment.

It is generally accepted that "domino effects" (i.e. the consequences of the failure of the structure or equipment are worse than the consequences of the initial explosion) have been very rare in industrial accidents because of the scale of the initial explosion.

Domino Effects in the Process Industries. http://dx.doi.org/10.1016/B978-0-444-54323-3.00004-X

However, some authors claim that about 30–50% of the industrial accidents involving domino effects can be addressed to such phenomenon (Darbra et al., 2010; Cozzani and Salzano, 2004a). Domino effects during plant design are nevertheless intensely studied. The following two approaches are commonly proposed:

- The comprehensive approach in which for each accidental explosion scenario the effects (pressure, impulse) on the surrounding structures and equipment are calculated and the potential damage for these structures and equipment are assessed.
- The rational approach in which it is checked whether the equipment causing the maximum consequence can be impacted ("fail") by an explosion effect.

In both approaches it is necessary to establish the blast load parameters of the possible explosion and to calculate the (dynamic) response of the structure or equipment to the load.

The following sections give an overview of the main issues of domino effects. The first is related to the characterization of the pressure wave propagating from the source point, which depends on the physics of the explosion phenomenon and on the total energy exerted by the source. The second is related to the interaction of the pressure wave with the target equipment, which is clearly dependent on the design of equipment, i.e. the resilience of industrial structure with respect to pressure loading. The third section addresses simplified models for the evaluation of the intensity and modality of the loss of containment (LOC) (hazardous materials) from the target system, which may lead to secondary catastrophic events such as fire, explosion and toxic dispersion, or their combination.

4.2 Overview of the Basics

The most significant feature of an explosion is the sudden release of energy to the atmosphere, which results in a blast wave. Figure 4.1 shows an example of two blast waves measured inside a 50-m^3 module. The pressure rises to a maximum over-pressure (called peak free field, peak side-on or peak incident overpressure) and then

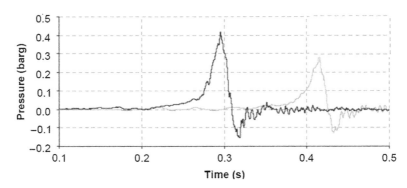

Figure 4.1 Typical blast waves measured in a 50-m^3 module. (For color version of this figure, the reader is referred to the online version of this book.)

returns gradually to ambient conditions with some highly damped pressure oscillations (positive and negative) around ambient pressure.

Commonly, two types of blast waves are distinguished:

- A shock wave that has a sudden, instantaneous rise in pressure above ambient conditions to the peak side-on overpressure followed by gradual pressure decay and a negative phase.
- A pressure wave that has a gradual pressure rise to the peak side-on overpressure followed by a gradual pressure decay and a negative phase similar to that for a shock wave.

Shock waves in the near field usually result from very energetic explosions. Vapor cloud deflagrations will give rise to pressure waves in the near field which may "shock-up" to a shock wave in the far field.

The negative phase of the blast loading is usually ignored in blast-resistant design in the hydrocarbon industry. The characterization and treatment of this loading can be studied with the methods described in UFC 3-340-02 (DoD, 2008).

The blast-induced loads on equipment and structures in the environment include

- Direct blast effects
  - Static overpressure
  - Dynamic (blast wind) pressure
- Indirect induced stress
  - Stress resulting from differential motion as a result of equipment being attached. A typical example is induced loads on piping due to the momentum induced by the motion of large equipments attached to the piping.
  - Shock and vibration loads transmitted through its mounting inside a structure.

4.2.1 Consequences of Blast Loads on Equipment

The consequences (damage pattern) of the blast load effects on a structure or equipment will depend not only on the characteristics of the blast load but also on the receptor characteristics such as its mass, its natural period and its dynamic resistance. Equipment may be sensitive to static overpressure, dynamic pressure or a combination of both. Deformation of outer surfaces of equipment can result from the variation of blast loading over the surface. Other types of equipment may be susceptible to rigid body motion. The response of small-diameter pipes will be primarily due to drag loadings (because the clearing time for the reflected pressure is very short, typically of the order of $5.9 \ 10^{-2} \times D$ ms (D = diameter in mm)). The response of larger pipes (>600 mm) will come more from the static overpressure.

The typical behavior of industrial equipment to pressure wave is sketched in Table 4.1.

4.2.2 Overall Approach of a Response Analysis

The overall approach to response modeling will be as follows:

- Step 1: Study the possible explosion scenarios
- Step 2: Perform explosion effect calculation for each of the scenarios
- Step 3: Determine for each receptor the blast load for each of the scenarios
- Step 4: Calculate the response of the receptor as function of the blast load

Table 4.1 The Behavior of Industrial Equipment to Pressure Waves

Equipment	Typical Failure Mode
Valves and controls	The most probable failure of a valve–actuator assembly is the loss of air line or other control connections due to relative motions and vibrations. Damage due to direct blast loading (static or dynamic overpressure) is unlikely because the potential load area is small and the actuator yokes as well as the valve are typically robust. Adequate protection is achieved by providing enough flexibility.
Piping and pipe racks	Typical failure mechanisms for piping on pipe racks are displacement of the support due to dynamic pressure or flange connection leakage. Most bolted connections will be adequate to ensure that pipes remains on the support. Shoe-type pip supports, however, rely on the weight of the pipe to keep it in the saddle. The equivalent static blast overpressure required to overcome the weight of a filled pipe and displace it from shoes of this type can be estimated with $0.06 \times D + 29$ (D in mm and overpressure in kPa). The allowable overpressure for pipe racks will depend on the design wind speed of the rack.
Flange connections	The typical mechanism for failure of flange connections is bending stresses at the flange gasket due to pipe deformations (indirect induced stress) induced by the bending moment created by the dynamic pressure. The response of a flange is very typical and has to be calculated on a case-by-case basis. A very important element is the support spacing. In a confidential study, it was calculated that for a 150-lb flange on a 3-in. pipe with a supporting spacing of 4 m, the required equivalent static overpressure for failure was at least 200 kPa.
Pressure vessels	The most common damage to vertical pressure vessels will be the yielding or pullout of anchor bolts. In case of high overpressure, the buckling of the skirt at the base of the vessel may occur. For horizontal pressure vessels and heat exchangers, the primary failure mechanism is failure of the supports.
Buildings and atmospheric storage equipment	The primary failure mechanism of buildings is static overpressure (and reflected pressure) in combination with the dynamic pressure.

4.2.2.1 Study of the Possible Explosion Scenarios

The most common type of explosion scenarios in the hydrocarbon industry are

- Decomposition of pure gas (e.g. acetylene can deflagrate or detonate with the oxygen in its molecule)
- Deflagration or detonation of a fuel–air mixture in unconfined or partially confined environment

- Deflagration and detonation of a fuel–air mixture in an enclosure
- Thermal explosion (runaway reaction)
- Physical explosions such as those resulting from a sudden failure of an equipment under pressure, steamtype explosion, BLEVE, and rapid phase transition
- Solid-phase explosion.

In the list, fuel can indicate vapor, gas, or dust. Details on these explosion phenomena are outside the scope of this contribution and can be found in many textbooks and recent scientific publications (Bartknecht, 1981; Baker et al., 1983; Crowl, 2003; Lees, 1996; Eckhoff, 2005; DoD, 2008; Abbasi et al., 2010; CCPS, 2010).

When occurring in unconfined environment, whatever the explosion phenomenon given above, the time profile of the generated pressure wave in the close vicinity of explosion source may be very complex and not amenable to standard shapes. However, as the wave moves outward, the effects of the nature of the explosion declines and the wave establishes a profile, which is common to all types of explosion. Hence, an ideal pressure wave may be defined and characterized – at any location in the far field – by the maximum pressure observed on a pressure–time plot reproducing the freely propagating (undisturbed) pressure wave (P_{max}); by the impulse (I_{exp}), i.e. the area under the pressure–time curve; by the total duration of pressure wave (t_{exp}); and by the drag pressure (P_d), which depends on dynamic pressure and represents the pressure correlated to the kinetic energy of fluid particles when loading an object with a given shape.

The static peak pressure is often defined as static pressure (P_k) or even side-on overpressure, if relative to the atmospheric pressure P°. Positive phase values for impulse and duration are also generally considered as they refer to the rising phase of pressure history only.

There is a lot of uncertainty involved in the selection of the most relevant scenarios and expertise will be needed in all cases. Consider as an example vapor cloud explosions (VCEs). The acronym VCE stands for a group of accidents that have in common that a large cloud, accidentally formed, explodes. The severity of the VCE depends largely on the rate at which the energy is released, which in turn depends on the propagation mechanism in the cloud. Two well-known propagation mechanisms are deflagration (from slow deflagration to very fast deflagration) and detonation (achieved by direct induced detonation or via a deflagration to detonation transition). Extensive research after an accident is not always conclusive (see e.g. HSE, 2009; for the Buncefield explosion).

When the explosions occur in a confined or partially confined enclosure, deflagration waves are very likely to cause secondary fire and explosions. However, it is apparently difficult to understand how a primary explosion can trigger domino effects. For the specific case of dust explosion, e.g. the pressure wave of a primary explosion, whether dust explosion or gas explosion, can stir and loft other settled dust, making an airborne dust cloud which can cause more destructive secondary fires and explosions (Amyotte and Eckhoff, 2010). In this regard, examples of primary nondust explosions which caused secondary dust explosions are the Ford River Rouge accident in 1999 (coal dust explosion) and the accidents at West Pharmaceutical Services in 2003 (polyethylene dust explosion) and CTA Acoustics, Inc in 2003 (phenolic resin dust explosion) (CSB, 2006). The Imperial Sugar Company explosion in 2008 (sugar dust explosion) was a primary–secondary dust explosion (CSB, 2009).

Dust explosions are more difficult than other types of explosions to model since many influential factors such as particle size, specific surface area and gravity effects are to be considered. However, this complexity should not prevent considering the significant role of dust explosion in the domino effect analysis. In 2002, a minor dust explosion in a bagging bin at Rouse Polymerics International, Inc, Vicksburg, Missouri, USA, caused a secondary atmospheric dust explosion followed by other fires and explosions through the plant, causing five deaths and seven injuries (Khakzad et al., 2012).

4.2.2.2   Explosion Effect Calculations

Few methods are available to calculate the effects for each of the explosion phenomena cited above. For point-source phenomena, as, e.g. solid explosions or BLEVE, TNT-like models are typically adopted. When more complex explosions must be analyzed, such as VCE, the Multi-Energy Model and the Baker–Strehlow method (see CCPS, 2010 for details) are adopted worldwide for the evaluation of peak pressure and positive impulse in the near and far field. More recently, the use of computational fluid dynamics (CFD) specifically developed for explosion analysis is increasing, particularly for the design of offshore units or complex industrial sites where confined, unconfined or partially confined gas explosion or VCEs are likely. The industrial version of CFD relies on the solution of Reynolds-averaged Navier–Stokes (RANS) equations. Several commercial CFD codes, and consultant firms, are available in the market. However, skilled operators are needed for obtaining physically acceptable results. The ability of RANS-based CFD model to reproduce the explosion phenomenon is limited by the adoption of submodels for the mathematical representation of turbulence and combustion: most commercial CFD codes are based on simplified laminar combustion models and on the derivation of empirical correlations for the description of turbulent combustion, such as eddy-dissipation or derived models and Bray or Gülder correlations for turbulent burning velocity, which have been proved to work in the scientific literature only in the laboratory scale (see Lea and Ledin, 2002; Bjerketvedt et al., 1993 for further details). The development of CFD codes based on large eddy simulation could be the future for numerical simulation of explosion phenomenon.

These methods can give significantly different results in the near field but quite similar results in the far field. This behavior can be easily observed in Figure 4.2, which gives the calculated pressure–distance relation for the same scenario studied by using different approaches with different levels of details (Hoorelbeke et al., 2005), starting from the release source calculated by simplified empirical methodologies to commercial codes such as Phast by Det Norske Veritas and a specialized CFD code such as FLACS® by Gexcon AS, Norway.

4.2.2.3   Definition of the Blast Load for Each Receptor

For each receptor that one wants to study, a well-defined blast load needs to be defined. Explosion blast loads can have very transient behavior. The blast load will at

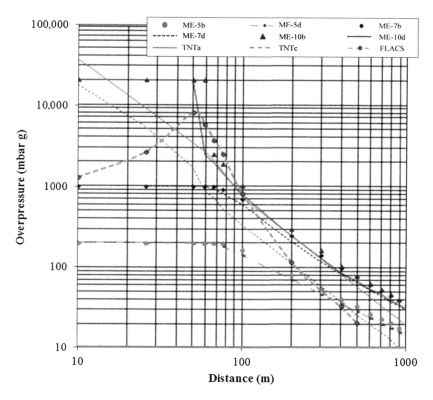

Figure 4.2 Overpressure vs. distance from explosion source calculated by different methodologies. (For color version of this figure, the reader is referred to the online version of this book.)

least include the side-on overpressure curve, the reflected pressure curve, the impulse curve and the drag–pressure curve.

4.2.2.4 Calculation of the Response of the Receptor as a Result of the Explosion Load

The last step is to calculate the behavior of the receptor when it is submitted to the calculated explosion load. This analysis should provide (ASCE, 1997) the following:

• Maximum relative deflections of each structural element
• Relative rotation angles at plastic hinge locations
• Dynamic reactions transmitted to the supporting elements
• Deflections and reactions due to rebound

Figure 4.3 shows an example from a response study. The figure reports the calculated displacements of the top of a propylene loop reactor due to the explosion load of a VCE, carried out with multi–degree of freedom (MDOF) systems (Hoorelbeke et al., 2005). The response (natural) period of the structure is about 1.4 s, which is

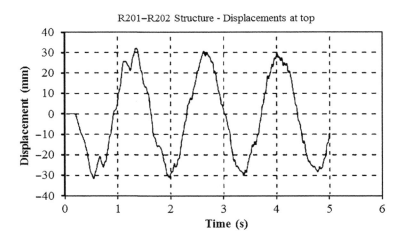

Figure 4.3 Calculated displacements of the top of a propylene loop reactor due to the explosion load of a VCE.

larger than the pressure load period (which is in the order of 150 ms), and hence the response will be in the impulsive regime (see also further section).

These deflections can be translated in, for instance, peak bending moments in the equipment and hence prediction can be made about elastic and plastic deformation and/or failure of the equipment.

The methods used for response analyses are equivalent static methods or simple analytical methods (see, for instance HSE, 1992), Single Degree Of Freedom (SDOF) systems (see, for instance ASCE, 1997) or Multi Degree Of Freedom (MDOF) (see, for instance Hoorelbeke, et al., 2005). SDOF, which is the structural modeling of uncoupled systems, neglects the deformation compatibility and equilibrium forces at contact points between the different elements of a structure or equipment. Details of these methods can be found in classical textbooks such as Biggs (1964).

In the past years, computational codes based on Finite Element Analysis (FEA) have been commonly used for detailed structural analyses. The next section shows a practical case of nonlinear dynamic response analysis, using FEA, for an existing installation. Some insights of the methodology are given.

4.3   Nonlinear Dynamic Response Analysis

The aerial view reported in Figure 4.4 shows a polypropylene unit (PP) and a poly-ethylene unit (PE) in a petrochemical complex. Between the two units there is a grass root land. The problem was whether this land could be used in the future for a new PP taking into account the potentiality of a domino effect in case of a major accident. The old PP would be located 60 m away from the new unit, while the PE would be located 55 m away from the new unit.

A major leak scenario in the existing PP unit was first analyzed. The calculated cloud engulfing the equipment of the existing unit would be of about 25,000 m^3.

Polypropylene unit (60 m)

Polyethylene unit (55 m)

Figure 4.4 The layout of the plant analyzed in the case study. (For color version of this figure, the reader is referred to the online version of this book.)

Calculations were performed with Multi-Energy and TNT equivalent method. A PP unit comprises several zones of high confinement and high congestion and it is reasonable to assume that flame acceleration up to fast deflagration would occur.

For the analyzed installation, the dispersion and explosion calculations were performed with the FLACS model taking into account the prevailing meteorological conditions and possible ignition sources. The FLACS model allowed to estimate all pressure load characteristics (overpressure, drag pressure, etc.) at all different locations in the new unit.

Hence, dynamic, nonlinear, FEAs were carried out using the LS-DYNA (Livermore, 2003 and CEAP-DYNA (CEAP-DYNA, 2004)) codes with postprocessing in the Oasys LS-DYNA Environment (Oasys, 2003). The blast loads information was prepared and delivered by TOTAL and Gexcon. The dynamic response analysis work was done by Ove Arup & Partners Ltd (Arup).

For the case reported here, a drum (D302) and the loop reactors (R201–R202) were chosen for the analysis as target equipment (Figure 4.5).

The primary structural elements of the R201–R202 and I201 structures were modeled using beam elements. The D202 tank and the D302 tank (with its associated support structure) were modeled using shell elements. A single layer of shells was used, representing the main tank structure. The insulation and the outer skin were only considered in terms of increased mass and surface area for blast loading. The contents of the reactor pipes and tanks were represented only as increased mass.

Due to the nature of the analyses, geometric nonlinearity (i.e. large displacements) is intrinsically included in the calculations. Most of the structural elements were modeled using elastic properties. However, where structural elements approached their elastic capacity (close to yielding), nonlinear properties (allowing the sections to yield) were introduced into the sections concerned.

Gravity loading was included in both the models. The loading was applied in a staged analysis where gravity was initially applied to the structures prior to the application of the blast load.

Figure 4.5 The target equipment analyzed in the case study. (For color version of this figure, the reader is referred to the online version of this book.)

The main tank structure with the steel saddles and the reinforced concrete base were modeled, as shown in Figure 4.6.

The reinforced concrete supports were modeled down to the foundation. The bolted connection details between the concrete supports and the steel saddle structures were not modeled explicitly (i.e. the saddles were fully connected to the tops of the concrete supports).

The steel plate thicknesses were 38 mm for the tank barrel and 45 mm for the ends. The components forming the saddle structure were generally 16 mm thick. The material used for all the primary steel components modeled was SA 516-Gr.70 carbon steel.

The R201–R202 reactor pipes consist of two concentric pipes: the inner (reactor) pipe carrying the polypropylene and the outer (jacket) primary pipe carrying the water. These pipes are fully connected at the top and bottom of the vertical portions (i.e. at the flange locations above and below the expansion joints) and at the level of the floor on a concrete support structure. In between these levels of full connection, there are guiding blocks, positioned at the same levels as the bracing members. The two reactor pipes were modeled as coincident beam elements. The outer stainless-steel skin covering the insulation was not modeled explicitly. The inner and outer pipes were fully connected together at the flange locations and at the level of the floor of the concrete support structure. The support ring detail was not modeled explicitly and the reactor pipes were considered to be rigidly

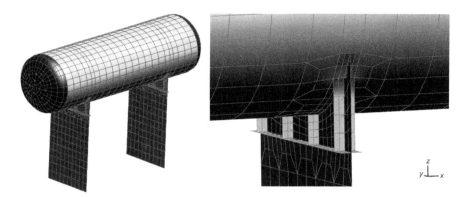

Figure 4.6 Geometrical model of the drum. (For color version of this figure, the reader is referred to the online version of this book.)

connected to the floor of the concrete structure. At the levels of the bracing members, the inner and outer reactor pipes were connected by nonlinear springs. These nonlinear springs effectively allowed ±5 mm relative horizontal movement between the pipes without resistance, representing the movement allowed by the guiding blocks. The bracing members were modeled as fully fixed to the outer pipes. Details toward the ends of the reactor pipes were only modeled in terms of the upper expansion joint.

Displacements, relative displacements and the peak bending moments in the equipment and structures were calculated. Figure 4.7 gives an example of the peak bending moments (in kNm) at the base of the reactor column.

Potential domino effect was evaluated and, when necessary, additional measures to make the structure resistant were proposed. The study was performed in 2004. The total (external) cost for the study (FLACS calculations, determination of the blast loadings, and the dynamic response study) was of the order of 150 k€.

The additional cost for implementing measures to make the structure blast resistant was estimated by the engineering department of the site. The Total cost of a new unit would be increased by a maximum of 2.5%.

4.4 Domino Effects: Simplified Analysis

Safety standards in the chemical and process industry require that a detailed assessment of safety is carried out on any operating plant, starting from the design phase up to the end of the plant life cycle. Quantitative risk assessment is usually the end point of safety assessment, in particular in the Oil and Gas sector. Thus, quantitative risk analysis (QRA) is routinely carried out on complex plants or even on extended industrial areas. Thus, a QRA generally requires simplifying assumptions both for the time evolution of the accidental phenomena and the consequence evaluation. Hence, the use of even simple "structural dynamics" codes and methodologies

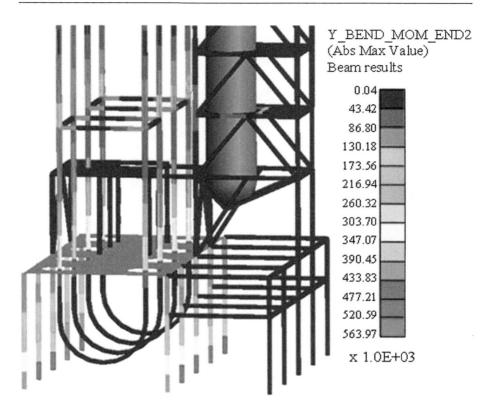

Figure 4.7 Peak bending moments (in kNm) at the base of the reactor column. (For color version of this figure, the reader is referred to the online version of this book.)

such as SDOF is too expensive and would not be practical. However, structural analysis for simple specimen can be still used as an "experimental" observation for postaccident analysis, in order to evaluate the destructive effectiveness of ideal blast waves on industrial equipment.

Several studies can be found in the open literature reporting data on damages to plant equipment caused by explosions. Most of the authors simply relate the observed system failures to the peak static overpressure. Also, several papers refer to the minimum value of static pressure with respect to specific equipment or, more generally, objects (e.g. window frames, concrete wall, rail tank).

An extensive review of these data has been reported in Cozzani and Salzano (2004b, 2004c) and Salzano and Cozzani (2005). An extract is reported in Table 4.2. The ambiguity in the definition of "damage" is evident: several types of damage (e.g. displacement, overturning, buckling, collapse) are reported without any distinction. Furthermore, details on structural or geometrical characteristics of the target equipment are not fully provided. As a conclusion, building a reliable model for overpressure damage to equipment required an accurate revision of the above data, based on univocal definition of damage, overpressure and equipment characteristics.

Table 4.2 Extract of Data Reported in the Literature for Damage to Process Equipment Due to Pressure Loads. See Cozzani and Salzano (2004a, 2004b, 2004c) for References and Details

P (kPa)	Damage	P (kPa)	Damage
1.72	Minor damage, cooling tower	39.12	Minor damage, pressure vessel horizontal
5.17	Minor damage, cone roof tank (100% filled)	42.00	Tubes failure
5.17	Minor damage, cone roof tank (50% filled)	42.00	Pressure vessel deformation
6.10	1% damage of equipment	42.51	Minor damage, floating roof tank (100% filled)
7.00	Failure of connection	42.51	Catastrophic failure, cone roof tank (100% filled)
10.00	Failure of atmospheric equipment	42.52	Minor damage, extraction column
10.00	5% damage of process plant	45.92	Catastrophic failure, fractionation column
10.00	50% damage of atmospheric tank	47.00	Failure of nonpressure equipment
14.00	Minor damage of cooling tower	49.32	Minor damage, heat exchanger
14.00	Minor damage of atmospheric tank	52.72	Minor damage, tank sphere
17.00	Minor damage, distillation tower	53.00	Pressure vessel failure
18.70	Minor damage, floating roof tank (50% filled)	53.00	Failure of spherical pressure vessel
18.70	Minor damage, reactor: cracking	55.00	20% damage of steel spherical steel petroleum tank
18.70	Catastrophic failure, cone roof tank (50% filled)	59.52	Catastrophic failure, reactor chemical
20.00	Displacement of steel supports	59.52	Catastrophic failure, heat exchanger
20.00	Tubes deformation	61.22	Catastrophic failure, pressure vessel horizontal
20.00	Deformation of atmospheric tank	69.00	Displacement and failure of heavy equipment
20.00	20% damage, process plant	69.73	Catastrophic failure, extraction column
20.00	100% damage, atmospheric tank	70.00	Structural damage of equipment
20.40	50% damage of equipment	70.00	Deformation of steel structures
22.10	Minor damage, pipe supports	70.00	100% damage, heavy machinery, process plant

(Continued)

Table 4.2 Extract of Data Reported in the Literature for Damage to Process Equipment Due to Pressure Loads. See Cozzani and Salzano (2004a, 2004b, 2004c) for References and Details—*cont'd*

P (kPa)	Damage	P (kPa)	Damage
22.11	Catastrophic failure, cooling tower	76.53	Catastrophic failure, reactor: cracking
24.00	20% damage of steel floating roof petroleum tank	81.63	Minor damage, pressure vessel vertical
25.00	Atmospheric tank destruction	81.63	Minor damage, pump
25.30	Minor damage, reactor chemical	83.00	20% damage of vertical cylindrical steel pressure vessel
27.00	Failure of steel vessel	88.44	Catastrophic failure, pressure vessel vertical
29.00	Distillation tower and cylindrical steel vertical structure	95.30	99% damage of vertical, steel pressure vessel
30.00	Failure of pressure vessel	97.00	99% damage of vertical cylindrical steel pressure vessel
34.00	99% damage of equipment	108.8	Catastrophic failure, tank sphere
35.00	80% damage of process plant	108.8	Catastrophic failure, pump
35.00	40% damage, heavy machinery	108.9	99% damage of spherical, pressure steel vessel
35.50	Structural damage of equipment	110.0	99% damage (total destruction) of spherical steel petroleum tank
35.71	Minor damage, fractionation column	136.0	Structural damage, low pressure vessel
37.42	Catastrophic failure, pipe supports	136.1	Catastrophic failure, floating roof tank (100% filled)
38.00	Deformation of nonpressure equipment	136.1	99% damage of floating roof tank

For these aims, in the framework of the quantitative assessment of domino hazards due to overpressure, the probit approach (Finney, 1971) is very attractive, due to its simplicity and the limited additional effort that is needed to implement the probit function in existing QRA algorithm. Moreover, probit models are not critically dependent on the definition of damage threshold values and may be easily modified to take into account specific categories of process equipment, when sufficient data are available. Therefore, probit analysis was applied both to revise existing models and to develop further the probabilistic models for the damage to specific categories of process equipment.

Eisenberg et al. (1975) first used a simplified model to assess the damage probability of process equipment caused by blast waves. The authors defined a probability function called "probit function" (Y) to relate equipment damage to the peak static overpressure $\Delta P°$

$$Y = k_1 + k_2 \ln(\Delta P°) \tag{4.1}$$

where Y is the probit for equipment damage, $\Delta P°$ is expressed in Pa, and k_1 and k_2 are the probit coefficients.

The model of Eisenberg et al. was based on "experimental" evaluation of equipment displacement with the subsequent deformation and breakage of connections, hence not considering the direct catastrophic failure of equipment. Results give $k_1 = -23.8$ and $k_2 = 2.92$.

The probit approach was then followed by other authors (see e.g. Khan and Abbasi, 1998), who proposed a probit function similar to the equation of Eisenberg, but substituting the static overpressure with the total pressure (the sum of static and dynamic pressure). Khan and Abbasi give the same probit coefficients of Eisenberg et al. In any case, the dynamic pressure is negligible for most industrial explosions, as discussed previously.

More recently, Cozzani and Salzano (2005) have elaborated an extended set of data regarding explosion in the process industry. With specific reference to explosions, the probability of damage of any equipment loaded by blast wave was considered as unity ($F_d = 1$) when the external pressure overcomes the allowable stress value of the material under consideration and any of the following events takes place:

- Catastrophic failure (catastrophic damage, total collapse, disintegration, fracture) of equipment
- Violent overturning or displacement of road, rail tank or heavy equipment
- Structural damage to the main system of containment for atmospheric and pressurized vessels.

On the other side, the probability of damage was considered as minimum ($F_d = 0.01$: "minor damage") when the pressure wave was sufficiently intense to produce very light damages which are unlikely to hinder the normal functioning of the system as a "buckling" of the equipment.

Quite clearly, probability values between the two limits are difficult to define without any arbitrary choice. In the context of QRA as a comparative tool, two further hypotheses were then introduced: a 10% failure probability was assumed to correspond to a partial failure, deformation, minor damage of the auxiliary equipment or minor structural damage of atmospheric equipment; a 30% failure probability was assumed for the complete rupture of connections or for minor structural damage of pressurized equipment.

Based on these considerations, the data were further divided, taking into account four equipment categories: (1) atmospheric vessels, (2) pressurized vessels, (3) elongated vessels, and (4) small equipment. Equipment not belonging to these categories were not considered in the analysis, whereas the data related to the piping failure were included in all categories. Final results in terms of probit analysis are given in Table 4.3.

Table 4.3 Probit Coefficients for the Damage of Different Equipment
Categories when Loaded by Blast Wave Characterized by Given Peak
Overpressure (Dose: Peak Overpressure in Pa)

Equipment	k_1	k_2
Atmospheric vessels	−18.96	2.44
Pressurized vessels	−42.44	4.33
Elongated equipment	−28.07	3.16
Small equipment	−17.79	2.18

Figure 4.8 shows a comparison of the probit values obtained for the different
process equipment categories as a function of overpressure. As expected, pressurized
vessels showed lower damage probabilities with respect to overpressure. Moreover,
higher overpressure are necessary in order to damage elongated vessels (as distillation
or absorption columns) compared with atmospheric storage vessels.

In the analysis reported, the structural (mechanical) damage of a process equipment item was roughly evaluated by defining two damage classes: light damage to the
structure or to the auxiliary equipment and intense or catastrophic damage.

In order to produce a domino effect, a further distinction should be, however,
related to the LOC for the damaged equipment. Giving a simple example on this
topic, a small amount of a low-risk hazardous materials such as diesel oil from the
damaged system of containment (the target) may be considered in terms of

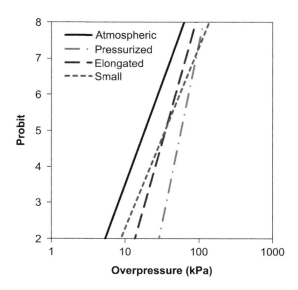

Figure 4.8 Comparison of probit models obtained for the damage of equipment due to
overpressure.

economical losses or in terms of return to service but can be loosely considered as the trigger for domino effect scenarios such as fire, explosion or more generally dispersion of toxic or flammable materials.

The first schematization for the importance of LOC in domino effects has been reported by Salzano and Cozzani (2006) and reported in Table 4.4. The data reported

Table 4.4 Credibility of Escalation as a Function of LOC Intensity for Different Equipment Category

Secondary Substance Hazard	Secondary Equipment	Expected Secondary Scenarios	Credibility of Escalation
		LOCI1	
Flammable	Atmospheric	Minor pool fire	Low
	Pressurized	Jet fire	High
	Elongated	Minor pool fire	Low
		Minor flash fire	
	Small/auxiliary	Minor pool fire	Low
Toxic	Atmospheric	Minor evaporating pool	Low
	Pressurized	Boiling pool	High
		Jet toxic dispersion	
	Elongated	Minor boiling pool	High
		Toxic dispersion	
	Small/auxiliary	Minor pool fire	low
		Minor flash fire	
		LOCI2	
Flammable	Atmospheric	Pool fire	High
		Flash fire	
		VCE	
	Pressurized	BLEVE/Fireball	High
		Jet fire	
		Flash fire	
		VCE	
	Elongated	Pool fire	High
		Flash fire	
		VCE	
	Small/auxiliary	Minor evaporating pool	Low
Toxic	Atmospheric	Evaporating pool	High
		Toxic dispersion	
	Pressurized	Boiling pool	High
		Toxic dispersion	
	Elongated	Boiling pool	High
		Toxic dispersion	
	Small/auxiliary	Evaporating pool	Low
		Minor toxic dispersion	

are only qualitative; however, the LOC intensity (LOCI) from damaged equipment should be also quantitatively defined. In this regard, in the framework of QRA, a main "standard" reference such as the "purple book" (TNO, 1999) suggests to consider at least two different LOC scenarios for any storage or process vessel: a release from a small-diameter pipe (10 mm equivalent diameter) and the total loss of inventory following severe damage (identified by the total loss of inventory in 10 min or by the "instantaneous" release of vessel content).

Thus, on the basis of the available damage data, taking into account the suggestions given by the "purple book", two categories have been defined by Salzano and Cozzani (2006) for the evaluation of LOCI due to damage following the impact of a blast wave (i.e. domino effects):

1. Damage state 1 (DS1): light damage to the structure or to the auxiliary equipment, associated with a 10 mm average release diameter, resulting in a moderate LOCI (LOCI1);
2. Damage state 2 (DS2): intense, or catastrophic damage, or even total collapse of structure, associated with an instantaneous release of the entire vessel content (LOCI2).

The above defined damage states (DSs) and the associated LOCI categories may then be used in a QRA framework to address the evaluation of the expected damage of the primary event and the possible consequences of the secondary scenario following the LOC.

Starting from these definitions, through fuzzy set analysis, Salzano and Cozzani (2006) have evaluated the probability of escalation effects for different types of equipment. The peak static overpressure was used as the input variable, while the damage probability was used as the output variable only in order to allow a straightforward clustering of the data. The general learning method by Hong and Lee (1996) was used to build a prototype fuzzy expert system and to develop a fuzzy inference procedure for process inputs. This general procedure was used to obtain probability values for the limit states of mechanical DS and LOCI.

In this framework, the conditional probability of domino effects may be calculated as

$$f_{\text{LOCI},i} = f_{\text{DS},i} = f_{\text{p}} \cdot P_{\text{d}} \cdot P_{\text{DS},i} \tag{4.2}$$

where $f_{\text{LOCI},i}$ is the overall expected frequency of the LOCI class of interest, $f_{\text{DS},i}$ is the overall expected frequency of the DS of interest, f_{p} is the expected frequency of the primary event, P_{d} is the expected damage probability following blast wave impact (as calculated by the probit models reported previously), and $P_{\text{DS},i}$ is the probability of the ith DS following the damage due to blast wave impact that needs to be evaluated.

Figure 4.9 reports the overall probability values for the above defined DSs or LOCI classes as a consequence of the blast wave impact on the different categories of process equipment considered (atmospheric and pressurized). As expected, the probability of the lower intensity LOC (LOCI1) is higher for the lower values of the peak static overpressure, whereas at higher overpressure, the expected structural damage is almost only associated with the probability of a high-intensity LOC (LOCI2). Moreover, the figure evidences the importance of considering different

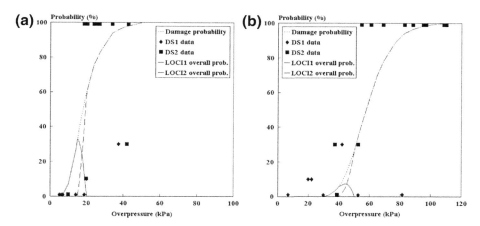

Figure 4.9 Damage probability and overall probability of LOC categories for atmospheric (a) and pressurized (b) equipment as evaluated by Salzano and Cozzani (2006).

categories of equipment. As a matter of fact, the LOCI1 release intensity prevails in different regions of peak overpressure values if different categories of equipment are considered. Atmospheric equipment are characterized by a LOCI1 release mode for impact overpressures between 7 and 15 kPa, while in the case of pressurized equipment, the same release mode is the more probable for overpressures up to 40 kPa (Salzano and Cozzani, 2004a, 2004c).

4.5 Response Regimes

Equipment or structures that are subjected to a dynamic transient load will vibrate. The magnitude of vibration will depend on the load of the forced vibration and on the amount of damping.

The plot of these functions, called "the frequency response of the system", presents one of the most important features in forced vibration. In a lightly damped system, when the forcing frequency nears the natural frequency, the amplitude of the vibration can get extremely high. This phenomenon is called resonance (subsequently the natural frequency of a system is often referred to as the resonant frequency). Whatever the methodology, the response of a structure to pressure loads may be characterized by the load ratio between the duration of pressure load τ_d and the fundamental natural period T of the structure. The fundamental natural period of a structure is the longest natural period at which the member will respond to any load, either seismic wave or explosion or any other impulse. As the overpressure interaction with any object is usually represented as a uniform load, this is the predominant mode of response in most explosion situations.

Depending on the value of the load ratio, response regimes (or realms) are typically defined, denoted as impulsive, dynamic and quasi-static or static. Conventionally, the

range for fully dynamic response is defined as, e.g. in the Interim Guidance Notes by
the Fire and Blast Information Group (FABIG) (Bowerman et al., 1992):

$$\frac{\tau_d}{T} < 0.4 \qquad \text{Impulsive}$$

$$0.4 \le \frac{\tau_d}{T} \approx 2.0 \quad \text{Dynamic} \tag{4.3}$$

$$\frac{\tau_d}{T} > 2.0 \qquad \text{Quasi Static}$$

Recently, NORSOK (2000) and Czujko (2001) proposed different limits, so that the
impulsive regime is recognized for time ratio lower than 0.3 and the static regime is
recognized for ratio larger than 5.

The regimes defined here reflect the behavior of any object when experiencing an
external load as a pressure wave. In general, when static or quasi-static regimes are of
concern, the interaction is only dependent on the overall static pressure exerted on the
object surface. Finally, for intermediate regimes such as the dynamic realm, no
simplified analyses are possible. A typical approximation regards the use of the
dynamic amplification factor (DAF), which is defined as the ratio of maximum
dynamic displacement over static displacement. The DAF transforms a dynamic peak
load into a static load with the same effect on the structure. For long explosion times
and in the case of an idealized triangle-shaped shock wave load, the value of DAF
approaches its boundary limit of 2, which means that the same damage is produced by
50% of the given static pressure.

When risk assessment or domino effect analyses are concerned, the cost of detailed
structural analysis may be too high and strong simplification is needed. To this aim,
pressure–impulse (P–I) diagrams (or isodamage plot), based on SDOF idealization,
are generally produced for any defined damage or failure defined in terms of
maximum displacement (Baker et al., 1983; Whitney et al., 1992; Schneider, 1997)
(Figure 4.10). Typical applications of P–I diagrams are, however, based on empirical
results and related to houses (Mercx et al., 1991; TNO, 1992), small office buildings,
light-framed industrial buildings, or human response to pressure waves.

Indeed, complex equipment are difficult to schematize and define by simplified
analysis if considering that different failure modes, damage and equipment section
may fail. In this regard, another difficulty is the definition of damage mode for knock-
on effects, which include the effects related to the LOC following structural damage
of equipment, as pointed out by Cozzani and Salzano (2004a).

An important assumption is related to the impulsive regime, where much higher
peak loads can be tolerated than the static capacity of the target (UKOAA, 2003). This
aspect is essential in risk assessment or domino effects analysis. Indeed, the evaluation
of the ability of any primary explosion to trigger secondary, catastrophic scenarios
(domino effects) is reliable, although conservative, if using only the minimum static
(side-on) overpressure parameter as observed in the static or in the quasi-static regime
on pressure–impulse plot. Quite clearly, this concept does not apply comfortably to
design phase, where ultraconservative options may be expensive.

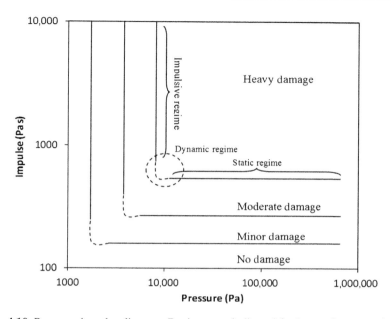

Figure 4.10 Pressure–impulse diagram. Regimes are indicated for heavy damage only.

Let us now consider the target equipment for explosion wave in the view of domino effects, as, e.g. horizontal and a vertical atmospheric (low pressure) storage tanks of different volume, containing fuel oil (either empty or half full), and pressurized horizontal cylinder containing propane gas only (see Table 4.5). For these equipment, explosion loading realm can be characterized by the results obtained by FEA model of equipment. In the following, an explicit-based, large deformation, dynamic FEA code LS-DYNA (2003) is used for the study. LS-DYNA is a commercial version of the public domain US Department of Energy code DYNA3D (Whirley and Englemann, 1993). The explicit formulation is ideally suited for analyzing the dynamic response of structures subjected to impulsive loading. It has a robust suite of constitutive material models and contact surface algorithms. The strain rate effects follow recommendations in TM 5-855-1 (1998). The finite element is deleted from the

Table 4.5 Sketch of Industrial Equipment Analyzed in this Work. Wall Thickness θ Ranges from Bottom to Top

N	Equipment		Fill	Vol. (m³)	Radius (m)	Wall θ (m)	Roof θ (m)
1	Atmospheric	V	Empty	250	6.6	0.005	0.005
2	Pressurized	H	Gas filled	100	2.8	0.018	0.018
3	Atmospheric	V	Empty	30,000	44	0.021–0.006	0.006
4	Atmospheric	V	Half	30,000	44	0.021–0.006	0.006

V = vertical; H = horizontal.

Table 4.6 Natural Period (ms) for the 10 Vibrational Modes, for Equipment Reported
in Table 4.5

n	1 (T)	2	3	4	5	6	7	8	9	10
1	82.37	82.37	82.25	82.19	76.17	76.13	74.22	74.22	67.23	67.23
2	530.70	121.96	87.80	83.65	68.17	53.92	47.09	40.06	38.91	35.22
3	492.51	453.64	342.71	341.85	335.82	314.85	302.46	277.52	269.81	263.16
4	711.69	492.10	469.15	453.14	427.50	391.05	387.19	373.32	369.81	349.65

calculation at that point in time after reaching its specified rupture criterion (i.e. ultimate effective plastic strain).

The geometrical model was discretized by using a variable-size mesh ranging from about 1 to 3 cm. A single shell element represents the wall and roof of the equipments. Due to the expected high-pressure loading that led to high strain rates and the possibility of plastic deformations, an isotropic, piecewise linear, elastic–plastic with failure material model (MAT 24, LS-DYNA) was used for the steel components. The steel material has an elastic modulus of 199,948 MPa, Poisson's ratio of 0.30, unit weight of 7850 kg/m^3 and a nominal rupture strain of 0.29.

For the equipment and filling level reported in Table 4.5, Table 4.6 reports the natural period for 10 vibrational modes, as calculated by LS-DYNA.

The data for equipment 3 and 4 are significantly greater than the data produced by Liu and Schubert (2002), which give an average natural period of about 150 ms. Table 4.7 reports the ratio of explosion duration to response time (hence allowing the identification of the explosion realm), by considering blast waves with triangular shape and total duration of, namely: (1) 200 ms, thus reproducing long-duration VCE; (2) 100 ms, i.e. a shorter VCE; (3) 10 ms, i.e. a short-duration explosion, which is characteristic of BLEVE, confined, partially confined and small-scale explosion; and (4) 1 ms, that represents strong condensed phase explosive and other point-source explosions.

Results clearly show that:

1. Empty large-scale vessels have lower values for time ratios with respect to partially or fully filled equipment
2. Nonempty large vessel are still characterized by impulsive realms unless very long-duration explosion such as VCE are considered, for which dynamic realm can be seen
3. Impulsive regime occurs always for short-duration explosions, whatever be the equipment or fill level or the scale of equipment
4. The quasi-static or the static regime is never reached if, following the NORSOK definition (NORSOK, 2000), the first fundamental period is considered.

Furthermore, these data clarify that static pressure is always suitable for the definition of escalation criteria unless large-scale VCEs are considered, for which half of the value of static pressure load may be conservatively adopted.

With respect to atmospheric tanks, the work done worldwide for seismic analysis can be applied, as the liquid content can add further load on the tank structure. To this

Table 4.7 Time Ratios τ_d/T for Equipment Reported in Table 4.5, for the First 10 Vibrational Modes, by Varying Explosion Duration

Equipment/ Modes	1	2	3	4	5	6	7	8	9	10
$t_d = 200$ ms										
1	2.43	2.43	2.43	2.43	2.63	2.63	2.69	2.69	2.97	2.97
2	0.38	1.64	2.28	2.39	2.93	**3.71**	**4.25**	**4.99**	**5.14**	**5.68**
3	0.41	0.44	0.58	0.59	0.60	0.64	0.66	0.72	0.74	0.76
4	0.28	0.41	0.43	0.44	0.47	0.51	0.52	0.54	0.54	0.57
$t_d = 100$ ms										
1	1.21	1.21	1.22	1.22	1.31	1.31	1.35	1.35	1.49	1.49
2	*0.19*	0.82	1.14	1.20	1.47	1.85	2.12	2.50	2.57	2.84
3	*0.20*	*0.22*	*0.29*	*0.29*	0.30	0.32	0.33	0.36	0.37	0.38
4	*0.14*	*0.20*	*0.21*	*0.22*	*0.23*	*0.26*	*0.26*	*0.27*	*0.27*	*0.29*
$t_d = 10$ ms										
1	*0.12*	*0.12*	*0.12*	*0.12*	*0.13*	*0.13*	*0.13*	*0.13*	*0.15*	*0.15*
2	*0.02*	*0.08*	*0.11*	*0.12*	*0.15*	*0.19*	*0.21*	*0.25*	*0.26*	*0.28*
3	*0.02*	*0.02*	*0.03*	*0.03*	*0.03*	*0.03*	*0.03*	*0.04*	*0.04*	*0.04*
4	*0.01*	*0.02*	*0.02*	*0.02*	*0.02*	*0.03*	*0.03*	*0.03*	*0.03*	*0.03*
$t_d = 1$ ms										
1	*0.01*	*0.01*	*0.01*	*0.01*	*0.01*	*0.01*	*0.01*	*0.01*	*0.01*	*0.01*
2	*0.00*	*0.01*	*0.01*	*0.01*	*0.01*	*0.02*	*0.02*	*0.02*	*0.03*	*0.03*
3	*0.00*	*0.00*	*0.00*	*0.00*	*0.00*	*0.00*	*0.00*	*0.00*	*0.00*	*0.00*
4	*0.00*	*0.00*	*0.00*	*0.00*	*0.00*	*0.00*	*0.00*	*0.00*	*0.00*	*0.00*

The line divides the impulsive realm (bottom) from the dynamic (top). Bold character defines static or quasi-static regime, italic character indicates impulsive realm.

aim, the work of Malhotra et al. (2000) defines two other natural periods related to the impulsive (T_{imp}) and convective (T_{conv}) responses of tank due to liquid movement (sloshing, overturning, due to liquid)

$$T_{imp} = C_i H \left(\frac{\rho r}{hE}\right)^{0.5} \tag{4.4}$$

$$T_{conv} = C_c r^{0.5} \tag{4.5}$$

where H is the total tank height, r is the tank radius, ρ is the density of liquid, E is modulus of elasticity of the material, h is the average thickness of shell, C_i (adimensional) and C_c [m s$^{0.5}$] are constants which depend on the ratio of H/r. For steel materials, Malhotra et al. (2000) gives simplified tables for C_i and C_c. The two effects (impulsive and convective) may produce damage to tanks. For a steel tank with radius r of 10 m and total height of 9.6 m, filled with water to a height H of 8 m ($H/r = 0.8$), the calculated values are $T_{imp} = 0.123$ s and $T_{conv} = 4.96$ s. If considering both effects conjunctly, the interaction may be considered essentially static or quasi-static unless very short-duration explosions are considered.

Table 4.8 Time Ratios τ_d/T for Tubes of Different Diameter and Materials, for the Fundamental Vibrational Mode, by Varying Explosion Duration

Equipment	Diameter (in.)	τ_d/T			
Tube		200 ms	100 ms	10 ms	1 ms
Threaded pipe	3/8	**2.032**	1.016	*0.102*	*0.010*
Threaded pipe	1	**7.418**	**3.709**	0.371	*0.037*
Conduit	2	**13.894**	**6.947**	0.695	*0.069*
Conduit	3/4	**4.706**	2.353	*0.235*	*0.024*
PVC	1	**5.842**	2.921	*0.292*	*0.029*
PVC	3/4	1.652	0.826	*0.083*	*0.008*

Bold character defines static or quasi-static regime, italic character indicates impulsive realm.
PVC: polyvinyl chloride

The work of Leal and Santiago (2004) can be usefully adopted for spheres with a variable filling level. For a 14.5-m-diameter sphere, natural periods of 2.90 and 4.44 s were, respectively, calculated for filling levels of 25% and 75%, thus demonstrating that the impulsive realm applies for any type of explosion.

The work of Roy and Antaki (2001) may be of help for pipelines, even if they refer to detonation. Their results have been adopted for the compilation of Table 4.8, which again reports the time ratios for several tubes and pipelines of different diameter and materials. Similar conclusions as for the equipment analyzed in the previous tables can be drawn. Unless very short impulsive explosions are considered, the realm is static or quasi-static for VCE only and dynamic for most accidental explosions.

Finally, typical values for the natural period for some element or equipment which may be of interest in the framework of industrial systems (e.g. as barriers for the escalation) are also reported in Table 4.9 for the sake of completeness.

Also in this case, the value of fundamental natural period allows prediction on the realm, which is impulsive for short-duration explosion and dynamic if the total duration of pressure loading is higher than 100 ms.

The conclusions of this analysis is that the impulsive realms, which may be evaluated with good approximation by one degree of freedom methodologies (e.g. the values of incident static pressure), may be always adopted for explosions having a very short duration ($\ll 10$ ms), providing conservative results.

Table 4.9 Time Ratios τ_d/T for Some Typical Industrial Element and Equipment (Fabig, 1999)

Equipment	T (ms)	τ_d/T			
		200 ms	100 ms	10 ms	1 ms
Concrete walls	10	**20**	**10**	1	*0.1*
Brick wall	20	**10**	**5**	0.5	*0.05*
Blast wall	35 (unstiffened)	**5.75**	2.85	*0.28*	*0.03*
Blast wall	31 (stiffened)	**6.45**	**3.23**	0.32	*0.03*

Pressurized equipment (cylinder, spheres) are under the impulsive regime even for explosions having longer durations (up to 100 ms). However, all structural elements as, e.g. saddles for pressurized equipment, should be considered for a proper escalation analysis.

When duration of explosions larger than 100 ms are considered, pipelines, blast wall, brick walls and the effect of liquid movement for large-scale equipment (sloshing, overturning) may be evaluated under the static or quasi-static realms.

For explosion duration between 1 and 100 ms the analysis should be addressed under dynamic realm for all equipment, for which the double of the static pressure should be considered under conservative assumption.

4.6 Conclusions

The effect of accidental explosions on industrial equipment is hardly predictable by deterministic approaches, and even the assessment of the resistance of a simple, flat blast wall for the protection of a residential unit on an offshore platform by considering an idealized triangular blast wave, in the design phase, is a matter of debate (Czujko, 2001; Louca and Boh, 2004) and requires complex numerical analyses, with large uncertainties in results. However, it is possible to carry out a reliable assessment of the possibility of an escalation either by simplified or by most complex methodologies.

Simple models (SDOF, single value criteria, P–I diagrams, etc.) can be used by any experienced risk engineer. The results of these simple approaches are conservative and enable the risk engineer to make a reliable statement concerning the possible exclusion of escalation. Probit models for damage to equipment caused by blast waves are available in the literature and can be usefully adopted for QRA and land use planning, if the damage probability data for different process equipment categories is separately considered. The use of probability functions and threshold values for the occurrence of LOC from damaged system is crucial for the analysis of escalation resulting in a domino effect.

When the risk engineer comes to the conclusion that an escalation due to blast load cannot be excluded, a more comprehensive dynamic response analysis is needed. This can only be performed with the aid of several experts (explosion modelers, finite element experts, etc.) each having extensive expertise in their own field. Substantial cost increase of the design engineering is, however, to be expected.

References

Abbasi, T., Pasman, H.J., Abbasi, S.A., 2010. A scheme for the classification of explosions in the chemical process industry. Journal of Hazardous Materials 174, 270–280.

Amyotte, P., Eckhoff, R.K., 2010. Dust explosion causation, prevention and mitigation: an overview. Journal of Chemical Health and Safety 17, 15–28.

ASCE, 1997. Design of Blast Resistant Buildings in Petrochemical Facilities. American Society of Civil Engineers.

Baker, W.E., Cox, P.A., Westine, P.S., Kulesz, J.J., Strehlow, R.A., 1983. Explosion Hazards and Evaluation. Elsevier, New York, NY.

Bartknecht, W., 1981. Explosions: Course, Prevention, Protection. Springer, Berlin.

Biggs, J.M., 1964. Introduction to Structural Dynamics. McGraw Hill.

Bjerketvedt, D., Bakke, J.R., van Wingerden, K., 1993. Gas Explosion Handbook. Journal of Hazardous Materials 52, 1–150.

Bowerman, H., Owens, G.W., Rumley, J.H., Tolloczko J.J.A. Interim Guidance Notes for the Design and Protection of Topside Structures Against Explosion and Fire, The Steel Construction Institute Document SCI-P-112/487, January 1992.

CCPS, 2010. Guidelines for Vapor Cloud Explosion, Pressure Vessel Burst, Bleve and Flash Fire Hazards, American Institute of Chemical Engineer (AIChE), The Center for Chemical Process Safety, second ed. Wiley, John & Sons, Inc.

CEAP-DYNA, 2004. LS-DYNA with Civil Engineering Applications Extensions by Oasys Ltd. Development version 8.0e. Oasys Limited, The Arup Campus, Blythe Gate, Blythe Valley Park, Solihull, West Midlands, B90 8AE.

Cozzani, V., Salzano, E., 2004a. The quantitative assessment of domino effects caused by overpressure. Part I: probit models. Journal of Hazardous Materials 107, 67–80.

Cozzani, V., Salzano, E., 2004b. The quantitative assessment of domino effects caused by overpressure. Part II: Case studies. Journal of Hazardous Materials 107, 81–94.

Cozzani, V., Salzano, E., 2004c. Threshold values for domino effects caused by blast wave interaction with process equipment. Journal of Loss Prevention in the Process Industries 17, 437–447.

Crowl, D.A., 2003. Understanding Explosions. Centre for Chemical Process Safety. American Institute of Chemical Engineers, New York.

CSB, 2006. Combustible Dust Hazard Study Investigation Report. Report No. 2006-H-1. <http://www.csb.gov/assets/document/Dust_Final_Report_Website_11-17-06.pdf> (last checked on 12.8.2012).

CSB, 2009. Imperial Sugar Dust Explosion and Fire Final Investigation Report. Report No. 2008-05-I-GA. <http://www.csb.gov/assets/document/Imperial_Sugar_Report_Final_updated.pdf> (last checked on 12.8.2012).

Czujko, J., 2001. Design of Offshore Facility to Resist Gas Explosion Hazard. CorrOcean ASA, Oslo.

Darbra, R.M., Palaciosa, A., Casal, J., 2010. Domino effect in chemical accidents: main features and accident sequences. Journal of Hazardous Materials 183, 565–573.

DoD, Department of Defense, 2008. Structures to Resist the Effects of Accidental Explosions, UFC 3-340-02.

Eckhoff, R.K., 2005. Explosion Hazards in the Process Industries. Gulf Publishing Company, Austin, TX.

Eisenberg, N.A., Lynch, C.J., Breeding, R.J., 1975. Vulnerability Model: A Simulation System for Assessing Damage Resulting from Marine Spills, Rep. CG-D-136-75, Enviro Control Inc., Rockville, MD.

Fabig, 1999. Technical Note 5: Design Guide for Stainless Steel Blast Walls. The Steel Construction Insitute.

Finney, D.J., 1971. Probit Analysis. Cambridge University Press.

Hong, T.P., Lee, C.Y., 1996. Fuzzy Sets and Systems 84, 33–47.

Hoorelbeke, P., Brewerton, B., Renoult, J. Explosion Analysis of an Onshore Plant with Worked Examples, Quantitative Risk Analysis, Probabilistic Explosion and MDOF Response Analysis, FABIG Technical Meeting on Protection of Onshore Oil and Gas Plants Against Fire and Explosion, 6–7, April 2005.

HSE, 1992. Explicit Analytical Methods for Determining Structural Response. OTI 92600.

HSE, 2009. Buncefield Explosion Mechanism Phase 1, Prepared by SCI, RR718 Research Report.

Khakzad, N., Khan, F., Amyotte, P., 2012. Dynamic risk analysis using bow-tie approach. Reliability Engineering and System Safety 104, 36–44.

Khan, F.I., Abbasi, S.A., 1998. Models for domino effect analysis in chemical process industries. Process Safety Progress 17, 107.

Lea, C.J., Ledin, H.S. A Review of the State-of-the-art in Gas Explosion Modelling, Health and Safety Laboratory Report CM/00/04, 19 February 2002.

Leal, C.A., Santiago, G.F., 2004. Do tree belts increase risk of explosion for LPG spheres? Journal of Loss Prevention in the Process Industries 17, 217–224.

Lees, F.P., 1996. Loss Prevention in the Process Industries, second ed. Butterworth-Heinemann, Oxford.

Liu, H., Schubert, D.H., 2002. Effects of Nonlinear Geometric and Material Properties on the Seismic Response of Fluid/Tank Systems, 10th ANSYS International Conference and Exhibition, Pittsburgh, PA.

Louca, L.A., Boh, J.W., 2004. Analysis and Design of Profiled Blast Walls, Health and Safety Executive 2004, Research Report 146.

LS-DYNA, 2003. User's Manual, Nonlinear Dynamic Analysis of Structures. Version 970. Livermore Software Technology Corporation.

Malhotra, P.K., Wenk, T., Wieland, M., 2000. Simple procedure for seismic analysis of liquid storage tanks. Structural Engineering 10, 197–201.

Mercx, W.P.M., Weerheijm, J., Verhagen, T.L.A., April 16–18, 1991. Some Considerations on the Damage Criteria and Safety Distances for Industrial Explosions, HAZARDS XI – New Directions in Process Safety. UMIST, Manchester, UK.

NORSOK Standard N-001, Structural Design, Rev 3, August 2000.

Oasys, 2003. LS-DYNA Environment. Version 9.0. Oasys Limited, The Arup Campus, Blythe Gate, Blythe Valley Park, Solihull, West Midlands, B90 8AE.

Roy, B.N., Antaki, G.A., 2001. Analysis of the Effects of External Detonations on Piping Systems. Smirt Transactions, Washington DC.

Salzano, E., Cozzani, V., 2005. The analysis of domino accidents triggered by vapour cloud explosions. Journal of Reliability Engineering and System Safety 90, 271–284.

Salzano, E., Cozzani, V., 2006. A fuzzy set analysis to estimate loss of containment relevance following blast wave interaction with process equipment. Journal of Loss Prevention 19, 343–352.

Schneider, P., 1997. Limit states of process equipment components loaded by a blast wave. Journal of Loss Prevention in the Process Industries 10, 185–190.

TM 5-855-1, 1998. Departments of the Army, the Navy and the Air force, Design and analysis of hardened structures to conventional weapons effects, Technical Manual for Army, Air Force AFPAM 32-1147(I), Navy NAVFAC P-1080, and Defense Special Weapons Agency DAHSCWEMAN-97.

TNO, 1992. Green Book, Methods for the Determination of Possible Damage to People and Objects Resulting from Releases of Hazardous Materials, Report CPR 16E, Voorburg, The Netherlands.

TNO, 1999. Department of Industrial Safety, Guidelines for Quantitative Risk Assessment (Purple Book). Committee for the Prevention of Disasters, The Hague (NL).

UKOAA, October 2003. UK offshore operators association, fire and explosion guidance – part 1: avoidance and mitigation of explosions ISSUE (1).

Whirley, R.G., Englemann, B.D. DYNA3D: A Nonlinear, Explicit, Three-Dimensional Finite Element Code for Solid and Structural Mechanics, User Manual, Report USRL-MA-107254, University of California, Lawrence Livermore National Laboratory, 1993.

Whitney, M.G., Barker, D.D., Spivey, K.H., 1992. Ultimate capacity of blast loaded structures common to chemical plants. Plant/Operations Progress 11, 205–212.

5 Heat Radiation Effects

Gabriele Landucci[*], Valerio Cozzani[†], Michael Birk[‡]

[*] Dipartimento di Ingegneria Civile e Industriale, Università di Pisa, Pisa, Italy, [†] LISES, Dipartimento di Ingegneria Civile, Chimica, Ambientale e dei Materiali, Alma Mater Studiorum - Università di Bologna, Bologna, Italy, [‡] Department of Mechanical and Materials Engineering, Queen's University, Kingston, Ontario, Canada

Domino Effects in the Process Industries. http://dx.doi.org/10.1016/B978-0-444-54323-3.00005-1

5.1 Introduction

This section is dedicated to the analysis of domino accidents triggered by fire, thus to the analysis of domino scenarios involving thermal radiation or fire engulfment as escalation vectors. This particular type of accidental scenarios may lead to extremely severe consequences, as evidenced by the analysis of past accidents (Chapter 2), which confirmed that more than half of the industrial "domino" accidents occurred between 1961 and 2010 involved fire as a primary event. Secondary targets affected were mostly pressurized tanks, atmospheric tanks, process vessels and pipelines (Lees, 1996; Roberts et al., 2000; Gòmez-Mares et al., 2008).

Since fire is the primary event, the escalation vector is the "heat load" due to the fire (Roberts et al., 2000). The heat load is a combination of the heat transferred from the fire to the target equipment by radiation and convection. While the target equipment receives the heat load, the shell of the target vessel heats up and heat is transferred into the liquid and vapor lading. Thus, the wall temperature increases and consequently the internal fluid temperature rises. This causes a strong negative effect on the resistance of the structure, since the internal pressure rises due to both the rapid heating of the gas phase and the increase in the vapor pressure of the liquid phase that may be present in the vessel. The wall temperature in the region of the gas phase rises very rapidly due to the poor cooling effect of the gas phase, while the wall temperature in the region wetted by the liquid remains near the liquid temperature because of the high heat transfer coefficient between the liquid and the wall.

The high temperature of the fire, typically ranging between 1000 and 1500 K (Lees, 1996; Roberts et al., 2000) can lead to gas phase wall temperatures higher than 700 K, which has a severe weakening effect on the shell materials, decreasing their resistance. For typical steel vessels the strength of the material drops rapidly at temperatures above 700 K. At the same time, the stress in the vessel shell increases, due to strong thermal dilatations and to inner pressure rise. Thus, conditions are created for a critical escalation due to vessel failure and consequent loss of containment.

The specific mechanism of escalation triggered by fire results in a peculiar aspect of fired domino events: the escalation is usually delayed with respect to the initiating event. In the case of domino effect caused by blast overpressure or fragment projection, the escalation rapidly occurs after the exposure of the target to the impact vector, while in the case of fired domino effect a time lapse is present between the start of the primary fire and the failure of the target equipment leading to a loss of containment (Landucci et al., 2009a). This time lapse can be minutes or hours in duration and is generally named "time to failure" (ttf) (Birk, 2006; Manu et al., 2009a; Landucci et al., 2009b; Paltrinieri et al., 2009). The ttf depends both on the type and geometry of the secondary equipment involved and on the exact details of the heat load. This heat load depends on the type of failure (i.e. spill, vapor jet, liquid jet, two-phase jet, etc.) that is generated by the loss of containment of the primary equipment.

Therefore, the key issue for the protection and prevention of escalation triggered by fire is the installation of systems able to prevent or, at least, delay the collapse or failure of the target equipment for a time lapse sufficient for effective emergency response. Therefore the assessment of the behavior of equipment items engulfed or impinged by fires is a key issue in order to understand on one side, the possibility of escalation of primary fires to severe domino scenarios (Cozzani and Zanelli, 2001; Delvosalle, 1998; Gledhill and Lines, 1998; Khan and Abbasi, 1998) and, on the other, to design effective protection measures (Landucci et al., 2009a; Cozzani et al., 2005).

Hence, this chapter is aimed on one side at the analysis of the fundamental elements of escalation triggered by fires (Section 5.2), reporting the results of specific studies on structural elements affected by fires. A detailed characterization of the primary fire scenarios potentially leading to severe damage to equipment is provided (Section 5.3). On the other side, numerical approaches for the prediction of ttf of equipment are presented and discussed (Section 5.4), in the perspective of the design and the selection of the more suitable techniques and safety systems for equipment protection (Section 5.5).

5.2   Mechanisms of Escalation Triggered by Fire

5.2.1   Experimental Analysis of Vessels Exposed to Fires

Experimental studies of fire effects on pressure vessels has been conducted by numerous groups (Birk and Cunningham, 1994a; Birk et al., 1997; Towsend et al., 1974; Droste et al., 1999; Moodie et al., 1988; Droste and Schoen, 1988; Persaud et al., 2001; Birk et al., 2006a; Birk et al., 2006b; Birk and Van der Steen, 2006; Faucher et al., 1993; Landucci et al., 2009c). However, the experimental work published over the last four decades was mainly undertaken in order to investigate the behavior of pressurized vessels engulfed in flames, while scarce attention has been dedicated to the experimental analysis of the behavior of the other types of vessels, and in particular of atmospheric tanks.

Tables 5.1 and 5.2 report a concise summary of the more significant experimental studies carried out on pressurized equipment exposed to fires. Table 5.1 focuses on experiments carried out on unprotected pressurized vessels (with pressure relief valves, PRVs), which allowed obtaining sufficiently detailed data to describe the behavior of such vessels in conditions similar to those corresponding to an involvement in an external fire event. Table 5.2 summarizes the experimental setup used for the analysis of passive fire protection (PFP) systems and the available results. Other studies are aimed at the analysis of active protection systems, such as water deluges or spray curtain systems, and their efficiency, without taking into account the analysis of the resistance of the vessel (Shoen and Droste, 1988; Planas-Cuchi et al., 1996; Lev, 1991; Shirvill, 2004; Davies and Nolan, 2004; Roberts, 2004a).

The typical setup of a vessel fire test is characterized by two key elements: a heat source and the target equipment. The heat source is aimed at the experimental simulation of a fire affecting the process equipment. In particular, confined pool fires with diesel or jet fuels are often employed to create a full engulfment in flames, while torch or jet fires are employed for partial or total engulfment. As shown in

Table 5.1 Summary of Relevant Literature Experiments on Unprotected Pressurized Vessels Exposed to the Fires

Test ID	EXP1	EXP2	EXP3	EXP4	EXP5	EXP6	EXP7
Data source	Towsend et al. (1974)	Droste et al. (1999)	Moodie et al. (1988)	Droste and Schoen (1988)	Persaud et al. (2001)	Birk et al. (2006a)	Birk and Van der Steen (2006)
Tank Specifications							
Tank geometry	Hor. Cyl.	Hor. Cyl.	Hor. Cyl.	Hor. Cyl.	Hor. Cyl.	Hor. Cyl.	Hor. Cyl.
External diameter (m)	3.05	2.9	1.7	1.25	1.2	0.953	0.953
Total length (m)	18.3	7.6	4.88	4.3	4.0	3.07	3.07
Minimum wall thickness (mm)	15.9	14.9	12.1	5.9	7.1	7.4	7.1
Type of material	TC-128 LCS*	LCS*	BS1500 LCS*	STE-36 LCS*	LCS*	SA455 LCS*	SA455 LCS*
Stored substance	Propane	Propane	LPG	Propane	Propane	Propane	Propane
Tested filling level(s) (%)	96	22	22,36,38,58,72	50	20,41,60,85	78	80
Pressure Relief Device Data							
Set pressure (MPag)	1.86	2.70	1.43	1.56	1.82	2.63	1.9
Pressure relief device nominal diameter (mm)	80	–	100	25.4	90	15[‡]	21–24[‡]
Number	1	0	2	1	1	1	1

(Continued)

Table 5.1 Summary of Relevant Literature Experiments on Unprotected Pressurized Vessels Exposed to the Fires—*cont'd*

Test ID	EXP1	EXP2	EXP3	EXP4	EXP5	EXP6	EXP7
External Heat Source							
Type of source	JP-4 pool fire	Fuel oil pool fire	Kerosene pool fire	Fuel oil pool fire	Propane jet fire	Butane burners	Butane burners
Ambient temperature (°C)	21.1	N.D.	1 ÷ 6.4	10, 2, −3	2	17, 20	10−20
Exposure mode	Full engulfment	Full engulfment†	Full engulfment	Surrounding fire	Full engulfment	Partial engulfment	Partial engulfment
Flame temperature (°C)	650−990	1200	−	843	900−1100	815−927	815−927
Measured heat flux (kW/m²)	123−138	−	100	−	180−200	−	−

*LCS = low-carbon steel.
†Strong wind effects did not allow the effective engulfment of the vessel surface.
‡Computer operated.

Table 5.2 Summary of Relevant Literature Experiments on Pressurized Vessels Exposed to the Fires in Presence of Thermal Protection

Test ID	EXP8	EXP9	EXP10	EXP11	EXP12	EXP13	EXP14
Data source	Faucher et al. (1993)	Faucher et al. (1993)	Faucher et al. (1993)	Droste and Schoen (1988)	Landucci et al. (2009c)	Towsend et al. (1974)	Birk et al. (2006b)
Tank Specifications							
Tank geometry	Spherical	Spherical	Spherical	Hor. Cyl.	Hor. Cyl.	Hor. Cyl.	Hor. Cyl.
External diameter (m)	1.7	1.7	1.7	1.25	1.25	3.05	0.953
Total length (m)	–	–	–	4.3	2.68	18.3	3.07
Minimum wall thickness (mm)	10	10	10	6.4	5.1	15.9	7.4
Type of material	LCS*	LCS*	LCS*	STE-36 LCS*	P355NH LCS*	TC-128 LCS*	SA455 LCS*
Stored substance	Propane	Propane	Propane	Propane	LPG grade A	Propane	Propane
Tested filling level (%)	20	20	20	20	50	84	71,78
Pressure Relief Device Data							
Set pressure (MPag)	1.6	1.6	1.6	1.46	1.46	1.86	2.63
Pressure relief device nominal diameter (mm)	–	–	–	32	32	80	15
Number	1	1	1	1	1	1	1

(Continued)

Table 5.2 Summary of Relevant Literature Experiments on Pressurized Vessels Exposed to the Fires in Presence of Thermal Protection—*cont'd*

Test ID	EXP8	EXP9	EXP10	EXP11	EXP12	EXP13	EXP14
Insulating Coating Properties							
Coating type	Intumescent	Mineral cement	Mineral cement	Rock wool	Intumescent	Polyurethane	Ceramic fibers
Applied thickness (mm)	10	38	35	100[†]	10	3.2	13[‡]
External Heat Source							
Type of source	Butane burners	Butane burners	Butane burners	Fuel oil pool fire	Diesel pool fire	JP-4 pool fire	Butane burners
Ambient temperature (°C)	29	20	22	25	12	6	11–21
Exposure mode	Full engulfment	Full engulfment	Full engulfment	Surrounding fire	Full engulfment	Full engulfment	Partial engulfment
Flame temperature (°C)	1200	1200	1200	–	700–1300	650–990	815–927
Measured heat flux (kW/m²)	–	–	–	–	–	123–138	–

[*]LCS, low-carbon steel.
[†]Coating was incapsulated in a watertight steel sheet coating 1 mm thick; 30 mm air gap was left between the sheet and the coating.
[‡]Coating was incapsulated in a steel jacket of 3 mm thick.

Tables 5.1 and 5.2, the typical heat load due to an engulfing pool fire is in the order of 100–140 kW/m^2 while for large jet fires it may be as high as 200 kW/m^2 or even higher. One of the critical issues related to the reproducibility of the fire test is the stability of the flame and the need of limiting flame tilting and/or disturbances due to the wind (Birk, 2012, 1995). When experiments are conducted to compare different fire protection systems it is critical that fire conditions are consistent between tests. Figure 5.1(a) and (b) shows the evolution of an experimental setup required in order to limit wind effects (EXP12 in Table 5.2): the influence of wind on the test is evident in Figure 5.1(a), while Figure 5.1(b) shows the sand wall built to limit such effect. However, controlling or managing the wind effects is a difficult task. Some researchers have stopped using open pool fires and have developed

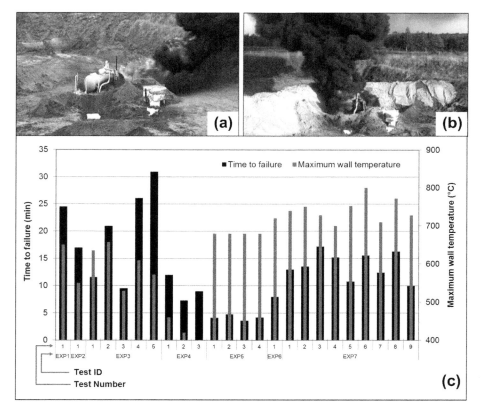

Figure 5.1 (a) Example of setup used for full engulfment fire tests on pressurized vessels in presence of wind. (b) Setup represented in panel (a) modified to limit the wind effect by a sand wall. (c) Time to failure (minutes) and maximum wall temperature (°C) obtained from the experimental fire tests on pressurized vessels available in the literature. See Table 5.1 for test number and ID. In test EXP4—Number 3, temperature data are not available. (For color version of this figure, the reader is referred to the online version of this book.)

multiburner arrays that can simulate full engulfing pool fires with high repeatability (Manu et al., 2009a; Faucher et al., 1993; Birk et al., 2006a).

The target equipment usually consists of small-scale cylindrical or spherical vessels featuring volumes ranging from 0.5 to 5 m^3 (Droste and Schoen, 1988; Moodie et al., 1988; Persaud et al., 2001; Birk et al., 2006a; Birk & Van der Steen, 2006), with liquefied petroleum gas (LPG) at different grades and featuring different filling levels. Only in the works of Townsend et al. (1974) and of BAM (Droste et al., 1999) large-scale vessels (respectively, 128 and 45 m^3) have been studied experimentally. These vessels are representative of both bulk storage and transport units. In these experiments the target equipment is usually equipped with sensors for the measurement of inner pressure, liquid and vapor temperatures and wall temperatures in various locations. In some cases radiometers were also provided for the evaluation of the heat flux received by the target equipment (Townsend et al., 1974; Persaud et al., 2001). These tests provide valuable information about how the vessel and its lading respond to accidental fire impingement. Key results from such tests include the following:

1. Wall temperature distribution vs time
2. Lading temperature distribution vs time
3. Tank fill level vs time
4. Tank pressure vs time
5. PRV behavior (if one is present)
6. ttf and conditions at failure, or time to empty
7. Modes of failure
8. Hazards due to blast, projectiles and resulting fire or explosion.

All these are needed to obtain an understanding of the key processes. These data are also needed to develop and validate detailed computer models of the process. Researchers need these data for baseline tanks without special protection. Then additional testing is needed with protected equipment under identical fire conditions to truly demonstrate the benefit of the protection.

The most significant result usually available from experimental tests is the determination of the ttf of equipment items exposed to fire, and the evaluation of the pressure and temperature conditions at the moment of failure. When protection systems are tested, as shown in Table 5.2, the aim is to verify the resistance of the vessel exposed to fire, thus the ability of the protection system to delay or, eventually, to avoid the failure.

The ttf is a key consideration for emergency response. Studies have shown that in very severe fires the ttf can be very short (minutes). This rapid failure can be predicted when the hoop stress exceeds the material's ultimate strength at the peak wall temperature. For this rapid failure the wall heating by fire must be very severe. However, with less severe fires failure is also possible but it may take more time (many minutes, or hours). With less severe fires creep rupture becomes a possibility. In creep rupture the hot material deforms with time under constant load and this may lead to failure after some time. This requires a specific and detailed calculation of failure as described by Birk and Yoon (2006).

Figure 5.1(c) collects the values of the ttf experienced during some experimental tests reported in the literature. The main features of the tests are reported in Table 5.1. Figure 5.1(c) also shows the maximum wall temperature recorded for the vessel shell during fire exposure. In the absence of any fire protection system (tanks were only provided with a pressure relief device), the ttf ranged between 3 and 35 min, depending on the type and size of equipment and on the fire exposure conditions. The ttf and the amount of product remaining change if the vessel is equipped with a PRV. If a vessel is equipped with such a PRV then there is a possibility that the PRV will empty the vessel before it fails. This depends on the details of the fire exposure.

The ttf depends on many factors including the tank design (wall thickness, diameter, length, head shape, etc.) and material, the setting of the PRV, the initial temperature and fill level and the fire conditions. Let us consider the case of a cylindrical vessel with a PRV set pressure of 2 MPa and a nominal (ambient temperature) burst pressure of 6 MPa. Let us consider the case where severe fire exposure causes the vessel to fail when the pressure rises to 2 MPa and the material degrades due to high wall temperatures (in vapor space) such that the burst pressure is reduced to 2 MPa (i.e. high temperature reduces material strength by 67%). The tank pressure is controlled to 2 MPa by the PRV. Therefore the ttf in this case would depend on how quickly the wall heats up to the failure temperature and the pressure builds up to 2 MPa. For the vessel to fail this must happen before the tank empties through the PRV.

The wall temperature in the vapor space depends mainly on the fire heat flux, the wall thickness and the liquid fill level in the tank. At high fill levels the vapor space wall area is small and is affected by the liquid. The liquid may not be in contact with the wall but it is close enough to affect the temperature in the vapor space by heat conduction and radiation. At lower fill levels (say below 60% full) the vapor space sees little cooling from the liquid. The tank wall thickness depends on the design pressure and the tank diameter. Therefore the larger the diameter, the thicker the wall, the longer it takes to heat-up. If the tank starts say at 80% full then some time will be needed for the PRV to expel product to reduce the fill level to expose more vapor space wall. If there is no PRV then the pressure will buildup without control until failure.

The rate of pressure rise in the tank depends on the fire-heated surface area that is wetted by liquid and this depends on the liquid fill level. At high fill levels there is a large wetted area for fire-heat input and the vapor space is very small. This results in very rapid pressurization. For example, a rail tank car initially filled to 94% will pressurize to the PRV set pressure in about 2 min when the tank is fully engulfed in a heavy hydrocarbon pool fire (Townsend et al., 1974). At lower fill levels the pressurization can be much slower (Balke et al., 1999).

The ttf can be changed with passive thermal protection. Thermal protection slows the rate of heating to the tank wall and lading. With thermal protection the rate of heat transfer can be reduced by 90% or more.

In some tests carried out with thermal protection, the failure was delayed and was not experienced even after more than 90 min of fire exposure. Only in trial EXP13

(Table 5.2), the protected tanker failed after 94 min of fire exposure (Townsend et al., 1974). Hence the tests provided indications on the resistance of the equipment given an impinging fire scenario, framing the timescale of the fired domino escalation in case of unprotected equipment, and giving information on the effectiveness of protection systems.

The tests yielded also useful indications on the dynamics of the main parameters governing the resistance of the vessels: inner fluid temperatures (Figure 5.2(a)), internal pressure (Figure 5.2(b)), and maximum wall temperature (Figure 5.2(c)). These data are of fundamental importance to support the theoretical analysis of vessel failure and to provide support for the development of vessel failure models. The effect of the above parameters on the structural resistance of a domino target will be discussed in the following sections.

Small-scale fire testing is a cost-effective procedure to obtain a preliminary assessment of fired domino effect and to gather information on the effectiveness of protection systems. However, some limitations are present in the scale-up of results. The reader is addressed to more specific publications for an in-depth discussion of this issue (Birk, 2012).

5.2.2   Increase in Vessel Wall Temperature

This is a key issue because it is the vessel wall temperature that determines the reduction in the pressure carrying ability of the vessel. The heat received at the vessel surface from the fire goes into heating the wall and the lading. The majority of heat from large luminous fires is transferred by thermal radiation (Keltner et al., 1990; Roberts et al., 2004). As the wall temperature increases the net heat flux to the tank wall will decrease because of convection and emission of radiation from the wall surface (both inner and outer surface).

The high convective heat transfer coefficient at the inner surface of the tank wetted by liquid keeps the wall cool and near the liquid temperature (Birk, 1983).

In the vapor space where the cooling effect from the vapor is small, the wall temperature rises rapidly to very high temperatures. Prediction of the liquid-wetted wall temperature is relatively simple because of the very high convective heat transfer coefficients. Radiation is negligible inside the vessel where the wall is wetted with liquid, except when the critical heat flux is exceeded (Collier and Thome, 1996). Such a case may be possible with some severe torch fires.

Accurate prediction of the vapor space wall temperature is much more difficult because several modes of cooling are possible including the following:

1. Free convection by saturated and superheated vapor
2. Forced convection with PRV open
3. Two-phase convection when PRV is open and two-phase swell takes place in liquid volume
4. Two-phase convection when PRV is open and liquid droplets are entrained at high fill levels
5. Thermal radiation from wall to wall and wall to liquid surface
6. Heat conduction through the wall from vapor-wetted (hot) to liquid-wetted (cool) wall at liquid/vapor interface.

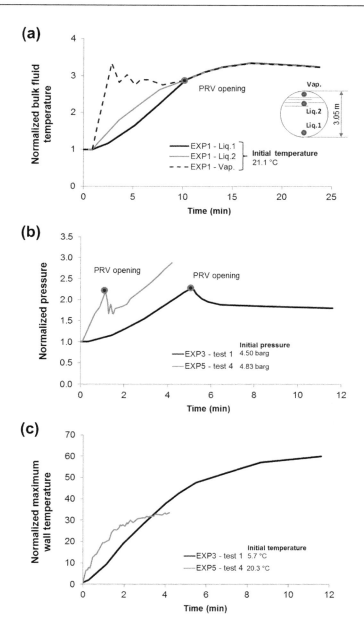

Figure 5.2 Sample results obtained in experimental tests on pressurized storage vessels (see Table 5.1 for details on test conditions): (a) normalized pressure; (b) bulk fluid (e.g. vapor and liquid) temperature; and (c) wall temperature. The variables are normalized with respect to the initial test conditions. (For color version of this figure, the reader is referred to the online version of this book.)

5.2.3   Increase of Internal Fluid Temperature

The heat load received by the target equipment, besides causing the temperature rise of the vessel shell, is also transferred to the inner fluid. Radiation and natural convection are the main heat transfer modes to the inner fluid. The inner fluid behavior is critical for the pressure rise in the vessel exposed to fire, since internal pressure is influenced by the fluid temperature. In vessels containing two-phase fluids (e.g. partially filled storage vessels, two- or three-phase separators, etc.), the internal pressure is also strongly affected by the liquid vapor pressure and thus by the temperature of the liquid (Birk and Cunningham, 1994b, 1996; Hadjisophocleous et al., 1990; Moodie, 1988). In the case of pressurized liquefied gases, the estimation of the temperature of the liquid is associated to the amount of energy that may be released in the case of vessel failure (Birk, 1995; Birk and Cunningham, 1996; Hadjisophocleous et al., 1990; Moodie, 1988; Reid, 1979). In particular, the liquid heat-up process affects the strength of a boiling liquid expanding vapor explosion (BLEVE), that is the instantaneous vaporization and expansion of a pressurized liquefied gas following the catastrophic rupture of the vessel. More details on BLEVEs and their consequences are reported in the literature (Reid, 1979; Roberts, 1981; Manas, 1984; Abbasi and Abbasi, 2007; American Institute of Chemical Engineers Center for Chemical Process Safety (AIChe-CCPS), 1996; Van Den Bosh and Weterings, 2005; Prugh, 1991).

In a vessel exposed to fire containing a liquid phase, as in a storage vessel, the temperature in the liquid phase is not uniform as evidenced in several experimental tests. The liquid temperature varies significantly from the bottom, where cooler liquid layers stratify, to the top of the vessel, where the hotter liquid accumulates. Figure 5.2(a) shows the liquid temperature behavior in two positions at different heights in the liquid phase (see sketch in Figure 5.2(a); EXP1 in Table 5.1; Townsend et al., 1974). Data show a 50% increase of the top temperature with respect to the bottom part of the liquid prior to the opening of the PRV. As remarked by several authors (Birk and Cunningham, 1996; Hadjisophocleous et al., 1990; Moodie, 1988; Venart, 1999; Lin et al., 2010), the stratification is due to a buoyancy-driven flow, caused by the more rapid temperature increase of the liquid in contact with the vessel walls heated by the fire with respect to the bulk of the liquid. Birk (1983) accounted for this process in his modeling of rail tank cars in fires based on data from Manda (1975). Moodie (1988) remarked that the heat transfer at the liquid–wall interface may be either convective, or boiling may occur (subcooled, saturated nucleate, or film boiling). Hence the heated liquid, less dense with respect to the bulk, rises to form a warm interfacial zone between the liquid and the vapor. This warm liquid layer drives the pressure in the vessel. This means that the bulk liquid is in a subcooled state. The vapor above the liquid is either saturated or superheated. This stratified temperature distribution will remain until other processes will dissipate the temperature gradients (such as bulk mixing, phase change and heat conduction, etc.). In particular, the opening of the PRV, installed on the storage vessels for overpressure protection, is likely to cause such intense mixing (Birk, 1995; Birk and Cunningham, 1996; Hadjisophocleous et al., 1990; Moodie, 1988).

When the PRV opens and releases vapor this results in a small pressure drop in the vessel which triggers boiling in the liquid. The release of vapor removes the superheated vapor and this is replaced by saturated vapor generated in the warm liquid. The bubbles rise up in the liquid generating convective currents that mix the warm upper layer with the liquid bulk. Over time with continued PRV action the liquid will become near isothermal and saturated. For a full-scale tank cars (128 m^3) engulfed in fire (Manda, 1975) it took about 10 min of continued PRV action to mix the liquid to isothermal conditions. Figure 5.2(a) shows how the opening of the PRV causes the evolution of the stratified liquid into a homogenous medium. Figure 5.2(a) also reports the vapor space temperature behavior registered in the experiment. As shown in the figure, a significant difference is registered with respect to the liquid side, since the vapor features lower heat transfer coefficients than the liquid phase. This temperature difference significantly affects the wall temperatures and may generate relevant thermal stresses, as discussed in Section 5.2.2.

After the PRV opening, liquid swelling may occur, affecting both the quality of the PRV venting (e.g. increasing the liquid content in the discharged vapor) and the average temperature. This is evident in Figure 5.2(a) that shows a uniform and constant liquid temperature in the final part of the test. The phenomenon of swelling and, more in general, a significant liquid carryover during the venting phase are experienced in all the experiments reported in Table 5.1 where a PRV was present. Hence after PRV opening the liquid level falls quickly, reducing the energy potentially released in case of vessel failure, thus mitigating the severity of escalation scenarios (Birk and Cunningham, 1994a).

5.2.4   Increase of Internal Pressure

As previously mentioned, during the heating process the liquid or the liquefied gas evaporates, causing a pressure increase inside the vessel. The pressure increase has a strong effect on the resistance of the structure, since it causes a higher mechanical hoop stress (American Society of Mechanical Engineers (ASME), 1989). Figure 5.2(b) shows the pressure increase due to fire exposure for two small-scale propane tanks in two different fire conditions. In the case of EXP3 (Table 5.1), a full engulfment in a pool fire is reproduced, while in EXP5 a propane torch fire partially engulfs the target vessel. The tests show a different behavior with pressure due to the action of the PRV, in both cases designed according to API standard RP 520 (American Petroleum Institute (API), 2000) and 521 (API, 2008). In the case of pool fire test, the pressure rise was limited due to the opening of the PRV. This behavior is in agreement with other results obtained by tests carried out with an engulfing pool fire, both on small- and large-scale vessels (Townsend et al., 1974; Droste et al., 1999; Droste and Schoen, 1988; Landucci et al., 2009c). On the contrary, due to the higher heat load (almost doubled with respect to EXP3, see Table 5.1), the pressure rise in the torch fire test (EXP5 in Table 5.1) was faster and not controlled by the PRV action, resulting in a lower ttf. Roberts et al. (2004) remarked on the criticality of the PRV design procedure, which might result in undersizing of the PRV for certain situations in which intense heat loads are experienced due to jet fires. The current sizing formulas are biased for large storage tanks and may not be suitable for smaller tanks in very severe fires. It should also be pointed out that all the tests summarized in Table 5.1, with the exception of EXP2 (Droste et al., 1999), ended in tank failure, even though the tanks used in the tests were were equipped with a PRV. This clearly shows that a PRV is a mitigation measure that alone is not sufficient to prevent tank failure under intense fire exposure conditions.

5.2.5 Direct Fire Damage: Weakening of Structural Elements

As shown by the experimental studies discussed in Section 5.2.1, one of the critical elements that may lead to the failure of equipment exposed to fire is the weakening of shell material due to the vessel wall heat-up. As evidenced in all the experiments reported in Tables 5.1 and 5.2, the temperature increase is more critical in the vessel section in contact with the vapor phase (e.g. the upper section of a storage vessel).

In order to represent the severity of the shell heat-up experienced by vessels exposed to fire, Figure 5.1(c) reports the maximum temperature, T_{max}, measured during the fire tests mentioned in Section 5.2.1, while Figure 5.2(c) shows the typical behavior of the temperature measured during the fire test of nonprotected vessel. Depending on the severity of the fire, the heat-up process may take only few minutes and the vessel shell temperature may rise up to 600–700 °C. Therefore, even if pressure relief devices are present, limiting the internal pressure buildup in the vessels, the strength of the vessel shell material is reduced by the temperature increase, jeopardizing the resistance of the whole structure and possibly leading to its failure.

Several approaches in the literature propose the estimation of the maximum temperature rise in order to directly predict the failure conditions (Khan and Abbasi, 1998). Such approach is not conservative and does not take into account the effective resistance of the vessels and specific geometrical factors. In order to carry out a more detailed stress analysis for failure prediction (Section 5.4.1), the effective resistance of the construction material is usually compared with parameters representative of the stresses acting on the equipment exposed to fire (ASME, 1989). The resistance may be synthetically represented by the maximum allowable stress, σ_{adm}, which is derived from the material mechanical properties as follows (ASME, 1989; Luecke et al., 2005):

$$\sigma_{adm} = (\text{yield or tensile strength})/(\text{saftey factor}) \qquad (5.1)$$

where the safety factor is a number higher than one.

Figure 5.3 reports a chart for the assessment of σ_{adm}, based on experimental results (Roberts et al., 2000; Persaud et al., 2001; Birk, 1995; ASME, 1989; Luecke et al., 2005) obtained for the more common types of construction materials employed for process, storage, and transportation equipment. In order to give a more immediate representation of the material weakening with respect to optimum operative

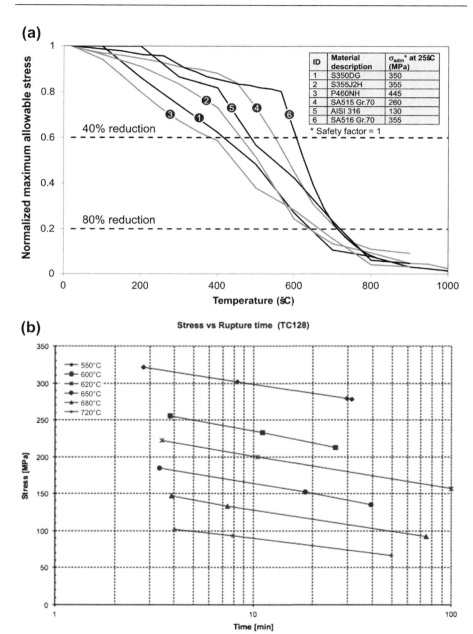

Figure 5.3 Effect of temperature on the normalized maximum admissible stress of construction steels used in the process industry. (a) The admissible stress is normalized with respect to the value at 25 °C reported in the right panel at the correspondent ID; (b) high-temperature stress rupture data for TC-128 Tank Car Steel (North America). (For color version of this figure, the reader is referred to the online version of this book.)
Source: From Birk and Yoon (2004).

conditions, Figure 5.3 shows the temperature variation of the normalized maximum allowable stress of the material ($\sigma_{\text{adm\_norm}}$) defined as follows:

$$\sigma_{\text{adm\_norm}} = \sigma_{\text{adm}}(T)/\sigma_{\text{adm}}(25\,^{\circ}\text{C}) \tag{5.2}$$

where $\sigma_{\text{adm}}(T)$ is the maximum allowable stress at the temperature of interest and $\sigma_{\text{adm}}(25\,^{\circ}\text{C})$ is evaluated at 25 °C (ambient temperature).

As shown in the chart reported in Figure 5.3(a), below 300 °C the strength of the materials considered is not drastically affected by the shell heat-up, while above this temperature limit a drop in the vessel resistance occurs, leading to the loss of 80–90% of its strength for temperatures higher than 600 °C (i.e. the temperature experienced by the steelwork in the case of severe fire exposure).

However, this type of analysis may not be sufficient for tanks that are exposed to less severe fires where wall temperatures may not reach 600 °C or higher. This could also be applied to tanks equipped with thermal protection if the protection has defects (Birk et al., 2006b). Creep rupture can be important at temperatures above 400 °C depending on the steel type and the level of stress (Boyer, 1988).

With high-temperature creep the ttf is a function of stress and temperature. For this analysis high-temperature stress rupture data are needed for the subject steel. These data are obtained by heating a tensile test sample to a certain temperature (and maintaining this temperature) and then stressing the sample under constant applied force. This force is then maintained until the sample fails and the time is recorded. Figure 5.3(b) is an example of high-temperature stress rupture data from Birk and Yoon (2004). This type of analysis may be important to consider with thermally protected tanks where we want to delay failure for a certain time (such as 90 min).

5.2.6   Mechanisms of Escalation for Vessels Containing Nonboiling Liquids

Process or storage vessels containing nonboiling liquid are usually designed to operate at nearly atmospheric pressure, but accounting for the liquid hydrostatic load (API, 2003). This type of vessel usually consists of a vertical cylinder with the base plate which can be either flat or conical, directly fixed on the ground. Both floating and fixed roofs may be used depending on the type of stored substance.

Several standard procedures for sizing and selection of these equipment items are available (API, 2003). The shell consists of different welded steel courses (1.8 or 2.4 m tall) where thickness decreases along the height, ranging between 5 mm at the top and 20 mm in the bottom course. Tank height rarely exceeds 20 m while tank diameter may reach up to 80 m. Hence, the thickness over diameter ratio is extremely small when compared to pressurized vessels (typically at least two orders of magnitude) (Landucci et al., 2012).

Therefore, under thermal loads, extremely thin-walled shells are expected to have large deflections that will develop buckling modes depending on the constraints due to boundary conditions and to the influence of other parts of the shell that have not been yet affected by fire, as evidenced in a recent work by Godoy and Batista-Abreu (2012). It is worth noting that due to the large size of these equipment items, a uniform fire exposure mode is not likely and partial engulfment with strong temperature gradients might occur leading to severe thermal dilatation of the structure (Lees, 1996; Landucci et al., 2009a; Cozzani et al., 2006; Godoy and Batista-Abreu, 2012). Therefore, the mechanism of failure in this case is related not only to the thermal weakening of the structure material (Section 5.2.5) but also to the structural instability related to buckling and strong dilatation. This was experienced in several past accidents involving atmospheric vessels, in which the collapsed tanks presented bulges and wave-type deformation among the shell (Lees, 1996; Gómez-Mares et al., 2008; Landucci et al., 2012; Godoy and Batista-Abreu, 2012; Chang and Lin, 2006 and references cited therein).

Another criticality related to atmospheric vessels exposed to fires is the possibility of boilover. This phenomenon occurs in storage tanks in which a separated layer of water may be present, such as in the case of crude oil storage (Shaluf and Abdullah, 2011). When the water, which is generally present at the bottom of storage tanks, vaporizes due to the fire heating, the resulting large volume of steam violently ejects a portion of the stored liquid and damages the tank itself, with possible rupture of the vessel. If the liquid is flammable, this results in an enormous fire enlargement, in the formation of a fireball, and in frothing over the entire tank content. More details on boilover experiments and numerical simulation are reported elsewhere (Shaluf and Abdullah, 2011; Argyropoulos et al., 2012; Fan et al., 1995).

5.3 Escalation Potential of Fire Scenarios

5.3.1 Industrial Fires Leading to Escalation

The experimental results discussed in Section 5.2 provided evidence that the severity of fires impacting on a target equipment is a critical issue that influences the dynamic temperature profile of involved vessels and the phenomena that may eventually lead to vessel failure. Industrial fires can have very different characteristics and encompass an extensive range of size (Lees, 1996; Roberts et al., 2000; Van Den Bosh and Weterings, 2005; Roberts et al., 2004; Steel Construction Institute (SCI), 1992; Cowley and Johnson, 1991). Fire properties are influenced by leakage rates and their time dependence, type of flammable substance burning, storage and discharge conditions, surrounding topside structures and equipment, and ambient wind conditions. Despite the large number of possible fire events, few categories of industrial fires are relevant for escalation leading to domino effect, as shown in Table 5.3. This section deals with the detailed characterization of fires able to trigger escalation, evidencing the relevant features of the industrial fires and the potential secondary effects due to the ignition of flammable material involved in domino accidents.

Table 5.3 Classification of Fires in the Process Industry, Evidencing Escalation Criteria Based on the Heat Load Received by the Target (Cozzani et al., 2006)

Features Relevant for Escalation		Type of Fire					
		Confined Jet Fire	Open Jet Fire	Confined Pool/Tank Fire	Open Pool Fire	Fireball	Flash Fire
Combustion mode		Diffusive	Diffusive	Diffusive	Diffusive	Diffusive	Premixed
Total heat load (kW/m²)		150–400	100–400	100–250	50–150	150–280	170–200
Radiative contribution (%)		66.7–75	50–62.5	92–100	100	100	100
Convective contribution (%)		25–33.3	37.5–50	0–8	0	0	0
Flame temperature range (K)		1200–1600	1200–1500	1200–1450	1000–1400	1400–1500	1500–1900
Escalation criteria for fire impingement	Atmospheric equipment	Escalation always possible	Escalation always possible	Escalation always possible	Escalation always possible	$Q_{HL} > 100$	Flammable vapors ignition*
	Pressurized equipment	Escalation always possible	Escalation always possible	Escalation always possible	Escalation always possible	Escalation unlikely	Escalation unlikely
Escalation criteria for distant source radiation	Atmospheric equipment	$Q_{HL} > 15$	$Q_{HL} > 15$	$Q_{HL} > 15$	$Q_{HL} > 15$	$Q_{HL} > 100$	Escalation unlikely
	Pressurized equipment	$Q_{HL} > 40$	$Q_{HL} > 40$	$Q_{HL} > 40$	$Q_{HL} > 40$	Escalation unlikely	Escalation unlikely

Q_{HL}: thermal flow received by the fire in kW/m².
*For floating roof tanks.

5.3.2 Jet Fires

Loss of containment from a pressurized vessel containing a flammable gas or a flashing liquid may result in a jet fire, if ignition takes place. A jet fire is a turbulent flame that may have a significant length in the direction of the release, due to the high kinetic energy of the jet (Van Den Bosh and Weterings, 2005). The jet may be vapor, liquid or two-phase and this of course affects the fire conditions and duration. The duration of the jet fire may be long (minutes or hours) or short (seconds) depending on the source of the jet. For this reason, a long-duration jet fire can be considered as a steady source of radiation (Lees, 1996; Roberts et al., 2000; Van Den Bosh and Weterings, 2005; Roberts et al., 2004). According to the features reported in Table 5.3, jet fires are considered among the more critical fire events with respect to escalation, due to the high heat loads and flame temperatures. Besides the contribution of radiation, the convective term is also considerable due to the high flame speed. Thus, as shown in Table 5.3, the heat load is higher than in low-velocity flames such as pool fires. Further details on jet fire features and jet fire modeling are reported elsewhere (Lees, 1996; Van Den Bosh and Weterings, 2005; AIChe-CCPS, 2000).

The relatively high frequencies of occurrence and the high damage radius cause the jet fire to be among the scenarios that more frequently result in escalation. A jet fire may cause an escalation as a result of two different events: direct flame impingement and steady radiation from the flame zone.

Jet fire impingement is a well-known cause of escalation, as shown by the analysis of past accidents where domino effects took place (Lees, 1996) and by the experiments previously discussed. Damage due to heat transfer caused by a distant stationary radiation source may result in vessel failure, although higher values of the ttf are expected and more time is available for active mitigation measures. Thus in the absence of direct flame impingement, the possibility of escalation needs to be specifically evaluated with models able to take into account the resistance of the target equipment according to the severity of the fire. This aspect will be discussed in Section 5.4. Table 5.3 summarizes the escalation criteria associated to jet fires derived in a previous study (Cozzani et al., 2006a). As shown in the table, in the case of fire impingement, escalation should always be considered possible, while for distant source radiation the proposed escalation criterion varies according to the target equipment, with atmospheric equipment having less resistance than pressurized equipment.

5.3.3 Pool Fires

A pool fire consists in the uncontrolled combustion of the vapors generated from a pool of a flammable liquid. The pool is usually formed as a consequence of a loss of containment from atmospheric or pressurized vessels (in this case, only the residual liquid after the flash and the entrainment forms the pool). Hence, the duration of pool fires may be relevant and higher than that of jet fires, resulting in a steady radiation source (Lees, 1996; Roberts et al., 2000; Van Den Bosh and Weterings, 2005; Roberts et al., 2004). Further details on pool fire description and modeling are reported in the literature (Lees, 1996; Van Den Bosh and Weterings, 2005).

Also in the case of pool fire, direct flame engulfment and steady radiation from a distant source may be responsible for escalation. Even if the heat load associated to pool fires is usually lower than that associated to jet fires, due to the limited convective term associated to the flame velocity (Table 5.3), an engulfment in flames may cause the failure of the target vessel, resulting in an escalation. This was confirmed by several past accidents and by the experimental tests discussed in Section 5.2.1. Therefore, the escalation should be considered possible for any target vessel located inside the pool area. In the case of a target vessel receiving a steady heat radiation but not engulfed in flames, the possibility of escalation should be addressed taking into account the intensity of heat radiation and the characteristics of the target vessel, as in the case of jet fires.

Table 5.3 reports escalation criteria derived for pool fires in a previous study (Cozzani et al., 2006). As shown in the table, the results are similar with respect to the ones obtained for jet fire escalation, since both types of flame are steady sources of heat, and thus are able to trigger similar failure mechanism.

5.3.4 Other Fire Scenarios

The possibility of other industrial fire scenarios and in particular of flash fires and fireballs to trigger escalation phenomena should also be considered. A typical scenario (Maremonti et al., 1999; Tacoma, Washington, USA, 2007; Toronto, Ontario, Canada, 2008; Bennet et al., 2008) could include a release of flammable product with delayed ignition that becomes a flash fire or local vapor cloud explosions (VCE) that works its way back to the source of the release resulting in a jet or pool fire. The subject of VCE was covered in Chapter 4.

A flash fire may be described as the "slow" laminar or low-turbulent combustion of a gas or vapor cloud, i.e. without the production of a blast wave due to the low confinement and/or congestion of the cloud, or to the low reactivity of the flammable mixture (e.g. a stratified cloud, a nonhomogeneous fuel–air mixture or a flammable cloud with average concentration close to the lower or upper flammability level). The flash fire phenomenon is characterized by a low flame speed, hence typical duration may range from few milliseconds to few seconds for large stratified flammable clouds (Lees, 1996; AIChe-CCPS, 1996; Landucci et al., 2011; Pontiggia et al., 2011). Therefore, these events have a characteristic duration that is a few orders of magnitude lower than the ttf due to heat radiation of any type of process vessels (e.g. 3–35 min). As a consequence, flash fires are not likely to result in the damage of a secondary vessel due to heat radiation. Nevertheless, an escalation may be caused by the direct ignition of flammable material due to flame impingement. A single case is usually of relevance within a process plant: the ignition of vapors above the roof of a floating roof tank, starting a tank fire.

Thus, as summarized in Table 5.3, it may be concluded that escalation due to a flash fire is unlikely and the secondary events due to flash fires are likely to involve only floating roof tanks containing high volatility flammable liquids. Nevertheless, it must be remarked that the above discussion does not consider the effect of stratification that may take place in the dispersion of heavy gas clouds in atmosphere.

Several events involving the dispersion of such type of clouds resulted in long-lasting flames, having a duration comparable to those present in a fireball, and capable of generating secondary fires due to the ignition of flammable material as wood, plastic and rubber (Lees, 1996; Landucci et al., 2011; Pontiggia et al., 2011; Venart, 2007; Cox, 1976).

Different considerations might be referred to fireballs, which are diffusive flames caused by the immediate ignition of a flammable vapor or gas cloud generated by a catastrophic loss of containment caused by the failure of a storage or process vessel. In this case a high-intensity heat radiation is associated with the combustion process. A fireball has a limited duration, although longer than the one of a flash fire (usually much less than 60 s) (Van Den Bosh and Weterings, 2005; AIChe-CCPS, 1996). In the assessment of escalation possibility, two different situations should be considered: (1) flame engulfment, if the target vessels are comprised within the cloud extension; and (2) radiation from distant source without flame impingement, for target vessels at distances higher than the flame radius.

The possibility of escalation following the damage of equipment items caused by fireball heat radiation, in the case of both full engulfment and distant source radiation, is a controversial point. As a matter of fact, the relatively short duration of the fireball makes questionable the possibility of radiation damage to process vessels. A specific assessment was carried out to shed some light on this point, evidencing significant differences between the behavior of atmospheric and pressurized vessels (Cozzani et al., 2006a). While the resistance of pressurized vessels allows for a ttf about an order of magnitude higher than the expected duration of the fireball, in the case of atmospheric vessels the fire duration might be sufficient to cause the failure of the equipment. Moreover, as discussed above, in the case of flame engulfment of floating roof storage tanks, the escalation may be caused by the ignition of flammable vapors above the roof sealing or by the failure of the roof sealing. Thus, even if the escalation due to fireball radiation involving atmospheric vessels seems to be credible only for a limited number of very severe scenarios, a specific assessment may be necessary.

5.4   Modeling the Behavior of Equipment Exposed to Fire

5.4.1   Approaching the Quantitative Assessment of Escalation

The analysis of fire scenarios relevant for escalation evidenced that in several situations the resistance of the target equipment needs to be specifically evaluated, taking into account the characteristics of the fire scenario and the actual mode of exposure to fire. Hence, reliable tools for the prediction of the ttf are required in order to determine the likelihood of escalation.

Modeling the failure of equipment exposed to fires is a very complex and multi-disciplinary task. This section aims at the description of the more critical aspects of the problem and at the discussion of tools available for the analysis of the problem, taking into account different levels of detail. Besides semiempirical correlations and simplified criteria for estimation of the vessel failure, more complex two-dimensional and three-dimensional (3D) models will be presented (Hadjisophocleous et al., 1990; Venart, 1986).

5.4.2 Theoretical Analysis of Vessel Heat-up and Failure Conditions

The detailed modeling of pressure vessels exposed to fire has been a topic of research around the world since the early 1970s. Several sophisticated models have been developed (Graves, 1973; Beynon et al., 1988; Ramskill, 1988; Johnson, 1998; Birk, 2000, 2004, 2005). In most cases these models were developed to help design thermal protection and pressure relief systems for rail tank cars in North America. These various models consisted of a number of submodels including fire heat transfer by radiation and convection, wall heat conduction, convection and radiation in the vapor space, convection and boiling in the liquid space, thermodynamic process in the liquid and vapor including temperature stratification, thermal protection degradation, wall stress and strength including high-temperature stress rupture and creep, PRV action and mass flow. The main objective of the models was to predict ttf for a range of fire scenarios and protection schemes.

Figure 5.4(a) reports the schematization of a target equipment engulfed by fire. When the target equipment is exposed to the fire, the outer surface receives the heat load (Q_{HL}, generally expressed in kilowatt per square meter). Part of the incoming heat flux is reflected and emitted from the external surface according to the material emissivity, and part is transmitted by conduction through the insulating coating layer (if present) and the equipment shell to the inner surface of the equipment wall. The heat is then transferred by radiation and convection into the lading. As evidenced in Section 5.2, if the equipment inner surface is in contact with a liquid phase, the heat exchange with the liquid is sufficiently high to have a strong cooling effect, especially if boiling takes place (e.g. if the target is a tank containing a pressurized liquefied gas). Where the equipment inner surface is in contact with the vapor phase, the internal heat exchange coefficient is so low that a very rapid rise in the equipment wall temperature is usually recorded.

In order to evaluate the vessel heat-up, the following energy balance can be written as follows (Birk, 2006; Landucci et al., 2009c; Hadjisophocleous et al., 1990; Moodie, 1988):

$$c\rho\frac{\partial T}{\partial t} = \frac{\partial}{\partial x}\left(k_x\frac{\partial T}{\partial x}\right) + \frac{\partial}{\partial y}\left(k_y\frac{\partial T}{\partial y}\right) + \frac{\partial}{\partial z}\left(k_z\frac{\partial T}{\partial z}\right) \tag{5.3}$$

where T is the temperature, t, the time, c, the heat capacity, ρ, the density, and k, the thermal conductivity. If the thermal conductivity is supposed to be uniform and isotropic, the following expression is obtained:

$$c\rho\frac{\partial T}{\partial t} = k\left(\frac{\partial^2 T}{\partial x^2} + \frac{\partial^2 T}{\partial y^2} + \frac{\partial^2 T}{\partial z^2}\right) \tag{5.4}$$

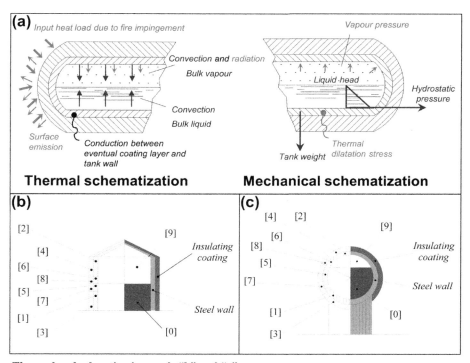

Thermal nodes location in panels "b" and "c"

[0]	Liquid space
[1]	Vessels wall in contact with the liquid phase but not in the zone impinged by the flame
[2]	Vessels wall in contact with the vapour phase but not in the zone impinged by the flame
[3]	Insulated material layer (if present) at the level of the liquid phase but not in contact with the flame
[4]	Insulated material layer (if present) at the level of the vapour phase but not in contact with the flame
[5]	Vessels wall in contact with the liquid phase and in the zone impinged by the flame
[6]	Vessels wall in contact with the vapour phase and in the zone impinged by the flame
[7]	Insulated material layer (if present) at the level of the liquid phase and in contact with the flame
[8]	Insulated material layer (if present) at the level of the liquid phase and in contact with the flame
[9]	Vapour space

Figure 5.4 Analysis of vessels exposed to fire: (a) schematization of thermal and mechanical balances; (b) schematization of thermal nodes for lumped parameters analysis of atmospheric vessels; (c) schematization of thermal nodes for lumped parameters analysis of pressurized vessels. (For color version of this figure, the reader is referred to the online version of this book.)

In order to solve Eqn (5.4) in each point of the structure an initial temperature should be assumed (e.g. the ambient temperature prior to the exposure to fire). Several further boundary conditions are needed, according to the schematization reported in Figure 5.4(a). First of all, the heat load Q_{HL} must be evaluated in each point of the external surface of the equipment on the basis of the analysis of the primary fire, thus determining the impact vector. Three simplifying limit conditions may be identified (Landucci et al., 2009a; Persaud et al., 2001): (1) full or partial fire engulfment; (2) distant source radiation; and (3) jet fire flame partial impingement.

In the first case the equipment surface receives a uniform heat flux distribution, thus in order to simplify the problem Q_{HL} can be estimated on the basis of the strength of the primary fire and may be imposed as a boundary condition at all the points of the external surface.

In the second case the equipment target receives the heat only by radiation, thus the heat is redistributed on each point of the target equipment surface as follows:

$$Q_{HL} = \tau_a \ F_w \ Q_{flame} \tag{5.5}$$

where F_w is the geometrical view factor (Lees, 1996; Van Den Bosh and Weterings, 2005), τ_a is the atmospheric transmissivity (Lees, 1996; Van Den Bosh and Weterings, 2005) and Q_{flame} is the total heat flux (kilowatt per square meter) generated by the flame and exchanged by radiation.

The geometrical view factor is the ratio between the received and the emitted radiation energy per unit area. The factor is determined by the flame dimensions and shape, and by the relative position and orientation of the receiving target equipment. The atmospheric transmissivity (τ_a) accounts for the fact that the radiation emitted by the fire is partly absorbed by the air present between the radiator and the target equipment, due to the presence of water vapor and carbon dioxide in the atmosphere, which are the main absorbing components within the wavelength area of the heat radiation. More details for the calculation of F_w and τ_a are extensively reported elsewhere (Lees, 1996; Van Den Bosh and Weterings, 2005).

The third possible type of exposure, e.g. partial engulfment due to fire impact on part of the target equipment surface, is characterized by the combined convection and radiation heat flux in the flame impact zone and to the radiation in the nearby zones.

Once the incoming heat load Q_{HL} is evaluated, the internal heat exchange with the inner fluid in contact with the vessel wall needs to be assessed. Hence a variable heat load Q_{conv} on the inner tank shell surface, due to the convective heat transfer to the fluid (gas or liquid phase), can be expressed as follows:

$$Q_{conv} = h \ (T - T_B) \tag{5.6}$$

The value of Q_{conv} depends on the inner wall temperature and on the bulk fluid temperature (T_B, which can be the liquid or the vapor temperature, depending on the position). The values of the heat transfer coefficient h between the equipment wall and the gas or liquid may be estimated on the basis of conventional semiempirical correlations for natural convection heat transfer (Kern, 1950; Knudsen et al., 1999).

Table 5.4 Energy Balance Equations Implemented in the Lumped Parameters Model and Heat Transfer Coefficient Correlations with Corresponding Reference

	Equation	Description
(A)	$-H_V \dfrac{dm_L}{dt} = q_L + q_{LV}$	Mass and energy balance on the liquid node
(B)	$q_{LV} = h_{LV} A_{LV}(T_V - T_L)$	Heat exchanged between liquid and vapor
(C)	$q_L = h_L A_{1,\text{int}}(T_1 - T_L) + h_L A_{5,\text{int}}(T_5 - T_L)$ $+ \sigma F_{w,L}(T_6^4 - T_L^4) A_{6,\text{int}} + \sigma F_{w,L}(T_2^4 - T_1^4) A_{2,\text{int}}$	Heat exchanged by the liquid with the vessel wall
(D)	$\dfrac{dL}{dt} = \dfrac{dm_L}{dt}\left(\dfrac{1}{\rho_L \frac{\pi}{4} D^2} \right)$	Dynamic behavior of the liquid level
(E)	$m_V c_V \dfrac{dT_V}{dt} = q_V + q_{LV} + \dfrac{dm_L}{dt}\left(c_V(T_V - T_L) - RT_L\right)$	Energy and mass balance on the vapor node
(F)	$\dfrac{dP}{dt} = \dfrac{\rho_V}{m_V}\left(\dfrac{P}{\rho_L}\dfrac{dm_L}{dt} + \dfrac{Rm_V}{M_w}\dfrac{dT_V}{dt} \right)$	Dynamic behavior of internal pressure
(G)	$q_V = h_V(T_2, T_L) A_{2,\text{int}}(T_2 - T_V) + h_V(T_6, T_V) A_{6,\text{int}}(T_6 - T_V)$	Heat exchanged by the vapor with the vessel wall
(H)	$d_s \rho_s c_s \left(\dfrac{A_{j,\text{int}} + A_{j,\text{ext}}}{2} \right) \dfrac{dT_j}{dt} = 2\dfrac{k_i}{d_i}(T_{j+2} - T_j) A_{j,\text{ext}}$ (a) in contact with liquid phase or	Energy balance on the j-th vessel wall node ($j = 1, 2, 5, 6$)
	$\quad -h_L(T_j - T_L) A_{j,\text{int}}$	
	$\quad -h_V(T_j - T_V) A_{j,\text{int}}$ (b) in contact with vapor phase	

(Continued)

Table 5.4 Energy Balance Equations Implemented in the Lumped Parameters Model and Heat Transfer Coefficient Correlations with Corresponding Reference—*cont'd*

	Equation	Description
(I)	$$d_i\rho_i c_i\left(\frac{A_{j,\text{int}}+A_{j,\text{ext}}}{2}\right)\frac{dT_j}{dt}=I_{\text{ext}}A_{j,\text{ext}}-2\frac{k_i}{d_i}(T_j-T_{j-2})A_{j,\text{int}}$$ $$-h_a(2T_j-T_{j-2}-T_a)A_{j,\text{est}}-\sigma\varepsilon_i[(2T_j-T_{j-2})^4-T_a^4]A_{j,\text{ext}}$$	Energy balance on the j-th insulating coating node in contact with the flame ($j=7,8$)
(J)	$$d_i\rho_i c_i\left(\frac{A_{j,\text{int}}+A_{j,\text{ext}}}{2}\right)\frac{dT_j}{dt}=(I_f+h_f(T_f-2T_j+T_{j-2}))A_{j,\text{ext}}$$ $$-2\frac{k_i}{d_i}(T_j-T_{j-2})A_{j,\text{int}}-\sigma\varepsilon_i[(2T_j-T_{j-2})^4-T_a^4]A_{j,\text{ext}}$$	Energy balance on the insulating coating node not in contact with the flame ($j=3,4$)
(K)	$$h_V(T_n,T_V)=3.41\ (T_n-T_V)^{0.25}$$	Heat transfer coefficient wall-vapor (W/m²K) (Kern, 1950)
(L)	$$h_L^I(T_n,T_L)=0.138\frac{k_L}{Z_L}\left(\frac{Z_L^3\rho_L^2 g\beta(T_n-T_L)}{\mu_L^2}\right)^{0.36}\left(\left(\frac{c_L\mu_L}{k_L}\right)^{0.175}-0.55\right)\quad(I)$$ $$h_L^{II}(T_n,T_L)=0.225\left(\frac{h_L^I(T_n-T*(P))c_L}{H_V}\right)^{0.69}\left(\frac{Pk_L}{\sigma_t}\right)^{0.31}\left(\frac{\rho_L}{\rho_V}-1\right)^{0.33}\quad(II)$$	Heat transfer coefficient wall-liquid (W/m²K) $h_L=\text{MAX}(h_L^I;h_L^{II})$ (Knudsen et al., 1999).
(M)	$$h_{LV}(T_L,T_V)=0.138\frac{k_V}{Z_V}\left(\frac{Z_V^3\rho_V^2 g\beta(T_V-T_L)}{\mu_V^2}\right)^{0.36}\left(\left(\frac{c_V\mu_V}{k_V}\right)^{0.175}-0.55\right)$$	Heat transfer coefficient between liquid and vapor (W/m²K) (Knudsen et al., 1999).

If there is a PRV and it opens to release vapor then there may be a significant forced convection component as well.

The interface between the liquid and vapor phases plays an important role for the thermal balance and significantly affects the heat transfer coefficient. Table 5.4 reports some examples of correlations that take into account the heat transfer between the target equipment and the internal fluid. Very high wall temperatures are possible in the vapor space and in this case thermal radiation may dominate. The vapor space wall sees the liquid surface and other parts of the vapor space wall. The vapor may also intervene by absorbing and emitting radiation.

$$Q_{rad\_i} = \sigma \varepsilon_i \left(T_i^4 - T_j^4 \right) \quad i \neq j \tag{5.7}$$

where Q_{rad_i} is the radiant heat flux received or emitted in a single portion (i) of the inner surface of the target equipment from/to a different portion (j), $\sigma = 5.6703 \times 10^{-8}$ W m^{-2} K^{-4} is the Stephan–Boltzmann constant and ε_i is the internal emissivity of the material.

As discussed in Section 5.2.2, the increase of the wall temperature depletes the resistance of the shell material, reducing its strength. At the same time, as evidenced in Figure 5.4(a), the increasing temperature gradients have a negative influence on the mechanical force balance leading to distortion and thermal stress (i.e. hot vessel top expands causing the vessel to bend). While the temperature rises, the thermal and hoop stresses on the vessel wall become more critical. Moreover, the heat-up of the vessel lading results in an increase of the internal pressure P, thus affecting the hoop stress on vessel wall (see detailed discussion in Section 5.2.4). The other mechanical loads present on the tank are due to the weight of the structure and of the inner fluid. In the case of atmospheric equipment the liquid hold up due to hydrostatic pressure plays a major role in the determination of the stress together with thermal dilatation (Section 5.2.6).

In order to predict the failure of the structure, the detailed evaluation of the temperature and stress profile is needed. The failure conditions are strictly dependent on structural design: geometry, material, and boundary conditions. For the process equipment of interest (horizontal cylindrical vessels, vertical cylindrical vessels, etc.) exposed to an intense heat flux, the failure conditions may be reached by wall thinning due to hoop stress and high-temperature material degradation (Roberts et al., 2000; Manu et al., 2009a; Birk, 1995; ASME, 1989). Hence the failure may be driven by plastic deformation following the yielding of the construction material. As recently investigated by Manu (2008) and Manu et al. (2009a,b), failure may also follow the occurrence of creep in the construction steel due to the extremely severe temperature conditions. Another failure possibility is related to vessel wall buckling: following a little deformation, the structure may change its initial configuration assuming another equilibrium configuration losing its integrity.

In order to simplify the problem, a mathematical representation of the failure condition can be introduced: the failure criterion. A failure criterion is generally derived by a direct comparison between a parameter representative of the stress field over the equipment shell and a parameter representative of the mechanical properties of the structure. More details on definition and application of failure criteria are reported in previous publications (Roberts et al., 2000; Birk, 1995; Moodie, 1988; ASME, 1989; Graves, 1973; Forrest, 1985; Beynon et al., 1988; Birk, 1988; Ramskill, 1988; Birk and Leslie, 1991; Johnson, 1998; Shebeko et al., 2000; Salzano et al., 2003; Gong et al., 2004). It is worth to remark that the application of each failure criterion requires specific information which are a function of the analysis level of detail. Hence in the following the available approaches for the evaluation of the vessel resistance to fire, and thus of the ttf, will be presented and associated to the corresponding failure criterion.

5.4.3   Lumped Parameters Modeling Approaches

Several lumped parameters models are available in the literature for the assessment of the thermal response of both the vessel and its content when subjected to fire (Landucci et al., 2009a; Birk, 2006; Persaud et al., 2001; Hadjisophocleous et al., 1990; Moodie, 1988; Graves, 1973; Forrest, 1985; Beynon et al., 1988; Birk, 1988; Ramskill, 1988; Birk and Leslie, 1991; Johnson, 1998; Shebeko et al., 2000; Salzano et al., 2003; Gong et al., 2004). These models are in general dedicated to the assessment of the response of horizontal cylindrical LPG vessels to engulfing fire, by predicting the wall temperature rise and internal pressure. This type of models (Birk, 1983) can give reasonable predictions but it is worth noticing that for high values of wall temperature, due to the strong weakening of structural steel (Birk and Yoon, 2006) advanced tools capable of more accurate predictions should be considered for more precise determination of vessel ttf (Section 5.4.4).

Nevertheless, simplified models usually provide support to the analysis of fire protection system influence or to the PRV sizing. Only few models manage different vessel geometries and different vessel categories, e.g. ENGULF (Ramskill, 1988), SAFIRE (Forrest, 1985) (currently implemented in DIERS), SuperChems® code (available at http://www.ioiq.com/superchems/features.aspx) and few others are able to predict correctly the influence of the PRV action, e.g. HEAT-UP (Beynon et al., 1988) and the model developed by Salzano et al. (2003).

More recently, an extended model was developed in order to evaluate the ttf of insulated or unprotected vessels of any type, both atmospheric and pressurized, undergoing different modes of fire exposure, taking into account the influence of the PRV actions (Landucci et al., 2009a; Gubinelli, 2005). The model is based on a lumped approach for the calculation of the time–temperature and time–pressure profiles, but is based on an extended validation based on experimental data. The approach attempts to divide the equipment in different zones (or nodes), defined on the basis of the fire scenario. Each node can be described by a simple set of parameters (Landucci et al., 2009a; Persaud et al., 2001; Hadjisophocleous et al., 1990; Moodie, 1988). Figure 5.4(b) and (c) shows the nodes defined respectively for atmospheric and pressurized vessels. Table 5.4 summarizes the equations used to model the thermal nodes defined in the lumped model (Landucci and Cozzani, 2009). Quite obviously, depending on the fire scenario, some nodes may be neglected in the simulations. As shown in Table 5.4, the thermal balance equations are solved in each node in order to reproduce the dynamic behavior of the vessel.

The parameters represent physical quantities (e.g. temperature, pressure, thermal conductivity, etc.) averaged over each node. Boundary conditions together with global conservation laws lead to a system of equations which determines the parameters of interest and, in particular, the temperature in each node. This allows the calculation of temperature–time profiles as a function of the radiation mode and intensity on the vessel. The estimation of these parameters is used for the calculation of the hoop stress generated in each zone of the vessel (namely, σ_{eq}) and to compare it with the maximum allowable tensile strength of the vessels material (σ_{adm}, see Section 5.2.5).

The ttf of the vessel is the time at which the following condition occurs at least in one of the nodes representative of the vessel wall (e.g. the nodes labeled with 1, 2, 5 and 6 in Figure 5.4(b) and (c)):

$$\sigma_{eq} = \sigma_{adm} \tag{5.8}$$

5.4.4 Simplified Correlations for Vessel Failure Prediction

Since in the risk assessment of complex industrial areas a huge number of possible targets of escalation triggered by fire may be identified, a very high number of simulations may be required to calculate the ttf for any exposed vessel in any possible fire scenario. Even if lumped models or other similar tools are characterized by a low computational time, their use may require a relevant effort in the analysis of extended industrial clusters. Moreover, even lumped models, although simplified, require the definition of several input parameters for each simulation (e.g. vessel geometrical data, properties of vessel content, radiation mode, etc.).

Thus, a further simplified approach might be preferred in the domino risk assessment of complex and extended industrial sites. In conventional analysis, empirical correlations were used in the attempt to quantitatively approach the problem. Rules of thumb showing different degrees of simplification were adopted to directly predict the failure conditions: escalation is considered likely if a given value of damage threshold for radiation intensity is exceeded (e.g. 37.5 kW/m^2) (Cozzani et al., 2006; Health and Safety Executive, 1978; British Standards Institution (BSI), 1990; Mecklenburgh, 1985) or if a critical rise in the vessel wall temperature is obtained (taking into account only the radiative contribution to the heat-up of the vessel) (Khan and Abbasi, 1998). In the probabilistic method proposed by Bagster and Pitblado (1991), both the intensity of the radiation and the distance between the primary fire and the potentially involved secondary unit are considered in the evaluation of escalation probability, which is carried out by the adoption of an empirical correlation describing the decay of the physical effect intensity as a quadratic function of the distance from the center.

Such failure criteria were often applied without any reference to the actual expected duration of the fire and to vessel structural features, and in most cases their validation was carried out through simplified simulations neglecting the actual geometry of the target vessel. As a consequence of their empirical nature that disregards the complexity of the phenomena leading to the failure of equipment exposed to fire, these approaches should be applied with caution, since, depending on the actual fire scenario and vessel geometry, they may be underconservative or strongly overconservative.

More recently, simplified correlations based on the results of validated model were introduced for the calculation of ttf (Cozzani et al., 2006a; Landucci et al., 2009a; Landucci and Cozzani, 2009). A specific approach was introduced to define simple analytical functions for the assessment of vessel ttf in different fire scenarios. The correlations are based on empirical functions correlating an extended dataset of ttf values obtained applying the lumped parameters model described in Section 5.4.3 (Landucci et al., 2009a). The dataset was obtained identifying the more important categories of secondary equipment involved in domino accidents, defining reference geometrical characteristics on the basis of typical design data used by engineering companies in the oil and gas sector. The more credible primary fire scenarios were selected on the basis of the discussion in Section 5.3. Further details concerning the reference vessels and the scenarios considered in the analysis are reported elsewhere (Cozzani et al., 2005, 2006b). Reference charts were obtained for ttf, such as the one shown in Figure 5.5 for pressurized equipment having nominal design pressure ranging between 1.5 and 2.5 MPa. In these plots, the ttf is plotted against the heat load, Q_{HF} for different types of vessels, whose geometry is resumed in the figure.

Table 5.5 reports the simplified correlations obtained for both atmospheric and pressurized equipment from the analysis of the dataset. Correlations in Table 5.5 yield conservative data for the ttf of vessels having volumes and operating pressures within the range specified in the table. Clearly enough, the data obtained from the correlations should be considered only a conservative estimate of the actual ttf, but are based on a physical model of the phenomena involved in vessel failure. Thus, the simplified correlation may be useful at least for a preliminary but more sound assessment of escalation credibility, taking into account at least some features of the target vessels, as the volume and the range of design pressure.

5.4.5   Advanced Tools for the Assessment of Escalation Triggered by Fire

As mentioned in Section 5.4.2, the detailed calculation of the ttf of process equipment requires a precise description of vessel geometry, design data, flame geometry and fire exposure modes. These input data are needed to evaluate in detail the behavior of the wall temperature and the consequent mechanical response. Complicating factors may be, e.g. the nonhomogeneous heat load on the external vessels shell, the presence of a thermal protection (e.g. an insulating layer), and the presence of thermal gradients in the fluid contained in the vessel. Therefore, if a more precise assessment of ttf is required for a specific fire scenario, a detailed description of these phenomena is required. Specific 3D approaches may be introduced to carry out a detailed analysis. Finite Element Modeling (FEM) is the main technique proposed to approach the problem (Roberts et al., 2000; Landucci et al., 2009a; Birk, 2006; Manu et al., 2009a; Landucci

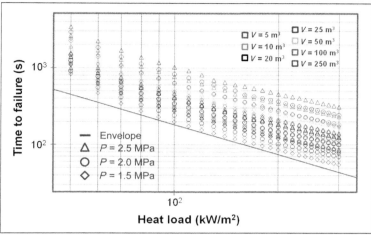

Nom. Volume	Diameter	Length	Wall thickness (mm)		
V (m³)	(m)	(m)	P_{des} = 1.5 MPa	P_{des} = 2.0 MPa	P_{des} = 2.5 MPa
5	1.0	6.1	11	14	17
10	1.2	7.7	11	14	17
20	1.5	9.7	12	16	20
25	1.7	10.5	15	20	24
50	2.1	13.2	17	23	29
100	2.8	18.0	18	24	30
250	3.8	24.0	24	32	40

Figure 5.5 Example of results obtained with the lumped parameters model described in Section 5.4.3: evaluation of time to failure (s) as a function of the heat load (kW/m²) for pressurized vessels. Vessels features are reported in the panel; V = nominal volume (m³) ranging between 5 and 250 m³; P_{des} = design pressure = 1.5, 2.0 and 2.5 MPa. (For color version of this figure, the reader is referred to the online version of this book.)

Table 5.5 Envelope Correlations for Distant Source Radiation. Time to Failure (ttf) is Expressed in Seconds, Volume (V) in Cubic Meter, and Heat Flux (Q_{HL}) in Kilowatt per Square meter

Equipment	Atmospheric Equipment	Pressurized
Volume range (m³)	25–17500	5–250
Design pressure range (MPa)	0.1	1.5–2.5
Envelope correlation	$\ln(\text{ttf}) = -1.13 \cdot \ln(Q_{HL})$ $- 2.67 \cdot 10^{-5} V + 9.9$	$\ln(\text{ttf}) = -0.95 \cdot \ln(Q_{HL})$ $+ 8.845 \, V^{0.032}$

et al., 2009b; Paltrinieri et al., 2009; Landucci et al., 2009c; Tan et al., 2011; Bi et al., 2011) from a structural standpoint. Advances in computational fluid dynamics (CFD) promises to provide techniques to study the thermodynamics and fluid dynamics in the vessel with and without two-phase effects and pressure relief (Yoon and Birk, 2004).

Figure 5.6 shows the examples of FEM results. In particular, Figure 5.6(b) reports the maps representing the stress intensity field acting on the equipment shell obtained from the detailed temperature simulations in Figure 5.6(a), obtained from the simulation of EXP1 in Table 5.1. The calculation of the detailed temperature and stress intensity maps allows the application of the correct failure conditions and thus an accurate calculation of the equipment ttf (Landucci et al., 2009a,b,c; Paltrinieri et al., 2009). Figure 5.7 shows the sample results for CFD modeling results for the internal heat transfer and temperature stratification (Yoon and Birk, 2004). In particular, Figure 5.7(a) shows the transient temperature distributions and the development of liquid stratification in an LPG vessel engulfed by the fire. Figure 5.7(b) shows the temperature distributions on liquid walls and free surfaces of various tanks with defective insulation after 180 s of engulfing fire exposure.

It is worth to mention that the higher complexity of the models, allowing on one side a deeper investigation of the behavior of the target equipment, needs advanced expertise and skills for the model setup and detailed input data, which might not be available for the analysis. A more extended discussion of the potentialities of FEM and CFD simulations for both fire and overpressure damage is reported in Chapter 11.

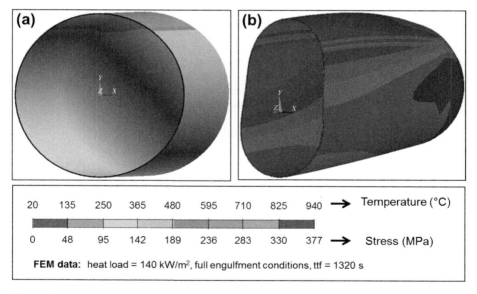

Figure 5.6 Examples of result obtained with the FEM simulations: (a) temperature map (in °C) obtained for the vessel tested in EXP1; (b) stress intensity map calculated by imposing the temperature profile shown in panel (a). (For color version of this figure, the reader is referred to the online version of this book.)

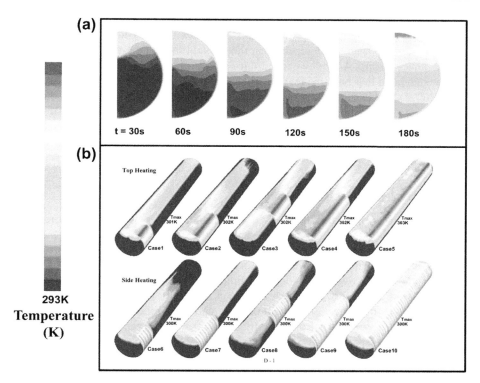

Figure 5.7 Sample results from CFD simulation showing (a) temperature stratification in an LPG tank before PRV opens (Yoon and Birk, 2004) simulated with heat fluxes of 108 kW/m^2 on liquid space wall and 50 kW/m^2 on vapor space wall. (b) Temperature buildup and convective currents inside rail tank car with localized heating of vessel by fire (Yoon and Birk, 2004). (For color version of this figure, the reader is referred to the online version of this book.)

5.5   Prevention of Escalation Caused by Fire

5.5.1   Time to Effective Mitigation

The analysis of the behavior of vessels exposed to fire shows that the key issue in the evaluation of the credibility of escalation by fire is the determination of the ttf of the target equipment. This represents the available time lapse for the activation of emergency procedures and of mitigation devices, as well as for the deployment of emergency teams aimed at the mitigation and/or suppression of the primary fire. It may be reasonably assumed, on the basis of experience, that the full and correct activation of the planned specific emergency measures should prevent, in general, escalation triggered by fire.

Therefore, besides the prevention of primary events, all the available strategies aimed at the prevention of escalation caused by fires prescribe measures to delay or eliminate secondary vessel failure, in order to guarantee a sufficient time for effective mitigation (tem), that is assumed as the time needed to put in place emergency measures that will effectively prevent the escalation caused by fire.

The tem is site specific, depending on the availability and the time of deployment of emergency teams. A survey carried out on several chemical and process sites in Europe indicated that baseline tem may be assumed to be of about 30 min (Landucci et al., 2009a). In the case of transportation and shipment of dangerous goods, the tem can be up to 3 times higher, due to the unavailability of mitigation systems if a fire accident occurs in nonindustrial areas (Hobert and Molag, 2006). In this section, mitigation measures that may be adopted to increase the ttf of a vessel up to values higher than the site tem will be discussed.

5.5.2 Passive Protection Systems

A generic passive protection device is a system or a barrier which does not require either power or external activation to trigger the protection action (Lees, 1996; AIChE-CCPS, 2001). PFP systems are usually based on the implementation of a set of barriers aimed at delaying the vessel failure, hence providing additional time for the implementation of active protections (e.g. firefighting) or of mitigation measures (e.g. blowdown and depressurization) (Di Padova et al., 2011; Tugnoli et al., 2012).

Usually PFPs of process equipment include a pressure relief device. This is aimed at limiting the vessel internal pressure by the control of the vapor pressure increase due to the rise of the liquid temperature (Lees, 1996; Gómez-Mares et al., 2008; Roberts et al., 2000; Landucci et al., 2009a; Birk, 2006; Towsend et al., 1974; Droste and Schoen, 1988; Faucher et al., 1993; Landucci et al., 2009c; Birk et al., 2006b; Shebeko et al., 2000; Roberts et al., 2010). A further mitigation barrier may be provided by the application of fireproofing materials (cementitious or vermiculite sprays, intumescent, mineral or ceramic fibers, etc.) (SCI, 1992; American Petroleum Institute (API), 1999) which are able to resist to the fire exposure delaying the temperature rise of the protected structural elements (SCI, 1992; Di Padova et al., 2011; Tugnoli et al., 2012).

Cost and maintenance issues require the identification of fire protection zones where the risk reduction justifies the application of PFP systems. Technical standards provide criteria for the application of PFP in both onshore (API, 1999; API, 2011; Health and Safety Executive, 1996) and offshore facilities (Less, 1996; AIChE-CCPS, 2003; Tugnoli et al., 2012; NORSOK, 2008; API, 2006; API, 2007; Det Norske Veritas (DNV), 2008; International Organization for Standardization (ISO), 1999, 2007).

However, an intense work was devoted in the past fourty years to the detailed to the detailed assessment of the performance of fireproofing materials and pressure relief devices in the protection of process equipment in fixed installations, as well as of tanks used for the road and rail transportation of liquefied pressurized gases. Table 5.2 reports the relevant large- or medium-scale fire tests for the analysis of protected vessels exposed to fires with different fireproofing materials. Birk et al. (Birk, 1999; Birk et al., 2006b; Vander Steen and Birk, 2003) specifically assessed the influence of defects in the insulating layer on the behavior of pressurized tanks engulfed by fires.

Table 5.6 Physical Properties of the More Common Fire-Proofing Materials. FSI: Flame Spread Index; SDI: Smoke Development Index According to ASTM E84 (American Society for Testing Materials (ASTM), 1994)

Coating Characteristics				
Commercial name	Chartek 7	Fendolite M-II	ISOVER MD2	Pyrocrete 241
Definition	Epoxy intumescent	Vermiculite spray	Fibrous mineral wool	Cementitious inorganic formulation
Fire resistance indexes (ASTM E84)	FSI = 25 SDI = 130.9	FSI = 0 SDI = 0	FSI = 0 SDI = 0	FSI = 0 SDI = 0
Thermal conductivity (W/mK)	0.066	0.03–0.2*	0.38	0.9
Heat capacity (J/kgK)	1172	970	920	1507
Surface emissivity	0.9–0.95	0.65–0.75	0.65–0.75	0.65–0.75
Density (kg/m^3)	1000	680	100	850

*Low value for thicknesses >35 mm; high value for thicknesses <35 mm.

Table 5.6 reports the characteristics of the more common commercial fireproofing materials.

Figure 5.8 reports some representative examples of application. Figure 5.8(a) shows an example of epoxy intumescent material applied on an LPG tank. During the fire exposure this type of coating expands as a result of the specific thermal degradation mechanism activated by the exposure to fire. A "foaming" effect takes place, which increases the insulating properties of the material, due to the decomposition of the volatile products and to the charring process (Landucci et al., 2009c; Roberts et al., 2010; API, 1999). Figure 5.8(a) shows the coating after a 90 min fire exposure. More details on the behavior of this category of materials are reported elsewhere (Gomez-Mares et al., 2012a,b; Jimenez et al., 2006).

Figure 5.8(b) shows a board of vermiculite spray material. These coating materials exploit the water content to limit the temperature of the nonexposed side (Jin et al., 2000). Water evaporation keeps the temperature to a constant plateau (close to the saturated steam temperature) for a time that depends on the coating thickness, which is related to its water content (Faucher et al., 1993; Roberts et al., 2010; Jin et al., 2000). On the basis of experimental outcomes of Faucher et al. (1993), Roberts et al. (2010), and Jin et al. (2000), a minimum required thickness of 35 mm is considered necessary for the activation of this peculiar behavior. An effective value of thermal conductivity is reported in Table 5.6, obtained from experimental data fitting. Figure 5.8(b) shows the material after 90 min fire exposure.

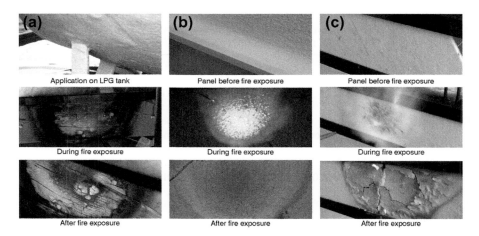

Figure 5.8 Different coating materials exposed to fire: (a) epoxy intumescent; (b) vermiculite spray; (c) cementitious inorganic formulation. (For color version of this figure, the reader is referred to the online version of this book.)

Mineral wool is widely used for conventional thermal insulation applications (Malloy, 1969). Some applications are the fire protection of building structures or the heat insulation of steam generators, reactors, and pipes. The term "mineral wool" typically refers to several types of insulation material (Malloy, 1969):

1. Stone wool: material consisting of natural minerals like basalt or diabase;
2. Slag wool: material from blast furnace slag (the layer that forms on the surface of molten metal);
3. Glass wool: material consisting of fiberglass wool;
4. Ceramic fiber: accumulation of fibers of different lengths and diameters, produced synthetically from mineral raw materials, typically silicates.

Mineral wool can be used in two different insulation forms: blanket (batts and rolls) and loose fill. The low density of this material (one order of magnitude lower than other products, as reported in Table 5.6) makes it attractive, but high thicknesses are needed for an adequate protection. Thus, different layers of material are usually applied, and are incapsulated with a watertight steel sheet leaving an air gap between the external surface of the coating and the steel itself. This enhances the thermal inertia of the system and it acts as a radiation shield. For example in North America rail tank cars carrying certain dangerous goods must have a PRV and PFP (US Federal Government, 2008). The most common system used for LPG and propane tank cars is a 13 mm layer of ceramic blanket covered by a 3 mm steel jacket.

Cementitious inorganic fireproofing formulations are also recommended for the fire protection of structural steel and bulkheads. These materials may also be used to upgrade the fire resistance of existing concrete (Malloy, 1969). Nevertheless, as shown in Figure 5.8(c), the exposed surfaces may be strongly damaged by the fire.

Moreover this material has a higher thermal conductivity with respect to other fireproofing materials (Table 5.6), thus the application of high thicknesses is required.

The application of fireproofing materials and/or of other passive protection devices such as a PRV may significantly delay the failure of a vessel exposed to fire. The PRV has also the further advantage of reducing the inventory of hazardous substances present in the vessel, conveying the vent to a safe disposal system. However, the application of passive protection always needs to be combined to other mitigation or emergency strategies, since a PRV or a fireproofing layer is only able to delay and not to avoid the failure of the protected vessel (Lees, 1996; Gómez-Mares et al., 2008; Roberts et al., 2000; Landucci et al., 2009a; Birk, 2006; Towsend et al., 1974; Droste and Schoen, 1988; Faucher et al., 1993; Landucci et al., 2009c; Birk et al., 2006b; API, 1999; AIChE-CCPS, 2003; Society of Fire Protection Engineers (SFPE) & National Fire Protection Association (NFPA), 2002).

5.5.3   Active Protection Systems

A generic active protection device is a system or a barrier which requires an external automatic activation (Lees, 1996; Roberts et al., 2000; Shoen and Droste, 1988; Planas-Cuchi et al., 1996; Lev, 1991; Shirvill, 2004; Davies and Nolan, 2004; Roberts, 2004a,b; Roberts et al., 2004). Thus, a common architecture is featured by these systems to achieve their function. In particular, they are composed of three subsystems in chain: a fire or gas detection system, a treatment system (logic solver, releasing panel or alarm advising operator) and an actuation system (mechanical, instrumented, human, etc.) (Lees, 1996; Roberts et al., 2000; Jin et al., 2000; NFPA, 2009).

The detection system has the primary function of alerting personnel of the existence of a fire condition and it may also be used to automatically activate emergency alarms, initiate Emergency Shutdown (ESD), isolate fuel sources, start fire water pumps and activate fire extinguishing systems. Typical detection systems are fusible loop systems and electrical fire detectors (Lees, 1996; Roberts et al., 2000; Roberts et al., 2004; Jin et al., 2000; NFPA, 2009; De Dianous and Fievez, 2006).

Actuation systems can be divided into two different categories: (1) systems for the delivery of firefighting agents (such as water or water-based foam), in particular spray systems, sprinklers and water deluge systems (Lees, 1996; Roberts et al., 2000; Shoen and Droste, 1988; Planas-Cuchi et al., 1996; Lev, 1991; Shirvill, 2004; Davies and Nolan, 2004; Roberts, 2004a; De Dianous and Fievez, 2006); (2) ESD Systems and Emergency Depressurization Systems (API, 2000; API, 2008; Roberts et al., 2004). Both categories are aimed to provide protection for storage tanks and vessels having large flammable gas or liquid inventories and for specific process vessels potentially exposed to primary fires.

According to the mentioned architecture of the typical active protection systems, the performance analysis may be focused on three main aspects: availability, effectiveness and response time. Dealing with complex systems, unavailability is expressed as probability of failure on demand (PFD) and can be evaluated using fault tree analysis (FTA) (Lees, 1996; AIChe-CCPS, 2000). Several studies are available in the technical literature concerning water-based fire protection systems performances presenting the required value for reliability (>98%) or PFD value resulting from FTA varying in the range 10^{-3}–10^{-2} (Manyatt, 1988; Finucane, 1988; Hauptmanns et al., 2008). Considering the effectiveness of the active protection systems, two different approaches are available in the literature, depending on the type of protection. In the case of sprinkler systems, statistical data are available from different sources (Bukowski et al., 2002), reporting effectiveness as the ratio between the number of cases in which the sprinkler system successfully controlled the fire and total number of cases in which system activates, presenting an overall success rates value of at least 95%. In the case of water deluge systems, experimental researches were carried out by Roberts (2004a,b) on LPG storage vessels, while Shoen and Droste (1988), Shirvill (2004), and Hankinson and Lowesmith (2004) deeply investigated offshore facilities in order to evaluate effectiveness by studying the variation introduced by protection system action in some physical parameters of the fire attack, such as the incident thermal radiation levels or the decrease in the rate of temperature rise of the target vessel.

Also in this case further protection or mitigation measures should be considered to avoid the failure of the protected vessel. In particular, additional cooling from emergency teams should be assured to provide a more effective protection of the vessel and a wider safety margin.

5.6 Conclusions

This chapter was devoted to the analysis of domino effect induced by fires. The failure mechanism of process and storage equipment induced by fires was described considering both the available experimental studies and the significant theoretical aspects. The analysis allowed the identification of the main parameters affecting the possible escalation leading to the failure of vessels or process equipment exposed to the fire, evidencing the complexity of the problem.

On one side, the analysis of the available experimental studies provided the typical dynamic of vessels parameters leading to escalation triggered by fire. In particular, the pressure rise and the wall temperature increase were recognized as the key factors affecting the integrity of the target structure. On the other side, the type and severity of the fire, as well as the type of vessel exposure, determine the credible fire scenarios leading to the escalation. Modeling approaches of different complexity were discussed for the calculation of the ttf of the target vessel resulting from the exposure to fire.

Atmospheric vessels show a strong vulnerability to fire and ttf may be very short. Pressurized vessels may have a higher resistance, but they also can fail in minutes if exposed to severe fires. Escalation triggered by fire should always be considered possible. Active and passive protections may be crucial to reduce escalation probability, but a detailed analysis of their reliability and performance is needed to quantify the robustness of such mitigation barriers in the prevention of escalation.

References

Abbasi, T., Abbasi, S.A., 2007. The boiling liquid expanding vapour explosion (BLEVE): mechanism, consequence assessment, management. Journal of Hazardous Materials 141 (3), 489–519.

American Institute of Chemical Engineers Center for Chemical Process Safety AIChe-CCPS, 2000. Guidelines for Chemical Process Quantitative Risk Analysis, second ed. AIChE, New York, NY.

American Institute of Chemical Engineers Center for Chemical Process Safety AIChE-CCPS, 2003. Guidelines for Fire Protection in Chemical, Petrochemical, and Hydrocarbon Processing Facilities. AIChE, New York, NY.

American Institute of Chemical Engineers Center for Chemical Process Safety AIChE-CCPS, 2001. Guidelines for Engineering Design for Process Safety. AIChE, New York, NY.

American Institute of Chemical Engineers Center for Chemical Process Safety, AIChE-CCPS, 1996. Guidelines for Evaluating the Characteristics of Vapour Cloud Explosions, Flash Fires and BLEVEs. AIChE, New York, NY.

American Petroleum Institute, 2000. API Recommended Practice 520-Sizing, Selection, and Installation of Pressure-Relieving Devices in Refineries, seventh ed. API, Washington, DC.

American Petroleum Institute, 2008. API Standard 521-Guide for Pressure-Relieving and Depressuring Systems: Petroleum Petrochemical and Natural Gas Industries-Pressure Relieving and Depressuring Systems, fifth ed. API, Washington, DC.

American Petroleum Institute API, 1999. Fireproofing Practices in Petroleum and Petrochemical Processing Plants, API Publication 2218, second ed. API, Washington, DC.

American Petroleum Institute API, 2001. Design and Construction of LPG Installations, API Publication 2510, eighth ed. API, Washington, DC.

American Petroleum Institute API, 2003. Welded Steel Tanks for Oil Storage – API Standard 650, eighth ed. API, Washington, DC.

American Petroleum Institute API, 2006. Recommended Practice for Design of Offshore Facilities against Fire and Blast Loading, API RP 2FB, first ed. API, Washington, DC.

American Petroleum Institute API, 2007. Recommended Practice for Fire Prevention and Control on Fixed Open-Type Offshore Production Platforms, API RP 14G, fourth ed. API, Washington, DC.

American Society for Testing Materials ASTM, 1994. ASTM E84-94. Standard Test Method for Surface Burning Characteristics of Building Materials. ASTM International, West Conshohocken.

American Society of Mechanical Engineers ASME-Boiler and Pressure Vessel Committee, 1989. Boiler and Pressure Vessel Code, Section VIII, Division 2. ASME, New York, NY.

Argyropoulos, D., Christolis, M.N., Nivolianitou, Z., Markatos, N.C., 2012. A hazards assessment methodology for large liquid hydrocarbon fuel. Journal of Loss Prevention in the Process Industries 25 (2), 329–335.

Bagster, D.F., Pitblado, R.M., 1991. The estimation of domino incident frequencies – an approach. Process Safety and Environment Protection 69/B, 195–199.

Balke, C., Heller, W., Konersmann, R., Ludwig, J., 1999. Study of the Failure Limits of a Tank Car Filled with Liquefied Petroleum Gas Subjected to an Open Pool Fire Test. Federal Institute for Materials Research and Testing (BAM), Berlin, Germany.

Bennet, J., Kassabian, A., Randsalu, E., Silvestri, B., 2008. Fire Protection Engineer Report of Investigation No. 212-007-2008. Fire Investigation Services, Toronto.

Beynon, G.V., Cowley, L.T., Small, L.M., Williams, I., 1988. Fire engulfment of LPG tanks: HEAT-UP a predictive model. Journal of Hazardous Materials 20, 227–238.

Bi, M., Ren, J., Zhao, B., Che, W., 2011. Effect of fire engulfment on thermal response of LPG tanks. Journal of Hazardous Materials 192, 874–879.

Birk, A.M., 1983. Development and Validation of a Mathematical Model of a Rail Tank-Car Engulfed in Fire. Department of Mechanical Engineering, Queen's University, Kingston, Ontario, Canada.

Birk, A.M., 1988. Modelling the response of tankers exposed to external fire impingement. Journal of Hazardous Materials 20, 197–225.

Birk, A.M., 1995. Scale effects with fire exposure of pressure-liquefied gas tanks. Journal of Loss Prevention in the Process Industries 8 (5), 275–290.

Birk, A.M., 1999. Tank-Car Insulation Defect Assessment Criteria: Thermal Analysis of Defects (TP 13518E report). Transport Canada.

Birk, A.M., 2000. Review of AFFTAC Thermal Model. Transport Canada.

Birk, A.M., 2004. Tank2004-A Computer Code for Modelling Pressure Vessels Containing Pressure Liquefied Gases in Fires. User Guide. Thermdyne Technologies Ltd.

Birk, A.M., 2005. Thermal Model Upgrade for the Analysis of Defective Thermal Protection Systems. Transportation Development Centre, Transport Canada.

Birk, A.M., 2006. Fire Testing and Computer Modelling of Rail Tank-Cars Engulfed in Fires: Literature Review. (TP 14561E).

Birk, A.M., 2012. Scale considerations for fire testing of pressure vessels used for dangerous goods transportation. Journal of Loss Prevention in the Process Industries 25, 623–630.

Birk, A.M., Cunningham, M.H., 1994a. A Medium Scale Experimental Study of the Boiling Liquid Expanding Vapour Explosion. Transport Canada.

Birk, A.M., Cunningham, M.H., 1994b. The boiling liquid expanding vapour explosion. Journal of Loss Prevention in the Process Industries 7 (6), 474–480.

Birk, A.M., Cunningham, M.H., 1996. Liquid temperature stratification and its effect on BLEVEs and their hazard. Journal of Hazardous Materials 48 (1–3), 219–237.

Birk, A.M., Leslie, I.R.M., 1991. State of the art review of pressure liquefied gas container failure modes and associated projectile hazards. Journal of Hazardous Materials 28, 329–365.

Birk, A.M., Van der Steen, J.D.J., 2006. On the transition from Non-BLEVE to BLEVE failure for a 1.8 m^3 Propane tank. Transactions of the ASME 128, 648–665.

Birk, A.M., Yoon, K.T., 2004. High Temperature Stress Rupture Testing of Sample Tank-Car Steels. Transportation Development Centre, Transport Canada.

Birk, A.M., Yoon, K.T., 2006. High temperature stress-rupture data for the analysis of dangerous goods tank-cars exposed to fire. Journal of Loss Prevention in the Process Industries 19, 442–451.

Birk, A.M., Cunningham, M.H., Ostic, P., Hiscoke, B., 1997. Fire Tests of Propane Tanks to Study BLEVEs and Other Thermal Ruptures: Detailed Analysis of Medium Scale Test Results. Transport Canada.

Birk, A.M., Poirier, D., Davison, C., 2006a. On the response of 500 gal propane tanks to a 25% engulfing fire. Journal of Loss Prevention in the Process Industries 19, 527–541.

Birk, A.M., Poirier, D., Davison, C., 2006b. On the thermal rupture of 1.9 m^3 propane pressure vessels with defects in their thermal protection system. Journal of Loss Prevention in the Process Industries 19, 582–597.

Boyer, H.E., 1988. Atlas of Creep and Stress Rupture Curves. ASM International, Ohio.

British Standards Institution, 1990. BS 5908-Code of Practice for Fire Precautions in Chemical Plant. British Standards Institution, London.

Bukowski, R.W., Budnick, E.K., Schemel, C.F., 2002. Estimates of the operational reliability of fire protection systems. Fire Protection Strategies for 21st Century. In: Proc. of Building and Fire Codes Symposium, Baltimore, MD, pp.111–124.

Chang, J.I., Lin, C., 2006. A study of storage tank accidents. Journal of Loss Prevention in the Process Industries 19, 51–59.

Collier, J.G., Thome, J.R., 1996. Convective Boiling and Condensation, third edition. Oxford University Press, Oxford, United Kingdom.

Cowley, L.T., Johnson, A.D., 1991. Blast and Fire Engineering Project for Topside Structures. Oil and Gas Fires: Characteristics and Impact. Health and Safety Executive, United Kingdom.

Cox, J., 1976. Flixborough-some additional lessons. Chemical Engineering 387, 353–358.

Cozzani, V., Zanelli, S., 2001. An approach to the assessment of domino accidents hazard in quantitative area risk analysis. In: Proceedings of the 10th International Symposium on Loss Prevention and Safety Promotion in the Process Industries, Amsterdam, p. 1263.

Cozzani, V., Gubinelli, G., Salzano, E., 2006a. Escalation thresholds in the assessment of domino accidental events. Journal of Hazardous Material 129 (1–3), 1–21.

Cozzani, V., Gubinelli, G., Salzano, E., 2006b. Criteria for the escalation of fires and explosions. In: Proceedings of the 7th Process Plant Safety Symposium. American Institute of Chemical Engineers, New York, NY, p. 225.

Cozzani, V., Gubinelli, G., Antonioni, G., Spadoni, G., Zanelli, S., 2005. The assessment of risk caused by domino effect in quantitative area risk analysis. Journal of Hazardous Materials A127, 14–30.

Davies, G.F., Nolan, P.F., 2004. Characterisation of two industrial deluge systems designed for the protection of large horizontal, cylindrical LPG vessels. Journal of Loss Prevention in the Process Industries 17, 141–150.

De Dianous, V., Fievez, C., 2006. ARAMIS project: a more explicit demonstration of risk control through the use of bow–tie diagrams and the evaluation of safety barrier performance. Journal of Hazardous Materials 130, 220–233.

Delvosalle, C., 1998. A Methodology for the Identification and Evaluation of Domino Effects. Belgian Ministry of Employment and Labour, Bruxelles. (Rep. CRC/MT/003).

Det Norske Veritas DNV, 2008. Fire Protection. DNV, Høvik, Norway. (Offshore Standard DNV-OS-D301).

Di Padova, A., Tugnoli, A., Cozzani, V., Barbaresi, T., Tallone, F., 2011. Identification of fireproofing zones in Oil & Gas facilities by a risk-based procedure. Journal of Hazardous Materials 191 (1–3), 83–93.

Droste, B., Schoen, W., 1988. Full scale fire tests with unprotected and thermal insulated LPG storage tanks. Journal of Hazardous Materials 20, 41–53.

Droste, B., Probst, U., Heller, W., 1999. Impact of an exploding LPG rail tank car onto a castor spent fuel cask. RAMTRANS Nuclear Technology Publishing 10 (4), 231–240. http://www.tes.bam.de/de/umschliessungen/behaelter_radioaktive_stoffe/dokumente_veranstaltungen/pdf/rmtp1999104231.pdf.

Fan, W.C., Hua, J.S., Liao, G.X., 1995. Experimental study on the premonitory phenomena of boilover in liquid pool fires supported on water. Journal of Loss Prevention in the Process Industries 8 (4), 221–227.

Faucher, M., Giquel, D., Guillemet, R., Kruppa, J., Le Botlan, Y., Le Duff, Y., Londiche, H., Mahier, C., Oghia, I., Py, J.L., Wiedemann, P., 1993. Fire Protection Study on Fire Proofing Tanks Containing Pressurized Combustible Liquefied Gases (GASAFE Program Report). Groupement Europèen d'Intèret Economique-GEIE, Paris, DC.

Finucane, M., 1988. Information and tools required for a fire PSA. In: Ballard, G.M. (Ed.), Nuclear Safety after Three Mile Island and Chernobyl. Elsevier Applied Science, London, UK, pp. 294–305. (ISBN 1-85 166-235-9).

Forrest, H.S., 1985. Emergency relief vent sizing for fire exposure when two-phase flows must be considered. In: Proceedings of the 19th Loss Prevention Symposium. American Institute of Chemical Engineers, Houston.

Gledhill, J., Lines, I., 1998. Development of Methods to Assess the Significance of Domino Effects from Major Hazard Sites (CR Report 183). Health and Safety Executive.

Godoy, L.A., Batista-Abreu, J., 2012. Buckling of fixed roof above ground oil storage tanks under heat induced by an external fire. Thin-Walled Structures 52, 90–101.

Gomez-Mares, M., Tugnoli, A., Landucci, G., Cozzani, V., 2012a. Performance assessment of passive fire protection materials. Industrial Engineering and Chemical Research 51, 7679–7689.

Gomez-Mares, M., Tugnoli, A., Landucci, G., Barontini, F., Cozzani, V., 2012b. Behavior of intumescent epoxy resins in fireproofing applications. Journal of Analytical and Applied Pyrolysis 97, 99–108.

Gómez-Mares, M., Zárate, L., Casal, J., 2008. Jet fires and the domino effect. Fire Safety Journal 43 (8), 583–588. http://dx.doi.org/10.1016/j.firesaf.2008.01.002.

Gong, Y.W., Lin, W.S., Gu, A.Z., Lu, X.S., 2004. A simplified model to predict the thermal response of PLG and its influence on BLEVE. Journal of Hazardous Materials A108, 21–26.

Graves, K.W., 1973. Development of a Computer Model for Modeling the Heat Effects on a Tank Car (Report FRA-OR&D 75–33). US Department of Transportation.

Gubinelli, G., 2005. Models for the Assessment of Domino Accidents in the Process Industry. Doctoral Dissertation in Chemical Engineering, University of Pisa.

Hadjisophocleous, G.V., Sousa, A.C.M., Venart, J.E.S., 1990. A study of the effect of the tank diameter on the thermal stratification in LPG tanks subjected to fire engulfment. Journal of Hazardous Materials 25 (1–2), 19–31.

Hankinson, G., Lowesmith, B.J., 2004. Effectiveness of area and dedicated water deluge in protecting objects impacted by crude oil/gas jet fires on offshore installations. Journal of Loss Prevention in the Process Industries 17, 119–125.

Hauptmanns, U., Marx, M., Grunbeck, S., 2008. Availability analysis for a fixed wet sprinkler system. Fire Safety Journal 43, 468–476.

Health and Safety Executive, 1978. Canvey: An Investigation of Potential Hazards from Operations in the Canvey Island/Thurrock Area. Health and Safety Executive, London.

Health and Safety Executive, 1996. Jet-Fire Resistance Test of Passive Fire Protection Materials (Report Oti 95 634). Health and Safety Executive, Merseyside, UK.

Hobert, J.F.A., Molag, M., 2006. Fire Brigade Response and Deployment Scenarios to Avoid a Hot BLEVE of a LPG Tank Vehicle or LPG Tank Wagon (Report 2006-A-R0069/B). Netherlands Organisation for Applied Scientific Research TNO, Apeldoorn (NL).

International Organization for Standardization ISO, 1999. International Standard ISO 13702, petroleum and natural gas industries – control and mitigation of fires and explosions on offshore production installations – requirements and guidelines first ed. ISO.

International Organization for Standardization ISO, 2007. International Standard ISO 22899-1, determination of the resistance to jet fires of passive fire protection materials—part 1: general requirements first ed. ISO.

Jimenez, M., Duquesne, S., Bourbigot, S., 2006. Intumescent fire protective coating: toward a better understanding of their mechanism of action. Thermochimica Acta 449, 16–26.

Jin, Z., Asako, Y., Yamaguchi, Y., Harada, M., 2000. Fire resistance test for fire protection materials with high water content. International Journal of Heat and Mass Transfer 43, 4395–4404.

Johnson, M.R., 1998. Tank Car Thermal Analysis, Volume 1: User Manual, Volume 2: Technical Documentation. (Reports DOT/FRA/ORD-98/09A and 09B). Federal Railroad Administration, U.S. Department of Transportation.

Keltner, N.R., Nicolette, V.F., Brown, N.N., Bainbridge, B.L., 1990. Test unit effects on heat transfer in large fires. Journal of Hazardous Materials 25, 33–47.

Kern, D.Q., 1950. Process Heat Transfer. McGraw Hill, New York, CA.

Khan, F.I., Abbasi, S.A., 1998. Models for domino effect analysis in chemical process industries. Process Safety Progress 17 (2), 107–113.

Knudsen, J.G., Hottel, H.C., Sarofim, A.F., Wankat, P.C., Knaebel, K.S., 1999. Heat and mass transfer, Section 5. In: McGraw-Hill (Ed.), Perry's Chemical Engineers' Handbook, seventh ed. McGraw-Hill, New York.

Landucci, G., Cozzani, V., 2009. Simplified assessment of protective measures to prevent fire escalation involving pressurized vessels. In: Søgaard, I., Krogh, H. (Eds.), Fire Safety. Nova Science Publishers Inc, Hauppauge, NY, pp. 65–83.

Landucci, G., Gubinelli, G., Antonioni, G., Cozzani, V., 2009a. The assessment of the damage probability of storage tanks in domino events triggered by fire. Accident Analysis and Prevention 41 (6), 1206–1215.

Landucci, G., Molag, M., Cozzani, V., 2009b. Modeling the performance of coated LPG tanks engulfed in fires. Journal of Hazardous Materials 172 (1), 447–456.

Landucci, G., Molag, M., Reinders, J., Cozzani, V., 2009c. Experimental and analytical investigation of thermal coating effectiveness for 3m3 LPG tanks engulfed by fire. Journal of Hazardous Materials 161 (2–3), 1182–1192.

Landucci, G., Tugnoli, A., Busini, V., Derudi, M., Rota, R., Cozzani, V., 2011. The Viareggio LPG accident: lessons learnt. Journal of Loss Prevention in the Process Industries 24 (4), 466–476.

Landucci, G., Antonioni, G., Tugnoli, A., Cozzani, V., 2012. Release of hazardous substances in flood events: damage model for atmospheric storage tanks. Reliability Engineering and System Safety 106, 200–216.

Lees, F.P., 1996. Loss Prevention in the Process Industries, second ed. Butterworth-Heinemann, Oxford.

Lev, Y., 1991. Water protection of surfaces exposed to impinging LPG jet-fires. Journal of Loss Prevention in the Process Industries 4, 252–259.

Lin, W., Gong, Y., Gao, T., Gu, A., Lu, X., 2010. Experimental studies on the thermal stratification and its influence on BLEVEs. Experimental Thermal and Fluid Science 34, 972–978.

Luecke, W.E., David McColskey, J., McCowan, C.N., Banovic, S.W., Fields, R.J., Foecke, T., Siewert, T.A., Gayle, F.W., 2005. Mechanical Properties of Structural Steels. National Institute of Standard and Technology (NIST). (NIST NCSTAR 1–3D).

Malloy, F., 1969. Thermal Insulation. Van Nostrand Reinhold Company, New York, NY.

Manas, J.L., 1984. BLEVEs: their nature and prevention. Fire International 87 (8), 27–39.

Manda, L.J., 1975. Phase II – Report on Full Scale Fire Tests (RPI/AAR). Tank Car Safety Research and Test Project. Association of American Railroads Research Center, Chicago.

Manu, C., 2008. Finite Elements Analysis of Stress Rupture in Pressure Vessels Exposed to Accidental Fire Loading. Department of Mechanical Engineering, Queen's University, Kingston, Ontario, Canada.

Manu, C.C., Birk, A.M., Kim, I.Y., 2009a. Stress rupture predictions of pressure vessels exposed to fully engulfing and local impingement accidental fire heat loads. Engineering Failure Analysis 16, 1141–1152.

Manu, C.C., Birk, A.M., Kim, I.Y., 2009b. Uniaxial high-temperature creep property predictions made by CDM and MPC omega techniques for ASME SA 455 steel. Engineering Failure Analysis 16, 1303–1313.

Manyatt, H.W., 1988. Fire, a Century of Automatic Sprinkler Protection in Australia and New Zealand 1886–1986 (ISBN 0-7316-4001-2). Australian Fire Protection Association, Blackburn North, VIC.

Maremonti, M., Russo, G., Salzano, E., Tufano, V., 1999. Post-accident analysis of vapour cloud explosions in fuel storage areas. Process Safety and Environmental Protection 77 (6), 360–365.

Mecklenburgh, J.C., 1985. Process Plant Layout. George Goodwin, London, UK.

Moodie, K., 1988. Experiments and modelling: an overview with particular reference to fire engulfment. Journal of Hazardous Materials 20, 149–175.

Moodie, K., Cowley, L.T., Denny, R.B., Small, L.M., Williams, I., 1988. Fire engulfment tests on a 5 tonne LPG tank. Journal of Hazardous Materials. 20, 55–71.

National Fire Protection Association NFPA, 2009. NFPA 15 Standard for Water Spray Fixed Systems for Fire Protection. Quincy MA: National Fire Protection Association.

NORSOK Standards Norway, 2008. NORSOK Standard S-001 Technical safety, fourth ed. Standards Norway, Lysaker, Norway. Available at http://www.standard.no/PageFiles/1055/S-001_e4.pdf.

Paltrinieri, N., Landucci, G., Molag, M., Bonvicini, S., Spadoni, G., Cozzani, V., 2009. Risk reduction in road and rail LPG transportation by passive fire protection. Journal of Hazardous Materials 167 (1–3), 332–344.

Persaud, M.A., Butler, C.J., Roberts, T.A., Shirvill, L.C., Wright S., 2001. Heat-up and failure of liquefied petroleum gas storage vessel exposed to a jet-fire. In: Proceedings of the 10th International Symposium on Loss Prevention in the Process Industries, Stockholm, pp.1069–1106.

Planas-Cuchi, E., Casal, J., Lancia, A., Bordignon, L., 1996. Protection of equipment engulfed in a pool fire. Journal of Loss Prevention in the Process Industries 9 (3), 231–240.

Pontiggia, M., Landucci, G., Busini, V., Derudi, M., Alba, M., Scaioni, M., Bonvicini, S., Cozzani, V., Rota, R., 2011. CFD model simulation of LPG dispersion in urban areas. Atmospheric Environment 45 (24), 3913–3923.

Prugh, R.W., 1991. Quantify BLEVE hazards. Chemical Engineering Progress 87 (2), 66–71.

Ramskill, P.K., 1988. A description of the "Engulf" computer code – codes to model the thermal response of an LPG tank either fully or partially engulfed by fire. Journal of Hazardous Materials 20, 177–196.

Reid, R.C., 1979. Possible mechanism for pressurized-liquid tank explosions or BLEVE's. Science 203, 1263–1265.

Roberts, A.F., 1981. Thermal radiation hazards from releases of LPG from pressurized storage. Fire Safety Journal 4 (3), 197–212.

Roberts, A., Medonos, S., Shirvill, L.C., 2000. Review of the Response of Pressurised Process Vessels and Equipment to Fire Attack (OFFSHORE TECHNOLOGY REPORT – OTO 2000 051, HSL 2000). Health and Safety Laboratory, Manchester, UK.

Roberts, T.A., 2004a. Directed deluge system designs and determination of the effectiveness of the currently recommended minimum deluge rate for the protection of LPG tanks. Journal of Loss Prevention in the Process Industries 17, 103–109.

Roberts, T.A., 2004b. Effectiveness of an enhanced deluge system to protect LPG tanks and sensitivity to blocked nozzles and delayed deluge initiation. Journal of Loss Prevention in the Process Industries 17, 151–158.

Roberts, T.A., Buckland, I., Shirvill, L.C., Lowesmith, B.J., Salater, P., 2004. Design and protection of pressure systems to withstand severe fires. Process Safety and Environment Protection 82 (B2), 89–96.

Roberts, T.A., Shirvill, L.C., Waterton, K., Buckland, I., 2010. Fire resistance of passive fire protection coatings after long-term weathering. Process Safety and Environmental Protection 88, 1–19.

Salzano, E., Picozzi, B., Vaccaro, S., Ciambelli, P., 2003. Hazard of pressurised tanks involved in fires. Industrial Engineering and Chemical Research 42, 1804–1812.

Schoen, W., Droste, B., 1988. Investigation of water spraying systems for LPG storage tanks by full scale fire tests. Journal of Hazardous Materials 20, 73–82.

Shaluf, I.M., Abdullah, S.A., 2011. Floating roof storage tank boilover. Journal of Loss Prevention in the Process Industries 24 (1), 1–7.

Shebeko, Yu. N., Bolodian, I.A., Filippov, V.N., Navzenya, V. Yu., Kostyuhin, A.K., Tokarev, P.M., Zamishevski, E.D., 2000. A study of the behaviour of a protected vessel containing LPG during pool fire engulfment. Journal of Hazardous Materials A77, 43–56.

Shirvill, L.C., 2004. Efficacy of water spray protection against propane and butane jet fires impinging on LPG storage tanks. Journal of Loss Prevention in the Process Industries 17 (2), 111–118.

Society of Fire Protection Engineers (SFPE), & National Fire Protection Association (NFPA), 2002. SFPE Handbook of Fire Protection Engineering, third ed. Quincy.

Steel Construction Institute-SCI, 1992. Availability and Properties of Passive and Active Fire Protection Systems. United Kingdom: Health and Safety Executive. (OTI 92 607).

Tan, D.M., Xu, J., Venart, J.E.S., 2011. Fire-induced failure of a propane tank: some lessons to be learnt. In: Proceedings of Instn Mechanical Engineers. Part. E: J. Process. Mech. Eng, vol. 217, pp. 79–91.

Townsend, W., Anderson, C.E., Zook, J., Cowgill, G., 1974. Comparison of Thermally Coated and Uninsulated Rail Tank-Cars Filled with LPG Subjected to a Fire Environment (Report FRA-OR&D 75–32). US Department of Transportation.

Tugnoli, A., Cozzani, V., Di Padova, A., Barbaresi, T., Tallone, F., 2012. Mitigation of fire damage and escalation by fireproofing: a risk-based strategy. Reliability Engineering and System Safety 105, 25–35.

US Federal Government, 2008. Code of Federal Regulations – CFR 49 Part 179 Specification for Tank Cars. US Federal Government (Ed.).

Van Den Bosh, C.J.H., Weterings, R.A.P.M., 2005. Methods for the Calculation of Physical Effects (Yellow Book), third ed. Committee for the Prevention of Disasters, The Hague.

VanderSteen, J.D.J., Birk, A.M., 2003. Fire tests on defective tank-car thermal protection system. Journal of Loss Prevention in the Process Industries 16, 417–425.

Venart, J.E.S., 1986. Tank Car Thermal Response Analysis – Phase II. Tank Car Safety Research and Test Project, Chicago.

Venart, J.E.S., 1999. Boiling Liquid Expanding Vapor Explosions (BLEVE): Possible Failure Mechanisms, vol. 1336. ASTM Special Technical Publication. pp. 112–134.

Venart, J.E.S., 2007. Flixborough: a final footnote. Journal of Loss Prevention in the Process Industries 20, 621–643.

Yoon, K.T., Birk, A.M., 2004. Computational Fluid Dynamics Analysis of Local Heating of Propane Tanks. Transportation Development Centre, Transport Canada.

6 Missile Projection Effects

Alessandro Tugnoli*, Valerio Cozzani*,
Faisal Khan†, Paul Amyotte‡

* LISES, Dipartimento di Ingegneria Civile, Chimica, Ambientale e dei
Materiali, Alma Mater Studiorum - Università di Bologna, Bologna, Italy,
† Safety and Risk Engineering Research Group, Faculty of Engineering and
Applied Science, Memorial University of Newfoundland, St. John's,
Newfoundland, Canada, ‡ Department of Process Engineering and Applied
Science, Dalhousie University, Halifax, Nova Scotia, Canada

6.1   Introduction

The projection of fragments following the catastrophic failure of equipment items is
a significant scenario for industrial accidents. It was recorded in several occurrences
as the cause of fatalities, injuries and of damage to process equipment (Baker et al.,

1983; Center for Chemical Process Safety (CCPS), 2000; Gubinelli and Cozzani, 2009a; Mannan, 2005). Moreover, the projection of fragments is among the more important causes of domino propagation of an industrial accident (Gledhill and Lines, 1998; Khan and Abbasi, 1998). The projection distances may be very high (up to 1 km), and the projected fragments are capable of generating secondary accidents at relevant distances from the primary scenario.

Fragment can be projected as a consequence of the catastrophic failure of equipment due to several accidental scenarios, including physical and confined explosions, boiling liquid expanding vapor explosions (BLEVEs), and runaway reactions (Bagster and Pitblado, 1991; Gledhill and Lines, 1998; Khan and Abbasi, 1998; Pettitt et al., 1993). Also rotating equipment, such as pumps and compressors, may result in fragment projection up to relevant distances.

This chapter provides a review of the assessment of escalation caused by fragment projection. In Section 6.2 the mechanism of fragments formation and projection is discussed. Section 6.3 provides criteria for the identification of possible fragment sources. Section 6.4 discusses the identification of reference shapers for the fragments. Sections 6.5 and 6.6 describe the methods currently available for the calculation of impact and damage probability.

6.2   Escalation Caused by Fragments

6.2.1   Scenarios Leading to Fragment Projection

Fragment projection is correlated with accident scenarios which both (1) can generate fragments, and (2) can transfer to the fragments a sufficient kinetic energy to make them a possible cause of damage to people or equipment.

The catastrophic rupture of a vessel is a typical accidental event in which fragment projection may occur. In fact, the catastrophic rupture is usually coupled with the release of the substances contained as well as of internal energy. While the released substances can cause other scenarios (e.g. fireballs, vapor cloud explosions, flash fires, and toxic dispersions) depending on the characteristics of the material in the vessel, the sudden release of energy can give rise to blast waves and high-velocity fragments. The fragments of the vessel can become missiles, which will be propelled for a long distance and may hit anything they find on their trajectory.

Gubinelli and Cozzani (2009a) identified a possible correlation among the vessel geometry, the accidental scenario causing vessel fragmentation and the shape and number of fragments generated. Their conclusion was validated by a survey of more than 180 past accidents. Table 6.1 shows the findings from the analysis of the different types of accidental scenarios resulting in the projection of missiles. In the table, a short definition of each category of scenarios that caused the projection of fragments in at least one of the accidental events recorded in the database is provided. Table 6.1 shows that BLEVE resulted as the more frequent scenario leading to fragment projection. However a significant number of events were recorded for the other scenarios considered. The higher number of events involving BLEVEs with respect to

Table 6.1 Scenarios Leading to Missile Projection from the Analysis of Past Accidents

Primary Scenario	Description	Percentage of Events
Fired BLEVE	Catastrophic failure of a vessel containing a liquid at temperature above its boiling temperature at atmospheric pressure, due to an external fire.	62%
Unfired BLEVE	Sudden loss of containment of a vessel containing a liquid at temperature above its boiling temperature at atmospheric pressure, not due to an external fire (e.g. due to corrosion, erosion, fatigue, and external impact).	12%
Physical explosion	Catastrophic failure of a vessel containing a compressed gas phase and/or a nonboiling liquid, due to an internal pressure increase not caused by fire or chemical reactions. Possible causes: overfilling, corrosion, etc.	10%
Confined explosion	Catastrophic vessel failure due to an internal pressure increase caused by the unwanted combustion of gases, vapors, or dust inside the vessel.	10%
Runaway reaction	Catastrophic vessel failure due to an internal pressure increase caused by the loss of control of a chemical reaction.	6%

Source: Adapted from Gubinelli and Cozzani (2009a).

other scenarios is related to the fact, on one hand, that BLEVE events may affect a larger number of installations (e.g. process plants, fuel transportation, domestic installations, and manufactory plants) and, on the other hand, that BLEVE accidents are usually more severe, and thus are more frequently reported in past accident databases.

All the scenarios considered are characterized by the availability of internal energy (usually in the form of internal pressure) before the vessel failure. This energy is sufficient both to propagate cracks in the vessel shell, leading to fragmentation, and to be partly converted into kinetic energy of the fragments.

Fragment projection can also follow the failure of rotating equipment (e.g. compressors and turbines). Fragments can be generated by the mechanical failure of moving parts of the machine (e.g. blade, section of rotor), which detach free from the unit assembly and pierce through the machine case. The kinetic energy is provided to the fragment by the mechanical energy associated with the original rotation/spinning of the components into the machine. The dynamic of the process makes it evident that only high-speed rotating machines may potentially generate high-speed missiles.

Fragments directly generated from equipment failures are named "primary frag-ments" or "primary missiles". The term "secondary missiles" refers instead to objects in the environment around the unit that are picked up by a blast wave (Van Den Bosh and Weterings, 1997). These can be potentially generated in any accident scenarios that create a blast wave with enough energy to lift/detach components or materials from their original location. In the following, primary missiles will be mostly considered, since these are known to be the more frequent cause of escalation.

6.2.2   Fragment Formation

Fragments originate when a unit is ruptured in sections that are physically detached from the original structure. Formation of fragments involves the formation and propagation of cracks through the structural material of the unit. In the applications of interest such material is usually metallic. The mechanics of the propagation of cracks in the metal is an important factor that may affect the equipment failure mode and the number and geometry of projected fragments. The different accident scenarios, together with the design of the equipment, were identified to influence the fracture mode and the fragmentation pattern (Gubinelli and Cozzani, 2009a).

Therefore, the actual mechanism of fracture propagation could be influenced by the type of material, the shell thickness, the shell temperature, and the loading rate. Considering the general features of pressure and thermal load in the different primary scenarios likely to cause vessel fragmentation, it is possible to draw basic information on the expected fracture mechanisms and on the consequent failure modes of the process vessels undergoing the event. Table 6.2 reports the conclusions of a qualita-tive analysis performed for the primary scenarios triggering vessel fragmentation (Gubinelli and Cozzani, 2009a). The table shows that very simple correlations could be drawn between the primary scenario and the expected number of fragments. These were derived considering the likely fracture mechanism and the credibility of crack branching and/or of crack arrest as a consequence of the typical behaviors of the shell loads due to internal pressure and wall temperature during the different scenarios. As shown in the table, a specific fragmentation mechanism and a qualitative evaluation of the fragment number may be associated to each primary scenario likely to cause vessel fragmentation, with the exception of runaway reactions. Ductile fractures resulting in a limited number of fragments are expected to be the prevailing frag-mentation mechanism in BLEVEs and in physical explosions. On the other hand, in the case of confined explosions, runaway reactions, and energetic material decom-positions, brittle fracture resulting in a high number of fragments is expected, although brittle–ductile transition is possible for high-toughness vessels. In the specific case of runaway reactions very different pressurization rates may take place (Design Institute for Emergency Relief Systems (DIERS), 1992), thus resulting in the different vessel fracture behaviors.

Important outlines on the mode of fragmentation may be drawn also considering the influence of the vessel shape on the crack propagation. The propagation of a crack on a vessel occurs in a normal direction to that of the maximum stress. Thus in cylindrical shells the cracks tend to start in the axial direction, since the

Table 6.2 Relations between the Type of the Primary Scenario, the Fracture Mechanics and the Fragmentation Characteristics

Primary Scenario	Load Parameters	Fracture Mechanics and Dynamics	Fragmentation Properties
Fired BLEVE (BLEVE-F)	Low dP/dt and low $d\sigma/dt$ High wall temperature	Ductile fracture Low fracture velocity: possible depressurization after crack initiation No branching	Low fragment number Possible crack arrest (due to depressurization and hot–cold zone transition)
Unfired BLEVE (BLEVE-NF) and Physical explosion (ME)	Low dP/dt and low $d\sigma/dt$ Low wall temperature	Ductile fracture (plastic deformation at the crack tip) Low fracture velocity but higher than (BLEVE-F): vessel depressurization not credible. Branching possible only for very low wall temperature (ME)	Low fragment number (high fragment number likely only for ME with very low wall temperature) Possible crack arrest for high-toughness material (brittle–ductile transition)
Confined explosion (CE)	High dP/dt and high $d\sigma/dt$ Low wall temperature	Brittle fracture High-fracture velocity: no vessel depressurization Branching	High fragment number
Runaway reaction (RR)	Low $d\sigma/dt$ Low wall temperature	For "low" pressure increase (depressurization possible, stress increase significantly lower than crack propagation speed), expected behavior similar to ME	
	High $d\sigma/dt$ Low wall temperature	For "high" pressure increase (depressurization not relevant, stress increase comparable to crack propagation speed), expected behavior similar to CE	
Energetic material decomposition (ENMAT)	High dP/dt and high $d\sigma/dt$ Low wall temperature	Brittle fracture High-fracture velocity: no vessel depressurization Branching	High-fragment number

Source: Adapted from Gubinelli and Cozzani (2009a).

circumferential stress is higher. The fracture may propagate in the circumferential direction only due to stress field changes caused by bends, and to stress intensification areas due to connections or to defects in the material (e.g. weldings). In spherical shells the fracture may start and propagate in any direction; however, the more likely initiation points are in those areas where material defects or stress intensification due to connections are present.

6.2.3 Fragment Impact

The fragments originating at the location of the primary accident are ejected and can occasionally impact a target unit. The behavior of a projected fragment during the flight depends on the interaction of gravitational and fluidodynamic forces (Baker et al., 1983; Mannan, 2005). The trajectory and velocity of the mass center of the fragment are usually used to describe the trajectory and velocity of the entire fragment, respectively. If the deviations of the trajectory of the center of mass due to the wind and to the possible oscillations caused by rotational movements of the fragment are neglected, the flight takes place on a vertical plane (Gubinelli et al., 2004). The velocity of projected fragments is usually higher of more than an order of magnitude than normal wind velocities, resulting in a limited influence of wind direction and speed on fragment trajectory. Moreover, the rotational movements of the fragment are likely to cause an oscillation of the mass center around its main direction rather than a deviation of the trajectory plane.

Several models were proposed in the literature for the description of the trajectory of projected fragments. A comprehensive review is given by Mannan (2005). Baker et al. (1983) developed a fundamental approach to the problem, based on the description of the fragment motion, considering the fragment acceleration and three types of forces acting on the fragment: gravitational, drag and lift forces. The last two were expressed as a function of the shape, the mass and the orientation of the fragment with respect to the trajectory of its mass center. However, the model of Baker et al. (1983), per se, requires providing deterministic input parameters for initial conditions (assigned initial velocity vector and fragment shape) and to solve by numerical methods the differential balance equations on which it is based. The more commonly used output of the method is a diagram which reports maximum fragment range. This kind of result is not compatible with a probabilistic approach typical of the Quantitative Risk Analysis (QRA) framework.

Hauptmanns (2001a,b) proposed a simplified approach based on the more simple equations generally used in mechanics to describe the ballistic motion of objects with velocities in the subsonic range:

$$\frac{d^2x}{dt^2} + k\left(\frac{dx}{dt}\right)^2 = 0 \tag{6.1}$$

$$\frac{d^2y}{dt^2} + (-1)^n k\left(\frac{dy}{dt}\right)^2 + g = 0 \tag{6.2}$$

where x and y are the coordinates of the position of the fragment at time t, g is the gravitational acceleration, k is the drag coefficient, and n equals 1 in the descending part of the trajectory and 2 in the ascending part. The terms multiplying k identify the drag forces exerted on the object in the x and y directions. In the subsonic range, these forces should be proportional, by a coefficient k independent of the direction, to the square of the velocity of the object. The coefficient k is a function of the mass and of the shape of the fragment. A more extensive discussion on the definition of the drag coefficient k as function of fragment characteristics is available in the literature (Gubinelli and Cozzani, 2009b).

Fragment impact occurs when the airborne fragment, during his trajectory, collides on a target of interest (piece of equipment, building, support structure, etc.). Scilly and Crowther (1992) introduced the concept of effective range interval (ERI) and postulated that missile impact is possible when the fragments fall within the ERI. With reference to the coordinate system shown in Figure 6.1, Pula et al. (2007) evidenced that two different scenarios can lead to the impact between a flying missile and a target:

- The impact results from the missile landing within the "vulnerable area" (VA) of the target (interval $\Delta\varphi_1$ in Figure 6.2(b)).
- The impact results from the missile colliding with the target object while in flight before reaching the final destination, that would have lead it to land beyond the target (interval $\Delta\varphi_2$ in Figure 6.2(b)).

The existence of these two impact scenarios is related to the fact that the target is a three-dimensional object, having nonnegligible dimensions also in the vertical direction (e.g. storage vessel, column, etc.). In particular, in the case of equipment having a relevant elevation above ground level, as columns, the shape of the target highly influences the impact probability.

In the case of domino accidents, the missile may have significant dimension compared to the target. Pula et al. (2007) introduced the concept of VA in order to account for the fragment size in the impact criteria. The VA is the probable impact zone around the target and is constructed as an envelope of the same shape of the

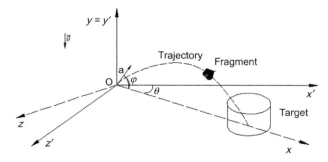

Figure 6.1 Schematization and reference coordinate systems adopted to represent the trajectory of a fragment and of the impact on a given target.

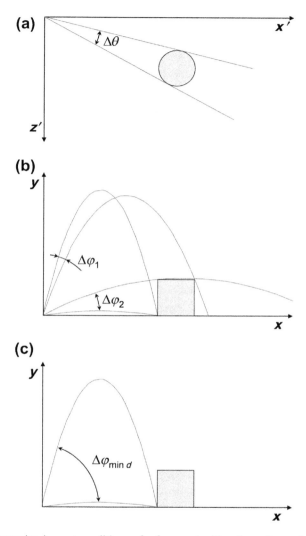

Figure 6.2 Alternative impact conditions of a fragment with a three-dimensional target: (a) range of horizontal angles ($\Delta\theta$) for which the missile hits the VA of the target; (b) range of elevation angles for which the missile hits the VA either by landing over it ($\Delta\varphi_1$) or by crossing the target during a flight trajectory that would have ended beyond the target ($\Delta\varphi_2$); (c) range of elevation angles ($\Delta\varphi_{\min\, d}$) for which the "minimum distance" assumption is satisfied.

actual target, but with a size that is the size of the actual target plus half the diameter of the smallest fragment generated during the primary accident. This definition ensures that any fragment whose center of mass is passing through the VA would definitely strike the target object. A similar approach in the definition of the impact criteria is used also by Mébarki et al. (2009b).

Figure 6.2 clearly evidences that, given a fragment (shape, dimensions and mass) and an initial velocity of projection, the impact with the target is realized by the trajectories that have initial angles within some limited intervals ($\Delta\theta$ for horizontal angles; $\Delta\varphi_1$ and $\Delta\varphi_2$ for elevation angles). Pula et al. (2007) introduced the concepts of effective trajectory interval and the effective orientation interval to describe these angles. The intervals of these angles that lead to an effective impact can be calculated from the geometry of the problem by solving the trajectory equations (Eqns (6.1) and (6.2)).

Gubinelli et al. (2004) also introduced a simplified impact criterion, named "minimum distance" assumption, aiming at the reduction of the calculation effort needed for the assessment of impact conditions. The criterion assumes that, given the θ angle, any fragment that can fly past the point of the target more near to the fragment source actually impacts the target (i.e. no missile can pass over the target without impacting it, Figure 6.2(c)). This condition is strictly valid only for "high" targets at a "sufficient" distance from the fragment origin, and makes it possible to neglect the actual target shape function in the calculation of impact criterion. The authors demonstrated that in all other cases the assumption leads to conservative results and that the difference in impact probability between the simplified and rigorous models was always below two orders of magnitude for a number of reference case studies.

6.2.4 Fragment Damage

A missile that hits a target may pierce through it (perforation), stop at some depth of penetration (embedment) or bounce back (ricochet), sometimes leaving a dent in the surface.

The impact of a fragment on a piece of equipment can damage the target either by penetration or by plastic collapse. Consequences on the target in terms of possible loss of containment are generally different. A penetrating fragment may pierce a hole through the shell of the target and initiate a continuous or semicontinuous release. A nonpenetrating fragment may cause significant deformations of the target, possibly leading to the catastrophic failure of the target and the instantaneous release of its entire content.

Penetration mechanics was extensively studied, primarily for military purposes (Backman and Goldsmith, 1978). The typical shapes and sizes of concern for impacting fragments in industrial accidents can be significantly different with respect to those of concern in military applications (e.g. small diameter cylinders, pointed chunks, etc.). However, some concepts and models from these studies can be extended to fragment penetration.

During target penetration, the fracture is caused by both superficially applied loads and stress waves through the impacted material. Different mechanisms of fracture may occur, depending on many parameters: characteristic dimensions of the missile and the target (diameter and length of the missile and thickness of the target), mechanical properties of the missile and the target (density, elastic module, toughness, yield stress, etc.), shape of the missile tip (angle of the nose, etc.), impact velocity, etc.

In most of the cases of domino propagation in industrial accidents, the target of concern is usually a piece of equipment or pipe work. The projected fragments are originated from the same kind of units, as discussed in Section 6.2.1. Therefore steel alloys, the more common materials used for plant construction, are expected to constitute both the missile and target material. Missile speed is generally low, if compared to sound speed in air. Under these conditions, penetration of the target may occur according to several plastic deformation mechanisms described in the literature (plugging, petaling rearward, petaling frontal, fragmentation, and ductile hole enlargement). Other mechanisms, such as fracture due to initial stress wave and radial fracture, can be reasonably excluded for ductile materials. Further details on fracture mechanisms are available in the literature (Backman and Goldsmith, 1978; Mannan, 2005).

The mechanism of plastic collapse can be considered credible for the impact of large-size fragments on targets having low-thickness walls or shells. This is for example the case of impact of fragments on atmospheric tanks.

The relative size of missile and target may span over a wide range. Missiles originated by vessel failure are usually quite large and can undergo strong deformation during the impact. Both larger and deformable missiles result in a lower penetration depth in the target. On the other side, relatively small fragments originated by the failure of rotating equipment or by the ejection of pipe fittings are expected to have larger potential for penetration.

Missiles originated by equipment failure have irregular shapes. The angle of presentation of the missile on the equipment is a random variable, since rotation is possible during flight. Since both the angle of the missile tip and the area of contact can have significant influence on the penetration depth, a sensitivity analysis could be required to assess target damage, considering different possible orientations. A conservative assumption can be to consider the impacting fragment configuration having the smaller area of projection on the target.

The scientific literature reports several models concerning perforation of metal and concrete panels. Some of these models are presented in Section 6.6. However, no specific models are reported in the literature on the plastic collapse of process equipment by fragment impact. The mechanism can be studied, though for specific geometries, by finite element simulation of impulsive loads on structures.

The general considerations reported in this section and the models later introduced in Section 6.6 refer to accidents where domino propagation occurs by direct damage of target equipment. For other mechanisms (damage of e.g. cables, marshalling cabinets, control rooms, support structures, etc.), specific consideration and damage models should apply, depending on target material and design (Mannan, 2005).

6.3   Identification and Characterization of Fragment Sources

6.3.1   Fragments Caused by Vessel Collapse

All vessels in which internal pressure can build up at values higher than their mechanical strength are potential sources of fragments. The possible accidental scenarios that may lead to such condition have been introduced in Section 6.2.1. The accidental scenario as well as the geometry and design of the vessel influence the pattern of fragment generation. Moreover a correlation exists between the possible accident scenarios and the taxonomy and service of the equipment. Altogether these elements define preferential patterns for vessel fragmentation. This aspect is extensively discussed in Section 6.4.1.

Gubinelli and Cozzani (2009a) analyzed a number of past accidents involving vessel fragmentation. The vessel category more frequently involved in fragmentation accidents was pressurized cylindrical vessels, since this type of vessel is more frequently used for the storage of liquefied compressed gases. In the case of accidents involving atmospheric vessels, overpressure was caused mainly by physical explosions, confined explosions and runaway reactions. All these scenarios lead to an internal pressure exceeding design pressure and to fragment projection.

Table 6.3 reports a heuristic criterion for the screening of possible sources of missiles from vessel fragmentation. The table points out the correlations between the accidental scenario, as identified from conventional hazard identification techniques, and the type of vessel that is likely to undergo fragmentation. As a matter of fact, BLEVE events leading to vessel fragmentation mainly affect cylindrical and spherical pressurized tanks, used for liquefied gas storage. On the other hand, atmospheric vessels including cone-roof tanks are likely to undergo fragmentation accidents mainly due to mechanical and confined explosions, or to runaway reactions.

The energy available for fragment projection depends on the design of the vessel and on the scenario. A fraction of the internal energy is transferred to the fragments as kinetic energy during vessel failure. Several methods were proposed for the evaluation of the kinetic energy transferred to the sample (Baker et al., 1983; Baum, 1984, 1987; Brode, 1959).

The first group of methods simply defines efficiency in the conversion of expansion energy in kinetic energy of the fragments. The internal energy is a function of the thermodynamic properties and mass of the vessel contents and is practically determined by two parameters: (1) the condition of the vessel at the moment of the rupture (volume, filling ratio, etc.), and (2) the conditions of the vessel contents (pressure, temperature, etc.). Well-accepted correlations can be found in the literature to calculate the available internal energy (Center for Chemical Process Safety (CCPS), 1994; Mannan, 2005; Van Den Bosh and Weterings, 1997).

Some methods consider that all or almost all the internal energy available is transferred as kinetic energy to the fragments. Other definitions prescribe that the fragments will receive only a part of the internal energy, based on experimental observations. Theoretically, these methods are applicable to all types of vessel bursts, except to vessels filled with energetic materials.

Other methods based on theoretical considerations define the initial velocity of fragments based on energy and momentum balances. The modeling has to account for complex aspects such as the gas flow through the ever-increasing gaps between fragments. Available solutions are usually limited only to specific vessel geometries and fragmentation patterns or to vessels filled with ideal gas. Therefore, one must be careful to generalize the results for other situations. The method originally proposed

Table 6.3 Possible Combinations of Accident Scenarios and Vessel Taxonomy Which May Generate Missiles

Vessel Taxonomy	BLEVE		Primary Scenario			
	Fired	Unfired	Physical Explosion	Confined Explosion	Runaway Reaction	Energetic Material Decomposition
Horizontal cylindrical atm. vessels			✓	✓	✓	✓
Vertical cylindrical atm. vessels			✓	✓	✓	✓
Cone-roof atmospheric tanks			✓	✓	(*)	(*)
Other sharp-edged atm. equipment			✓	✓	(*)	✓
Horizontal cylindrical press. vessels	✓	✓	✓	✓	✓	✓
Vertical cylindrical press. vessels	✓	✓	✓	✓	✓	✓
Spherical pressurized vessels	✓	✓	(*)	(*)	(*)	(*)
Other pressurized vessels	✓	✓	✓	✓	✓	✓

Combination marked as (*), though possible, have no typical application in industrial practice.

by Grodzovskii and Kukanov (1965) and further developed by Baker et al. (1983) belongs to this group. The method proposed by Gel'fand et al. (1989) comprises some dimensionless diagrams which can be used to determine the velocity of fragments generated in the rupture of spherical and cylindrical vessels containing either an inert or a reactive gas mixture.

The method of Gel'fand is based, first, on the assumption that nonuniform distribution of parameters with respect to the fragmentation effects may be ignored and, second, on the assumption of isentropic flow. These assumptions are allowed in the case of large fragments.

Empirical correlations for the initial velocity were proposed by Moore (1967). The relation is derived for fragments accelerated by high explosives packed in a casing. Therefore, this equation predicts velocities higher than actual, especially for low vessel pressures and few fragments. The method was later adapted by Gurney (Brown, 1985).

The empirical relations provided by Baum (1987) calculate an upper value for the initial fragment velocity from a spherical or cylindrical vessel, filled with an ideal gas or a flashing liquid (BLEVEs), that bursts into a small or, alternatively, a large number of fragments.

In some accidents rocketing effects were developed. In these cases large fragments were propelled for unexpectedly long distances. Baker et al. (1978) and Baum (1987) provided equations for a simplified rocketing problem. However the results of these methods are not considered, even by the same developers, sufficiently reliable for prediction purposes.

TNO's Yellow Book (Van Den Bosh and Weterings, 1997) suggests the use of kinetic energy models only for rough estimates and provides a guideline for the selection of the correlations for the initial velocity. Baker and/or Gel'fand's methods are suggested for physical explosions. Baum's formula for BLEVEs; Gel'fand's method for runaway reactions and internal explosions. Moore's relation is suggested for burst with high-scaled pressures and decomposition of energetic materials.

Gubinelli et al. (2004) used a ballistic model to retrofit data from the investigation of past accidents. The analysis suggested that a single mean value of the initial velocity of projection is sufficient to estimate the initial projection velocity of the fragments. The use of a kinetic energy model (Mannan, 2005) was found sufficiently precise for BLEVEs and mechanical explosions. In the case of BLEVE accidents involving liquefied petroleum gas (LPG) vessels, the analysis showed that an average value of about 4% of the Explosion Energy was actually transferred to the fragments as kinetic energy. Higher and more conservative values (10–40%) are proposed in the literature (Fingas, 2002; Holden and Reeves, 1985). Mébarki et al. (2009a) propose a probability distribution for the factor α that expresses the fraction of the explosion energy transferred to the fragments (Table 6.4). The mean value identified from the analysis of past experimental data is equal to 5.77%, not so far from the value of 4 proposed by Gubinelli et al. (2004).

The Baker model (Baker et al., 1983) yields the best result in the case of confined explosions. Conservative results were obtained using the Baker model in the case of runaway reactions. Table 6.4 reports some suggestions for the calculation of fragment velocity (Gubinelli et al., 2004).

Table 6.4 Suggested Models for the Initial Velocity of Fragments

Scenario	Model	Main Equations
BLEVEs	Kinetic energy model (Baker et al., 1983; Khan and Abbasi, 1998)	$u^2 = \alpha \left(\dfrac{2E_v}{M_v} \right)$
Mechanical explosion	Kinetic energy model (Baker et al., 1983; Khan and Abbasi, 1998)	$u^2 = \alpha \left(\dfrac{2E_v}{M_v} \right)$
Confined explosion	The Baker model (Baker et al., 1983; Brode, 1959; Holden, 1986)	$\log(u_s) = 0.56 \log(P_s) + 0.23$ (cylindrical vessels) $\log(u_s) = 0.6 \log(P_s) + 0.13$ (spherical vessels) $P_s = (P_1 - P_0)V/(M_V a_0\, 2);$ $u_s = u/(k\, a_0)$
Runaway reactions	The Baker model (Baker et al., 1983; Brode, 1959; Holden, 1986)	$\log(u_s) = 0.56 \log(P_s) + 0.23$ (cylindrical vessels) $\log(u_s) = 0.6 \log(P_s) + 0.13$ (spherical vessels) $P_s = (P_1 - P_0)V/(M_V a_0\, 2);$ $u_s = u/(k\, a_0)$

u: fragments initial velocity; E_v: explosion (expansion) energy; M_V: vessel mass; P_1: pressure inside vessel at failure; P_0: atmospheric pressure; a_0: speed of sound of the contained gas; V: volume of the vapor phase in the vessel subject to explosion.

6.3.2 Fragments Caused by the Collapse of Rotating Equipment

Fragments can be generated and projected following the failure of some types of rotating equipment. The analysis of accident dynamics for this scenario allowed the definition of the following heuristic criteria for the identification of rotating equipment units which may be possible sources of fragments:

1. assessment of the geometrical characteristics;
2. assessment of the specific operating conditions;
3. assessment of the average yearly working time.

With respect to the first criteria, rotating equipment able to cause relevant events should be able to generate, by the failure of the impeller, the projection of fragments having a nonnegligible geometrical size. This condition is necessary in order to have fragments with a sufficiently high inertia to cause significant damage to structures.

Specific operating conditions are relevant in order to assess if fragments having a sufficient kinetic energy to be projected at a significant distance from the source equipment are likely to be generated. It is clear enough that low-speed rotating equipment items are not able to cause the generation of fragments having a relevant kinetic energy. If so, even the casing may be a sufficient protection to avoid the projection of fragments. This criterion leads to exclude pumps and fans as relevant sources of fragments, while compressors and turbines usually have a sufficient kinetic energy to cause effective fragment projection.

A further element is the average yearly working time. This directly influences the expected frequency of an event leading to fragment projection from the equipment, since this is usually proportional to the number of worked hours. Thus, a low yearly average number of working hours may be a further criteria in order to exclude a rotating equipment as a relevant source of fragments.

Initial velocity of the fragment can be conservatively assumed equal to the maximum tangential velocity of the rotor before fragmentation. Alternative evaluations can be based on the conservation of the kinetic energy that the section constituting the fragment had before the detachment. A specific reduction of the initial velocity may be considered in order to take into account the energy required for casing perforation.

6.3.3 Secondary Fragments

Secondary fragments may be generated when the primary scenario involves blast waves. The energy from the wave should be such to lift or detach the projected items from their original location and make them airborne. Currently no extensive study considers secondary fragments in domino assessment. The difficulty is mainly into predicting the energy transferred by the blast wave to the fragment, since it is strongly dependent on the characteristics of the wave (wind forces, duration, etc.) and of the missile-to-be (shape, weight, initial location, etc.) (Baker et al., 1983). The analysis of past accidents carried out by Gubinelli & Cozzani evidences that no domino accident was caused by secondary fragments. This possibly suggests that the energies involved are lower than the ones associated with primary fragments and the blast waves itself (Gubinelli and Cozzani, 2009a). Particular forms of secondary fragments (debris, broken glasses, etc.) have been studied for their direct effect on humans (Mannan, 2005), but their potential as direct cause of domino is yet to be explored.

6.4   Assessment of Fragment Characteristics

6.4.1 Fragmentation Patterns

The definition of a limited number of reference fragmentation modes (i.e. fragmentation pattern), linked to accidental scenario and vessel geometry, allows for the assessment of the expected number and shape of fragments generated in different accidental scenarios. This is the basis for the definition of parameters necessary for the description of fragment trajectory, such as the fragment mass, velocity and drag factor.

Simplified approaches to the assessment of fragment number and properties rely on the definition of statistic distribution functions for the fragment size, weight and number (Mannan, 2005; Mébarki et al., 2009a,b; Van Den Bosh and Weterings, 1997). Since these simplifications result in a less detailed description of the fragment flight, these approaches are not extensively discussed here.

More detailed approaches may be based on the concept of "fragmentation pattern," introduced by Holden, Westin and Reeves (Holden and Reeves, 1985; Holden, 1986;

Westin, 1971, 1973). Westin (1971) recognized that a limited number of reference patterns were sufficient to describe the modes of fragmentation of cylindrical vessels containing LPG undergoing BLEVE events. Also in the work of Holden (1986) reference fragmentation patterns were used to classify the observed vessel fragmentation modes for horizontal cylindrical vessels subjected to BLEVE. Gubinelli and Cozzani (2009a) based the definition of a set of reference fragmentation patterns for different vessel categories on the analysis of fracture mechanics and of past accident data.

The definition of a fragmentation pattern also leads to the identification of the number of fragments originated in a failure event. Since the initial geometry of the primary equipment is known (shape, dimensions and density of the material), the identification of a fragmentation pattern paves the way to the definition of the expected dimensions and weight of the single fragments originated.

Some alternative approaches, as the "pragmatic approach" suggested in the TNO's Yellow Book (Van Den Bosh & Weterings, 1997), do not resort to the definition of a fragmentation pattern, but directly introduce simplified heuristic rules for the definition of the number of fragments and fragment mass.

6.4.2   Available Fragmentation Patterns for Vessels

Table 6.5 reports a graphical representation of the set of reference fragmentation patterns identified for metallic vessels by Gubinelli and Cozzani (2009a,b). For cylindrical shells no distinction was made between horizontal and vertical cylindrical vessels. The same table contains a brief description of the patterns and of the expected fragment number. The fragmentation patterns reported in the table were selected among all the ones that could be considered possible by considerations based on the fracture fundamentals and on the observation of past accident records (149 events). The reference patterns were defined taking into account the influence of a number of different factors: vessel shape, fracture initiation, preferred initial location, fracture propagation mechanism, credibility of branching and arrest mechanisms. The proposed fragmentation patterns refer to ideal vessel shells: the presence of chunky attachments (e.g. manholes, stirrer motors, nozzles) may generate other fragments that should be identified on a case by case basis.

The patterns resulting in the formation of a low number of fragments can be associated to BLEVEs and physical explosions. The observed average number of fragments for cylindrical vessels undergoing a fragmentation was 2 in the case of a BLEVE, and 3 for a physical explosion (Gubinelli and Cozzani, 2009a).

Patterns resulting in a higher number of fragments (CV21) are relevant mainly for events originated by confined explosions and runaway reactions. This kind of pattern is expected for physical explosions only in the case of cryogenic vessels. In the case of fired BLEVEs, the analysis of past accidents evidenced a limited number of events following patterns with a high number of fragments (e.g. CV21): besides the possible rupture of fragments due to the impact with the ground, the high number of fragments formed in some of these events (up to 9) also suggests the possibility of fragmentation by a brittle fracture mechanism in the cold zones of the vessel wall, outside the section engulfed in the fire.

Table 6.5 Likely Fragmentation Patterns Obtained from the Analysis of Fracture Mechanics Fundamentals and Past Accident Data

ID	Fragmentation Pattern	Description	Expec. N_f	Expected Fragment Shape
CV1		An axial fracture starts and propagates in two opposite directions. If the two tips do not meet (more probable), one fragment (the entire vessel) may be projected, but no detached piece is formed. No branching and no direction turn (no connections and no defects) should take place to obtain this pattern. The vessel may not be deformed (cylindrical fragment) or it may be flattened (plate fragment).	1	Configuration A ($P_{fs} = 0.5$): 1 cylindrical fragment (CE) or Configuration B ($P_{fs} = 0.5$): 1 flattened plate (PL)
CV2		The fracture, likely to start in the axial direction, may turn in the circumferential direction due to stress field changes (bending or stress intensification near connections), or to defects. If the axial crack propagates on the tube-end and stops, a flattened tube-end may be generated.	2	Configuration A ($P_{fs} = 0.28$): 2 tube-ends (PTE2, $\psi = 0$) or Configuration B ($P_{fs} = 0.72$): 1 tube-end (PTE2, $\psi = 0$), 1 flattened plate (PL)
CV3		Similar to CV2. Credible if the fracture starts on a pipe connection or if one of the two tube-ends impacts on a near object at the moment of the projection.	3	1 tube-end (PTE2, $\psi = 0$), 2 parts of tube-end (PTE2, $\psi > 0$ and PTE1)
CV4		Similar to CV2. Credible if the fracture starts on a pipe connection or if one of the two tube-ends impact on a near object at the moment of the projection. The axial fractures on the tube-end could arrest originating flattened tube-ends.	≥ 3 (R: 4)	1 tube-end (PTE2, $\psi = 0$), 2 parts of tube-end (PTE1) 1 plate (PL)

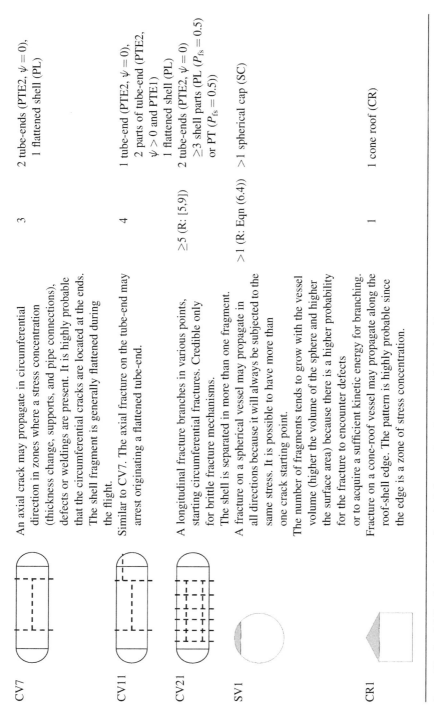

CV7	An axial crack may propagate in circumferential direction in zones where a stress concentration (thickness change, supports, and pipe connections), defects or weldings are present. It is highly probable that the circumferential cracks are located at the ends. The shell fragment is generally flattened during the flight.	3	2 tube-ends (PTE2, $\psi = 0$), 1 flattened shell (PL)
CV11	Similar to CV7. The axial fracture on the tube-end may arrest originating a flattened tube-end.	4	1 tube-end (PTE2, $\psi = 0$), 2 parts of tube-end (PTE2, $\psi > 0$ and PTE1) 1 flattened shell (PL)
CV21	A longitudinal fracture branches in various points, starting circumferential fractures. Credible only for brittle fracture mechanisms. The shell is separated in more than one fragment.	≥ 5 (R: [5,9])	2 tube-ends (PTE2, $\psi = 0$) ≥ 3 shell parts (PL ($P_{fs} = 0.5$) or PT ($P_{fs} = 0.5$))
SV1	A fracture on a spherical vessel may propagate in all directions because it will always be subjected to the same stress. It is possible to have more than one crack starting point. The number of fragments tends to grow with the vessel volume (higher the volume of the sphere and higher the surface area) because there is a higher probability for the fracture to encounter defects or to acquire a sufficient kinetic energy for branching.	>1 (R: Eqn (6.4))	>1 spherical cap (SC)
CR1	Fracture on a cone-roof vessel may propagate along the roof-shell edge. The pattern is highly probable since the edge is a zone of stress concentration.	1	1 cone roof (CR)

P_{fs}: conditional probability of alternative configuration in fragment shape; R: recommended number of fragments in QRA.

Source: Adapted from Gubinelli and Cozzani (2009a,b).

In the case of cone-roof tanks, only physical explosions, confined explosions and runaway reactions are likely to cause the fragmentation of the equipment. In the case of spherical vessels, only BLEVE events are credible, since these vessels in general are used only for the storage of liquefied pressurized gases. According to Gubinelli and Cozzani (2009a), spherical vessels and atmospheric cone-roof tanks evidence a single fragmentation mode (SV1 and CR1 respectively). In cone-roof tanks, as well as in other sharp-edged equipment, the fracture is likely to occur and propagate along zones where stress concentration is present. In roof tanks this is represented by the roof-shell edge. On the other hand, a fracture on a spherical vessel may propagate in all directions because it will always be subjected to the same stress. Therefore several segments of the size of the vessel can be generated. This conclusion, based on fracture theory, was validated by a limited number of available accident data (on fired BLEVE scenarios for spherical vessels and on confined explosion scenarios for conical tanks).

The definition of the fragmentation patterns allows for the identification of the expected number and shapes of the fragments. Table 6.5 reports the expected shape and number of fragments for some fragmentation patterns. It must be remarked that in this study the number of fragments should be intended as the final number of parts in which the vessel is fragmented, including the main body of the vessel. Moreover, only fragments having relevant sizes (more than 2% of empty vessel weight) were considered. Also in this case, the analysis of past accident by Gubinelli and Cozzani (2009a) provides data on the number of fragments formed as a consequence of the different fragmentation patterns, allowing a validation of the method.

The shape of the projected fragment does not always correspond to the shape the section had in the original vessel: during the tearing of the vessel, deformation (e.g. flattening) may occur. In particular, the shell fragments formed in the CV1, CV7 and CV21 fragmentation patterns can be flattened. Moreover, flattened tube-end fragments were formed in some fragmentation events following the CV3 pattern. These results are in agreement with those obtained by Birk (1996) that reports fragment flattening in several full-scale experimental tests.

Gubinelli and Cozzani (2009a) also provide a classification of a limited set of "ideal" reference shapes for fragments, validated by the analysis of past accidents (Table 6.6). The vessel fragmentation process obviously results in fragments having irregular shapes; however, on the basis of the actual fragment shapes it was possible to define a few reference model shapes for the fragments which may be used to represent the actual fragment shapes.

Table 6.5 reports the expected number of fragments for the fragmentation patterns considered. The observed data in past accidents correspond to the theoretical values reported in the table (Gubinelli and Cozzani, 2009a). Some fragmentation patterns may lead to the formation of a variable number of fragments (e.g. CV21 and SV1).

In the case of fragmentation pattern CV21 for cylindrical vessels, Gubinelli and Cozzani (2009b) suggest considering a number of fragments comprised between 5 and 9, and a uniform probability of distribution in the interval.

In the case of spherical vessels, a rough correlation is present among vessel volume and the number of fragments generated, since a higher probability for fracture

Table 6.6 Reference Shapes for Fragments

ID	Fragment Shape	Description	Geometric Parameters
CE		Cylinder: the fragment is constituted by the entire horizontal or vertical cylindrical vessels	t: wall thickness; ρ: density; l: length; r: end radius.
SC:		Spherical cap: the fragment is a section of a spherical surface, defined by a solid angle γ	t: wall thickness; ρ: density; r: radius; γ: arc angle.
CR		Cone roof: the fragment is a cone, generated by the detachment of the roof of an atmospheric tank	t: wall thickness; ρ: density; h: height; r: radius.
PL		Plate: flattened shell or pipe section end, or section of a sharp-edged atmospheric equipment	t: wall thickness; ρ: density; l: length; $l1$: width.
PT:		Tube section: fragment of a cylindrical shell. If $\xi = 2\pi$ the fragment is a tube.	t: wall thickness; ρ: density; l: length; r: radius; ξ: arc angle.
PTE1		Tube end section (Reference Shape #1): fragment of a cylindrical shell, generated by the propagation of a circumferential and of an axial crack	t: wall thickness; ρ: density; l: length; r: radius; ψ: sector angle.

(*Continued*)

Table 6.6 Reference Shapes for Fragments—*cont'd*

ID	Fragment Shape	Description	Geometric Parameters
PTE2		Tube end section (Reference Shape #2): fragment of a cylindrical shell, generated by the propagation of a circumferential and of an axial crack. If $\psi = 0$ the fragment is a tube end (e.g. generated from a cylindrical vessel by the propagation of a circumferential crack).	t: wall thickness; ρ: density; l: length; r: radius; ψ: sector angle.

Source: As proposed by Gubinelli and Cozzani (2009b).

branching corresponds to higher vessel volumes and higher vessel surface areas. Holden and Reeves (1985) proposed the following linear correlation:

$$N_f = -3.77 + 0.96 \cdot 10^{-2} \cdot V \tag{6.3}$$

where N_f is the number of fragments and V is the vessels volume in cubic meters.

Gubinelli and Cozzani (2009b) updated the correlation, based on a larger number of observations:

$$N_f = -0.425 + 6.115 \cdot 10^{-3} \cdot V \tag{6.4}$$

Equation (6.4) provides more reasonable results for low vessel volumes (number of fragments approaches zero for volumes close to zero). However, both correlations are based on a limited number of data and yield results affected by significant uncertainty. Nevertheless, in the framework of quantitative risk analysis, the simplicity of the approach makes it attractive for the preliminary estimation of the expected number of fragments.

6.4.3 Fragmentation Patterns for Rotating Equipment

Fragmentation patterns can be defined also in the case of rotating equipment. Fragments from rotating equipment are usually originated by failure of high-speed rotors, while the projection of fragments originated by the casing of the original unit, although

Table 6.7 Likely Fragmentation Patterns Obtained from the Analysis of Past Accidents and Fracture Propagation in Rotating Equipment

ID	Fragmentation Pattern	Description	Expec. N_f	Expected Fragment Shape
RE1		Detachment of a single blade. The rest of the rotor is not projected at relevant distances.	1	One fragment with the shape of the blade (PL or PT)
RE2		The rotor or impeller fractures and a section is detached. A circular sector with an angle of 120° is assumed as the detached fragment. The other sector of the rotor/impeller is not projected at relevant distances.	1	One fragment with the shape of a circular sector (PL)

possible, is rather unlikely. The analysis of past accidents evidenced the occurrence of two main fragment shapes in the catastrophic failure of rotating equipment:

RE1. fragments constituted by a single blade of the rotor;
RE2. fragment of the impeller/rotor, usually corresponding to a 120° section of the impeller itself.

The resulting fragmentation patterns thus considers the generation of a single fragment having one of the two shapes described above. The fragmentation patterns are summarized in Table 6.7. The casing of rotating equipment may provide a barrier to fragment projection. In particular, casing design may account for blade detachment (fragmentation pattern RE1 of the rotor), allowing a sufficient casing thickness to provide an effective passive barrier to fragment penetration.

6.4.4 Drag Factors for Fragments

The definition of reference model shapes for the fragments make possible the identification of the main characteristics influencing the flight of a missile. Among them, the fragment drag factors depend on fragment shape, size and weight. The reference shapes defined in Section 6.4.2 can be used for the evaluation of these parameters.

The following general expression may be used for the calculation of the drag coefficient (Gubinelli et al., 2004):

$$k(\overline{x}) = a \cdot \mathrm{DF}(\overline{x}) + b \tag{6.5}$$

where a and b are dimensional constants not dependent on the geometrical parameters of the fragment ($a = 0.69$ kg/m^3 and $b = 3.28.10^{-5}$ 1/m), and DF is the fragment drag factor. For chunky fragments, the fragment drag factor may be estimated as follows (Baker et al., 1983):

$$DF = \frac{C_D A_D}{M} \qquad (6.6)$$

where C_D is a drag coefficient, function of the fragment shape and of its orientation with respect to the flow direction, A_D is the section of the fragment on a plane perpendicular to the trajectory, and M is the mass of the fragment. Since the orientation of the fragment with respect to the trajectory is usually unknown when possible fragmentation accidents are assessed, Gubinelli and Cozzani (2009b) suggest that an average value of DF may be used in Eqn (6.5):

$$DF_a = \frac{DF_{max} + DF_{min}}{2} \qquad (6.7)$$

where DF_{min} and DF_{max} are, respectively, the minimum and the maximum values of DF that may be obtained considering all the possible orientations of the fragment with respect to the flight trajectory. Taking account of the effective fragment rotation is possible, as described by Mébarki et al. (2009a). However the angular velocities of the fragment are difficult to predict and may vary during the flight, making this approach quite cumbersome for practical use.

Gubinelli and Cozzani (2009b) proposed analytical functions for the drag factors of the reference shapes defined in Section 6.4.2. The equations, as a function of the geometrical parameters of the fragment defined in Table 6.6, are reported in Table 6.8.

The results of a sensitivity analysis carried out by Gubinelli and Cozzani (2009b) suggest that in the calculation of drag factors, only a limited number of parameters are critical and should be carefully assessed, while a default value may be used for other noncritical parameters. Table 6.8 lists the critical parameters and the proposed simplified functions for drag factors calculation. The table also reports the suggested value for the noncritical parameters in the development of the simplified functions. In the simplified revised functions, only the critical parameters listed are therefore present. With few exceptions, the simplified drag factor functions allow the calculation of the fragment drag factor only on the basis of the design details of the vessel undergoing fragmentation. On the other hand, the drag factor functions obtained for tube sections and for tube ends show a dependence on the ξ and ψ angles. Since the value of these parameters is unknown and is not predictable "a priori", a discrete distribution of the values of ξ and ψ was introduced, and different simplified drag factor functions were obtained for each value considered. A uniform probability distribution was assumed for each of the values considered, in coherence with the results coming from past accident analysis that did not evidence any preferential value of these angles.

Table 6.8 Critical Parameters Identified in Drag Factor Functions and Simplified Drag Factor Functions for the Evaluation of Fragment Impact Probability

Shape ID	Typical Range of k Values	Critical Parameters	Reference Values for Noncritical Parameter	Proposed Simplified Function
CE	$[1.9 \times 10^{-4}, 1.6 \times 10^{-3}]$	ρ, t	$l = 10\,m$ $r = 2.5\,m$	$DF''_{CE} = \dfrac{0.166}{\rho t}$
SC	$[5 \times 10^{-4}, 4.5 \times 10^{-3}]$	ρ, t	$\gamma = 0.6\,\pi$	$DF''_{SC} = \dfrac{0.460}{\rho t}$
CR	$[2 \times 10^{-4}, 1.1 \times 10^{-2}]$	r, h, t, ρ	–	$DF_{CR} = DF_{CR}(r, h, t)$
PL	$[1 \times 10^{-3}, 9 \times 10^{-3}]$	ρ, t	$l = 5\,m$ $l_1 = 5\,m$	$DF''_{PL} = \dfrac{1.17 + 0.41t}{\rho t}$
PT	$[4 \times 10^{-4}, 1 \times 10^{-2}]$	ξ, t, ρ	$l = 5\,m$ $r = 2.5\,m$	$\xi = \pi/2 \quad DF''_{PT} = \dfrac{1}{\rho t}\left(\dfrac{2.701}{5 - t} + 0.205t\right)$
				$\xi = \pi \quad DF''_{PT} = \dfrac{1}{\rho t}\left(\dfrac{1.910}{5 - t} + 0.205t\right)$
				$\xi = 3\pi/2 \quad DF''_{PT} = \dfrac{1}{\rho t}\left(\dfrac{1.273}{5 - t} + 0.205t\right)$
				$\xi = 2\pi \quad DF''_{PT} = \dfrac{1}{\rho t}\left(\dfrac{0.955}{5 - t} + 0.205t\right)$
PTE1	$[4 \times 10^{-4}, 1 \times 10^{-2}]$	ψ, t, ρ	$l = 5\,m$ $r = 2.5\,m$	$\psi = \pi/8 \quad DF''_{PTE1} = \dfrac{0.550}{\rho t}$
				$\psi = \pi/4 \quad DF''_{PTE1} = \dfrac{0.450}{\rho t}$
				$\psi = 3\pi/8 \quad DF''_{PTE1} = \dfrac{0.440}{\rho t}$
				$\psi = \pi/2 \quad DF''_{PTE1} = \dfrac{0.350}{\rho t}$
PTE2	$[3 \times 10^{-4}, 5 \times 10^{-3}]$	t, ρ	$l = 5\,m$ $r = 2.5\,m$ $\psi = \pi/4$	$DF''_{PTE2} = \dfrac{0.240}{\rho t}$

Source: Adapted from Gubinelli and Cozzani (2009b). Units: t (m), ρ (kg/m^3), h(m), r(m)

6.5   Calculation of Impact Probability

6.5.1   *Probability of Impact*

As underlined by Mannan (2005), the flight of a fragment is a standard problem in mechanics, for which a fundamental approach is described by Baker et al. (1983).

However, the method proposed by these authors is concerned with the accurate determination of the trajectory of a single fragment, and it is not directly suitable to derive the impact probability of a fragment. Hauptmanns (2001a,b) proposed a valuable and comprehensive approach to the calculation of impact probabilities of a fragment based on trajectory analysis. Impact probabilities in a given position with respect to the fragment origin were calculated by Monte Carlo methods, assuming probability distributions for the initial projection parameters (e.g. initial fragment velocity, number, mass and energy of fragments, etc.). Nevertheless, the study was mainly oriented to the determination of impact probabilities on exposed individuals and not on process equipment, thus the influence of the target geometry was not taken into account.

Gubinelli et al. (2004) presented a general method for the assessment of the probability of domino scenarios caused by fragment impact. The methodology is based on a ballistic analysis of all the possible trajectories of a fragment with a given mass, shape and initial velocity. A comparison with data from past accidents (Holden and Reeves, 1985) and from former studies (Baker et al., 1983) provided a validation of the model.

The frequency of a domino event caused by the impact of a fragment generated in a primary accident on a given secondary target may be expressed as:

$$f_F = f_p \cdot P_F \cong f_p \cdot \sum_F \left(P_{gen,F} \cdot P_{imp,F} \cdot P_{dam,F} \right) \tag{6.8}$$

where f_p is the expected frequency of the primary event and P_F corresponds to probabilities of the following event chain: (1) generation of several fragments of defined mass and shape during the primary event; (2) projection and impact of one (or more than one) of the fragments with the given target; and (3) probability of irreversible effects following target impact (i.e. the probability of damage of the target equipment).

The frequency of the primary event (scenarios discussed in Section 6.2.1) can be obtained by conventional methodologies (e.g. standard frequencies, parts count, fault tree analysis, failure frequency databases, etc.).

If the probability that two fragments impact on the same target is sufficiently low, as usual if a limited number of fragments is generated in the primary event, the expected probability of a domino event on a given secondary target (P_F) can be calculated as sum of the domino probabilities for single fragments (Eqn (6.8)). These, in turn, can be evaluated as a product of three contributions, corresponding to the main phases of the event chain described above: (1) the probability of each fragment to be generated ($P_{gen,F}$); (2) the probability of impact ($P_{imp,F}$) on a given target; and (3) the probability of irreversible effects following target impact ($P_{dam,F}$).

The probability of each fragment to be generated ($P_{gen,F}$) quantifies the conditional probability that a fragment with a given shape and mass is formed and projected following the primary event. It can be expressed, in turn, as the product of three contributors:

$$P_{gen,F} = P_{cp} \cdot P_{fp} \cdot P_{fs} \tag{6.9}$$

where P_{cp} is the probability of fragment generation after initial crack propagation; P_{fp} is the conditional probability of a fragmentation pattern; and P_{fs} is the conditional probability of fragment shape, that quantifies the probability of a given shape in the fragment, in cases where alternative shapes are possible for a given fragmentation pattern. Typical values for these parameters are discussed in Section 6.5.2.

The probability of impact of each fragment on a target, $P_{imp,F}$, is dependent on the initial conditions of the flight (direction and velocity), on the fragment characteristics (shape, dimensions, and weight) and on the characteristics of the target (dimension and location). As discussed in Section 6.2.2, the impact between a flying fragment and a target occurs under a limited range of initial projection angles. Defining $\Delta\theta$ and $\Delta\varphi$ as the intervals identifying all the directional angles θ and φ for which the impact takes place (in particular, according to Figure 6.2, $\Delta\varphi$ may be $\Delta\varphi = \Delta\varphi_1 \cup \Delta\varphi_2$), the total impact probability of the fragment F on the target or on its VA may be expressed as follows:

$$P_{imp,F} = \int_{\Delta\theta} \int_{\Delta\varphi} \wp_{dir}(\theta, \varphi) \cdot d\theta \cdot d\varphi \tag{6.10}$$

where $\wp(\theta, \varphi)$ is the probability distribution function of the fragment initial direction. The direction of fragment projection may depend on several factors, as the features of the ruptured unit, the position of the main connection pipes, the characteristics of the explosion causing the generation of the fragments, etc. Nevertheless, literature reports typical distributions, summarized in Section 6.5.2. In practice, the calculation of the value of $P_{imp,F}$ according to Eqn (6.10) can follow two alternative, but equivalent, approaches. Pula et al. (2007) and Mébarki et al. (2009b), starting from assumptions of probability distribution of projection angles, used a Monte Carlo approach to generate a probability map of trajectories. From this map the probability of impact is calculated by applying the relevant impact conditions (fragment trajectories falling within the VA of interest). On the other side, Gubinelli et al. (2004) solved Eqns (6.1) and (6.2) for the fragments of interest in order to identify the limits of the interval of initial direction angles that result in effective impact trajectories. Successively, they integrated the probability distribution functions for the identified initial angles by Eqn (6.10).

The probability of irreversible effects following target impact, $P_{dam,F}$, defines the vulnerability of the target. In the case of equipment or structures it can be assumed to be equal to the probability of effective penetration of the missile through the target structure. Fragment penetration can be analyzed according to the fragment shape and kinetic energy by suitable models (Section 6.6). Clearly enough target protection barriers may be accounted for if present. Though several models are available for the calculation of fragment penetration on a given target, no criteria are provided to date for the estimation of the actual damage probability. A common conservative hypothesis, in absence of reliable damage models, is to assume a unit value for the damage probability if the impact takes place (Mannan, 2005).

6.5.2 Typical Values for Fragment Probability

In the following typical values relevant for the calculation of domino probability as a consequence to fragment impact are reported, having as a reference the approach described in Section 6.5.1.

The probability of fragment generation after initial crack propagation (P_{cp}) is the first component of the probability of fragment generation ($P_{gen,F}$). In order to have fragment projection events, the crack, once formed, should propagate along the vessel enough to lead to the separation of at least one fragment that will be projected away from the equipment. Otherwise, a simple breach is created in the shell, allowing the release of the vessel content, but without the generation of missiles. The problem is particularly critical in the case of fired BLEVE scenarios, in which the fracture may stop outside the heated wall area, causing a loss of containment but not vessel fragmentation (Fingas, 2002). The analysis of past accident data allowed obtaining the data summarized in Table 6.9 (Gubinelli and Cozzani, 2009a). Also in accordance with the findings of Holden and Reeves (1985), past accident analysis evidenced that fragment projection following vessel fragmentations induced by fired BLEVEs has a conditional probability of 0.9 given vessel failure. On the basis of the discussion concerning the crack propagation mechanism observed for physical explosions and unfired BLEVEs, it seems reasonably conservative to assume that for these scenarios a fragment projection probability is equal to 0.9, in analogy with that estimated for fired BLEVEs. However, this conclusion lacks of confirmation by past accident data, probably due to underreporting of relevant events (Holden and Reeves, 1985). A conservative value of the fragment projection probability equal to 1 is proposed in the case of confined explosions and runaway reactions, since in these events the crack arrest is unlikely (Gubinelli and Cozzani, 2009a).

The conditional probability of a fragmentation pattern (P_{fp}) to occur following vessel fragmentation can be evaluated by past accident data. Table 6.10 shows the conditional probabilities obtained from the analysis of the accidental events recorded in the database collected by Gubinelli and Cozzani (2009a). The distribution of the events, reported in the table, shows that a quite limited number of different fragmentation patterns were sufficient to describe the fragmentation modes of the 141 vessels analyzed. The authors observe that other fragmentation patterns are possible in the presence of specific factors (e.g. very high-pressure increase rates, presence of

Table 6.9 Probability of Fragment Generation After Initial Crack
Propagation (P_{cp})

Type of Primary Event	P_{cp}
BLEVE, fired	0.9
BLEVE, unfired	0.9
Physical explosion	0.9
Confined explosion	1
Runaway reactions	1

Source: Gubinelli and Cozzani (2009a).

Table 6.10 Observed Conditional Probabilities of Credible Fragmentation Patterns (P_{fp}) for Different Primary Scenarios Leading to Vessel Fragmentation

ID	BLEVE (fired)	Physical Explosion, BLEVE (nonfired)	Confined Explosion	Runaway Reaction	All Scenarios
Cylindrical Vessel					
CV1	–	–	–	0.29	0.02
CV2	0.59	0.67	0.90	0.43	0.62
CV3	0.12	0.08	–	–	0.09
CV4	–	0.13	–	–	0.03
CV7	0.29	0.08	–	0.14	0.22
CV11	–	0.04	–	–	0.01
CV21	–	–	0.10	0.14	0.02
Spherical Vessel					
SV1	1.00	1.00	–	–	1.00
Cone Roof Tank					
CR1	–	–	1.00	–	1.00

BLEVE, boiling liquid expanding vapor explosion.
Source: Adapted from Gubinelli and Cozzani (2009a).

bending momentum due to pipe connections, and influence of vessel internals). This actually occurred only in about 5% of the events in their database.

Table 6.10 also reports the distribution of the fragmentation patterns for the different categories of primary scenarios. The data on events caused by physical explosions and unfired BLEVEs are presented together in the table, since the only difference between these primary events is the available explosion energy. For cylindrical vessels, CV2 is the prevailing fragmentation pattern for all the scenarios. An important number of events could be associated also to the CV3, CV7 and CV11 reference patterns. In a limited number of events also the CV1 fragmentation pattern was observed. However, it must be remarked that in the CV1 fragmentation pattern, the axial fracture may stop outside the heated zone of the vessel (in particular in the case of jet fire impingement resulting in a partial wall engulfment) causing only a loss of containment but not fragment projection. Such events may have been under-reported or not described as vessel fragmentation accidents (Holden and Reeves, 1985). Therefore, in the evaluation of the probability of fragmentation patterns, CV1 was excluded for BLEVE accidents, since the probability of fragment projection following this fragmentation pattern was found to be negligible. This may be easily explained, since when this fragmentation pattern takes place, the fragments are generally flattened on the ground due to the start of the axial crack in the upper zone of the vessel, where no liquid is in contact with the vessel walls. A single fragmentation pattern was observed for two equipment categories: CR1 in the case of cone-roof tanks and SV1 for spherical vessels. Thus, a conditional probability equal to 1 may be assumed for these fragmentation patterns.

The conditional probability of fragment shape (P_{fs}) is relevant for those fragmentation patterns where, due to deformation forces during equipment failure, one of the final fragments may assume different shapes (e.g. flattened or nonflattened). In the schematization of fragmentation patterns by Gubinelli & Cozzani (Gubinelli and Cozzani, 2009b) this is the case for patterns CV1, CV2, and CV21 of Table 6.5. The same table reports the conditional probability of the alternative configurations, as estimated on the basis of observation of past accidents.

The direction of fragment projection from the failure of process vessels may depend on several factors, as the features of the ruptured equipment, the position of the main pipes, the characteristics of the explosion causing the generation of the fragments, etc. Although the study of a specific layout may yield more precise information on the possible directions available for fragment projection, this approach is not feasible in a QRA framework of a complex plant.

Thus, unless precise information are available on the presence of preferential directions for fragments projection, a uniform probability distribution was proposed by several authors (Gubinelli et al., 2004; Mannan, 2005; Mébarki et al., 2009a; Pula et al., 2007). In this case, the probability distribution function of the initial missile direction can be expressed as:

$$\wp\left(\theta, \varphi\right) \cdot \mathrm{d}\theta \cdot \mathrm{d}\varphi \; = \; \frac{\cos \varphi \cdot \mathrm{d}\theta \cdot \mathrm{d}\varphi}{4\pi} \tag{6.11}$$

Thus, Eqn (6.10) may be rewritten as follows:

$$P_{\mathrm{imp,F}} \; = \; \frac{1}{4\pi} \int_{\Delta\theta} \int_{\Delta\varphi} \mathrm{d}\theta \cdot \cos \varphi \cdot \mathrm{d}\varphi \; \cong \frac{\Delta\theta}{4\pi} \int_{\Delta\varphi} \cos \varphi \cdot \mathrm{d}\varphi \tag{6.12}$$

where the last passage is valid under the assumption that the φ intervals that verify the impact condition are not dependent on θ intervals. Obviously, the intervals $\Delta\theta$ and $\Delta\varphi$ should be calculated on the basis of the impact condition and of the fragment trajectory (Sections 6.2.2 and 6.5.1).

If the specific geometry of the vessel is considered, the assumption of uniform distribution of the initial direction typically holds only in the case of spherical vessels. For cylindrical vessels, a preferential direction of projection along the axis of the vessel is usually expected. In the work of Holden and Reeves (1985), 11 incidents involving 15 cylindrical vessels containing mainly LPG were analyzed. It was found that about 50% of the total fragments were projected into one-third of the total area, in arcs of 30° to either side of the vessel's front and rear axial directions. Based on this study, Pula et al. (2007) proposed a probability distribution for horizontal angles in the fragmentation of horizontal cylindrical vessels where 50% of the fragments are projected, with uniform probability, within a horizontal angle of 15° per side from the axis of the vessel; the other 50% of the fragments are projected with uniform probability distribution on the other two angle intervals at the sides of the vessel. For elevation angles, Pula et al. suggest that projection of end caps in horizontal cylinders is quite likely with small elevation angles, and propose a uniform distribution of probability in the interval $\varphi \in [0°, 15°]$.

Mébarki et al. (2009a) proposes a similar approach for the description of initial angles in a horizontal vessel, but using different probability figures and interval limits: 60% of the fragments are projected with uniform probability distribution in the horizontal angles $\theta \in [330°, 30°] \cup [150°, 210°]$, while, for the elevation angle, a uniform probability distribution characterizes the sin of the angle, $\sin(\varphi) \in [-1, +1]$, over the full range of $\varphi \in [-90°, +90°]$.

For ejection of a fitting, the direction of the gas jet and, hence, the angle of departure of the fitting is usually defined by the initial orientation (Mannan, 2005). Likewise, this will generally be so for the two parts in the separation and rocketing of a gas-filled vessel.

In the case of fragments generated by rotating equipment, projection has a prevalent direction in a solid sector perpendicular to the spinning axis. Practically a uniform distribution of probabilities can be assumed for the horizontal angle within this sector, while negligible probability is assumed outside the range. For a horizontal spinning axis orientated along the z' axis of Figure 6.1, the solid sector can be defined by a limit angle θ_0 (typical values are about 5°). For vertical angle uniform probability can be assumed on all the angle interval $\varphi \in [-90°, +90°]$. The probability to project a fragment in a given initial direction may be therefore expressed as

$$
\wp(\theta, \varphi) \cdot d\theta \cdot d\varphi =
\begin{vmatrix}
0 & \text{if } \theta \in]\theta_0, \ 180° - \theta_0[\cup]180° + \theta_0, \ -\theta_0[\\
\dfrac{\cos \varphi \cdot d\theta \cdot d\varphi}{8 \cdot \theta_0} & \text{if } \theta \in [-\theta_0, \ \theta_0] \cup [180° - \theta_0, \ 180° + \theta_0]
\end{vmatrix}
$$

$$(6.13)$$

6.6   Calculation of Damage Probability

The damage by missile impact more frequently analyzed in the literature is the penetration of the target. The models available for fragment penetration in metal targets are mostly based on the fitting of experimental data, usually from tests on chunky missiles with sizes smaller than the typical fragments originated by vessel failure. Table 6.11 reports some examples of these models. The same table reports the applicable ranges of values for the parameters in the model, when available. Other approaches can be found in the literature (Mannan, 2005).

In general, the fragment penetration models provide the evaluation of a penetration parameter in function of the missile and target characteristics. Depending on the model, the penetration parameter can be the minimum kinetic energy of the missile for perforation of a target of a given thickness (E), or the maximum thickness of penetration (t), or the minimum missile impact velocity for perforation of a target of a given thickness (U), or the missile impact velocity resulting in 50% probability for perforation of a target of a given thickness (U_{50}). Under the assumption that surface effects do not influence penetration (that is reasonably true for targets which are thin

Table 6.11 Example of Models for Penetration of Metal Targets

Model Equations	Units	Range	Note	Ref
$E = \dfrac{d \cdot \sigma_R \cdot s^2}{10.29} \cdot \left(42.7 + \dfrac{L}{s}\right)$	E (J) d (m) σ_R (Pa) s (m) L (m)	4.7 mm $\leq s \leq$ 40 mm 4.7 mm $\leq d \leq$ 85 mm 300 MPa $\leq \sigma_R \leq$ 480 MPa	Target: flat, steel Missile: rigid, cylindrical, flat head	Neilson, 1985
$E = 1.4 \cdot 10^9 \cdot (d \cdot s)^{3/2}$	E (J) d (m) s (m)	N/A	Target: flat, mild steel	Neilson, 1985
$E = 2.94 \cdot 10^9 (D \cdot s)^{3/2}$ **Conical head missile:** $D = s \cdot [1 + 2.9 \cdot \tan(\Phi/2)^{2.1}]$ **Flat or spherical head missiles:** $D = d$	E (J) d (m) s (m) Φ	7 mm $\leq s \leq$ 38 mm 3 kg $\leq W \leq$ 50 kg 25 m/s $\leq u \leq$ 180 m/s $20° \leq \Phi \leq 90°$ 87.5 mm $\leq d \leq$ 160 mm	Target: SGV 49 $L = 1.5$ m Missile: AISI 304, flat head and conic head	Ohte et al., 1982
$E = 1.1 \cdot d^{1.48} \cdot s^{1.63}$	E (J) d (mm) s (mm)	1.23 mm $\leq s \leq$ 3 mm 6.35 mm $\leq d \leq$ 12.7 mm	Target: flat, mild steel	Stronge, 1985
$E = 8 \cdot 10^9 \cdot \sigma_R \cdot d^3 \cdot \left(\dfrac{s_t}{d}\right)^{1.7} \cdot \left(\dfrac{d}{D}\right)^{0.5}$	E (J) d (m) σ_R (Pa) s_t (m) D (m)	25 mm $\leq d \leq$ 170 mm 7 mm $\leq s_t \leq$ 18 mm 4 kg $\leq W \leq$ 50 kg	Target: pipe $D = 150$ mm	Neilson et al., 1987
$E = 3.28 \cdot d^{1.5} \cdot s^{1.4}$	E (J) d (mm) s (mm)	N/A	Target: steel	US-NOL, 1955

Equation	Variables	Range	Target/Missile	Reference
$t = 6 \cdot 10^{-5} \cdot W^{0.33} \cdot u$	t (m) W (kg) u (m/s)	$W < 1$ kg $u < 1000$ m/s	Target: mild steel Missile: steel, flat head	Cox and Saville, 1975
$t = \dfrac{C \cdot W}{A} \log_{10}(1 + 5 \cdot 10^{-5} \cdot u^2)$ $C = 3 \cdot 10^{-5}$ for steel alloys	t (m) W (kg) u (m/s) A (m^2)	$W > 1$ kg $u < 1000$ m/s	Target: steel Missile: steel	Cox and Saville, 1975
$t = \left(3.3 \cdot 10^{-8} \cdot \dfrac{W \cdot u^2}{d^3}\right)^{1/1.41}$	t (m) d (m) W (kg) u (m/s)	$u < 1000$ m/s	Target: mild steel Missile: rod shaped	Cox and Saville, 1975
$t = \dfrac{459}{\left(\sigma_R \cdot \dfrac{E_T}{E_{70}}\right)^{0.5}} \cdot N \cdot K_p \cdot \dfrac{W}{d^{1.43} \cdot d_e^{0.6}} \cdot \left(\dfrac{u}{1000}\right)^{1.65}$ **Flat missiles:** $N = 0.72$ **"Nosed" missiles:** $N = 0.72 + 0.25 \cdot \left(\dfrac{r}{d} - 0.25\right)^{0.5} \leq 1.17$ **Tubes or irregular fragments:** $N = 0.72 + 0.0306 \cdot \left[\left(\dfrac{d}{d_e}\right)^2 - 1\right] \leq 1$ $K_p = \dfrac{(0.632 \cdot BHN_p + 94.88)}{275} \cdot \dfrac{\rho_p}{\rho_R}$ $d_e = \sqrt{\dfrac{4 \cdot A}{\pi}}$	t (in) E_T E_{70} r (in) BHN_p ρ_R (lb/in^3) A (in^2) u (ft/s) σ_R (psi) ρ_p (lb/in^3) W (lb) d (in)	N/A	Target: steel	Kar, 1979

(Continued)

Table 6.11 Example of Models for Penetration of Metal Targets—cont'd

Model Equations	Units	Range	Note	Ref
$U = \dfrac{4 \cdot \Omega \cdot s^2 \cdot \psi \cdot \eta}{L \cdot d} \cdot (1 + b^{0.5})$ $b = \dfrac{L + \Omega \cdot s}{\Omega \cdot s}(1 + d/(4 \cdot \rho_t \cdot s \cdot \eta \cdot \psi^2))$ $\Omega = \dfrac{\rho_t}{\rho_p}; \ \psi = \dfrac{\rho_p \cdot a_p + \rho_t \cdot a_t}{\rho_p \cdot a_p \cdot \rho_t \cdot a_t}$ $\eta = 1.76 \cdot 10^6$ psi for mild steel	U (in/s) L (in) d (in) s (in) ρ_p (lb/in³) ρ_t (lb/in³) a_p(in/s) a_t (in/s)	$s/L < 0.5$, $s/d < 0.5$	Target: mild steel Missile: rigid, flat head Plugging perforation mechanism	Recht and Ipson, 1963
a) $\ U_{50} = 3.155 \cdot \dfrac{\sqrt{\sigma_t \cdot \rho_t} \cdot \dfrac{2 \cdot s}{d}}{\dfrac{\rho_p}{\sqrt{\sigma_t \cdot \rho_t} \cdot \dfrac{2 \cdot s}{d}}}$ b) $\ U_{50} = 2.941 \cdot \dfrac{\sqrt{\sigma_t \cdot \rho_t} \cdot \dfrac{2 \cdot s}{d}}{\rho_p}$ $d = 2 \cdot \left(\dfrac{W}{\rho_p \cdot \dfrac{4 \cdot \pi}{3}} \right)^{1/3}$	U_{50} d s ρ_p ρ_t σ_t	$s/d < 1.1$	Target: flat or low curvature Missile: spherical a) deformable missile b) rigid missile	Baker et al., 1975
$U_{50} = 2.05 \cdot 10^4 \cdot \left(\dfrac{s}{W^{1/3}} \right)^{3/4}$	U_{50} (ft/s) s (in) W (g)	$u < 2000$ ft/s $200 \le BHN_p \le 300$	Target: steel Missile: flat head	US-NOL, 1955
$U_{50} = 4.5 \cdot 10^4 \cdot \left(\dfrac{s^{0.906}}{W^{0.359}} \right)$	U_{50} (ft/s) s (in) W (g)	N/A	Target: steel	BAL, 1961

A: impact area of missile; a_p: sound velocity in the missile material; a_t: sound velocity in the target material; BHN$_p$: Brinell hardness of missile material; d: missile diameter; D: pipe external diameter; E: kinetic energy of projected missile; E_{70}: Charpy V test energy for the target material at ambient temperature (70 °F); E_T: Charpy V test energy for the target material at operating temperature; L: distance among supports; r: radius of the "nose" of the missile; s: maximum plate thickness; s_t: pipe wall thickness; t: maximum thickness of penetration; u: missile impact velocity; U: minimum missile impact velocity for perforation of the target; U_{50}: missile impact velocity resulting in 50% probability for perforation of the target; W: missile weight; ρ_p: density of missile material; ρ_R: reference density, 0.283 lb/in³; ρ_t: density of target material; σ_R: maximum allowable stress of the target material; Φ: angle of missile "nose".

compared to the missile size), all the correlation in the table can be mathematically adapted to express the maximum thickness that can be perforated by a fragment with a given velocity, mass and geometry.

Most of the available models refer to missiles impacting flat surfaces. A study of Baker et al. (1983) confirmed a negligible influence of the curvature of the target on missile penetration.

Several models assume the direction of the impacting missile perpendicular to the target surface. If this is not the case, Baker et al. (1975) suggest considering in the equation only the normal component of the missile velocity. The same concept is proposed by Recht (1974). The normal component of the velocity can be easily evaluated from the trajectory model, knowing the geometry of the target.

The missile diameter is a parameter required by many of the perforation models. In the case of irregular fragments, the definition of the diameter should be based on an equivalency criteria. Unless a specific criterion is provided by the penetration model, a general rule can be derived (Health and Safety Executive (HSE), 1992):

$$d = \frac{p}{\pi} \tag{6.14}$$

where p is the perimeter of the missile projection on the target surface and d is the equivalent diameter of the missile. The equation has been validated for different shapes of fragments, and seems reasonable when failure is due to shear stress.

Clearly enough, the orientation of the missile at the moment of the impact is usually unknown, since missile rotation can take place during the flight. If the missile shapes of Table 6.6 are considered, a large range of impact diameters is evidently possible for the same fragment, depending on the presentation angle. A conservative rule can be to consider the worst-case orientation, which is generally the one characterized by the smaller equivalent diameter.

As observed earlier, the penetration models generally allow the definition of the maximum thickness that can be perforated by the impact of given fragment. This provides a yes/no criteria on the perforation of the target that can be easily translated in a step function for damage probability. The models which provide values for 50% probability of penetration (Baker et al., 1975, US Naval Ordnance Laboratory (US-NOL), 1955, Ballistic Analysis Laboratory (BAL), 1961) provide only that single value, not allowing for a better definition of the damage probability functions. More recently Mébarki et al. (2008) proposed a penetration model associated with an error function, based on the maximum entropy fitting of a set of laboratory scale data. No probit models are however currently available in open literature for the damage probability as function of the penetration depth.

A few models are available for damage mechanisms different from fragment penetration. Ellinas and Walker (1983) proposed a model for calculating energy required for creating indents of different depth in pipes. However they do not propose any criteria for the plastic collapse of the pipe following the indent. Mannan (2005) reported results from a study on the different failure modes for a 4 in condensate line involved in the Piper Alpha accident. Westine and Vargas (1978) proposed an equation for verifying the failure by fragmentation mechanism in the case of impulse loading. Nevertheless, also in the case of damage mechanisms different from fragment penetration, a consolidated probit model for damage of targets is not available.

All the models presented above refer to the damage of metal targets. This is the most common pattern involved in domino propagation by fragments. Clearly enough, fragments can also lead to domino escalation by damaging control and communication structures, such as control rooms, marshalling cabinets, cable lines, etc. A few models are available for maximum thickness that can be perforated by chunky fragments in the case of concrete and brick structures. An account is given by Mannan (2005) and references cited therein. Though the mechanism of fracture is different in these cases, the general procedure and considerations reported in this chapter for the calculation of damage probability of metal targets can be applied also in the case of nonmetal targets.

6.7 Conclusions

Projection and impact of missiles is a credible cause of escalation resulting in domino accidents. The patterns underlying this escalation mechanism are complex, involving three main phases: fragment formation, fragment ejection and flight, and damage from fragment impact on a target. The analysis of the literature evidenced as a limited number of fundamental studies has been carried out on the topic. However, sufficiently validated models are available to describe the phases of fragment formation and flight. In particular, the approach based on the definition of fragmentation patterns provides a fundamental insight for the description of these phases and the quantification of the impact probability. On the other side, gaps still exist in modeling the damage of the target. Available models usually refer to specific damage modes (e.g. penetration) and have been developed for fragments with shape and size different than the typical missiles originated in industrial accidents.

References

Backman, M.E., Goldsmith, W., 1978. The mechanics of penetration of projectiles into targets. International Journal of Engineering Science 16, 1–99.

Bagster, D.F., Pitblado, RM., 1991. The estimation of domino incident frequencies—an approach. Process Safety and Environmental Protection 69, 196.

Baker, W.E., Kulesz, J.J., Ricker, R.E., Bessey, R.L., Westine, P.S., Parr, V.B., Oldham, G.A., 1975. Workbook for Predicting Pressure Wave and Fragment Effects of Exploding Propellant Tanks and Gas Storage Vessels. NASA CR-134906. NASA Lewis Research Center, San Antonio, TX.

Baker, W.E., Kulesz, J.J., Ricker, R.E., Westine, P.S., Parr, V.B., Vargas, L.M., Moseley, P.K., 1978. Workbook for Estimating the Effects of Accidental Explosions in Propellant Handling Systems. NASA Contractor Report No. 3023. NASA Scientific and Technical Information Office, Washington, D.C.

Baker, W.E., Cox, P.A., Westine, P.S., Kulesz, J.J., Strehlow, R.A., 1983. Explosion Hazards and Evaluation. Elsevier, Amsterdam, The Netherlands.

Ballistic Analysis Laboratory (BAL), 1961. The Resistance of Various Metallic Materials to Perforation by Steel Fragments, Empirical Relationships for Fragments Residual Velocity and Residual Weight. Project THOR Tech Report No. 47. John Hopkins University.

Baum, M.R., 1984. The velocity of missiles generated by the disintegration of gas pressurised vessel and pipes. Journal of Pressure Vessel Technology 106, 362–368.

Baum, M.R., 1987. Disruptive failure of pressure vessels: preliminary design guide lines for fragment velocity and the extent of the hazard zone. In: Advances in Impact, Blast Ballistics and Dynamic Analysis of Structures. ASME PVP, New York, NY, p. 124.

Birk, A.M., 1996. Hazards from propane BLEVEs: an update and proposal for emergency responders. Journal of Loss Prevention in the Process Industries 9 (2), 173–181.

Brode, H.L., 1959. Blast wave from a spherical charge. Physics of Fluids 2, 217.

Brown, S.J., 1985. Energy release protection for pressurized systems, Part I. review of studies into blast and fragmentation. Applied Mechanics Reviews 38 (12), 1625–1651.

Center for Chemical Process Safety (CCPS), 1994. Guidelines for Evaluating the Characteristics of Vapor Cloud Explosions, Flash Fires, and BLEVEs. CCPS–AIChE, New York, NY.

Center for Chemical Process Safety (CCPS), 2000. Guidelines for Chemical Process Quantitative Risk Analysis, second ed. AIChE, New York, NY.

Cox, B.G., Saville, G (Eds.), 1975. High Pressure Safety Code. High Pressure Technol. Ass., Imperial Coll, London, UK.

Design Institute for Emergency Relief Systems (DIERS), 1992. Emergency Relief System Design Using DIERS Technology. AIChE, New York, NY.

Ellinas, C.P., Walker, A.C., 1983. Effects of damage on offshore tubular bracing members. In: Ship Collision with Bridges and Offshore Structures. IABSE Colloquium, Copenhagen, Denmark.

Fingas, M., 2002. The Handbook of Hazardous Materials Spills Technology. McGraw-Hill, New York, NY.

Gel'fand, B.E., Frolov, S.M., Bartenev, A.M., 1989. Calculation of the rupture of a high-pressure reactor vessel. Combustion, Explosion and Shock Waves 24 (4), 488–496.

Gledhill, J., Lines, I., 1998. Development of Methods to Assess the Significance of Domino Effects from Major Hazard Sites. CR Report 183. Health and Safety Executive, Warrington, UK.

Grodzovskii, G.L., Kukanov, F.A., 1965. Motions of fragments of a vessel bursting in a vacuum. Inzhenemyi Zhumal 5 (2), 352–355.

Gubinelli, G., Zanelli, S., Cozzani, V., 2004. A simplified model for the assessment of the impact probability of fragments. Journal of Hazardous Materials A116, 175–187.

Gubinelli, G., Cozzani, V., 2009a. Assessment of missile hazards: identification of reference fragmentation patterns. Journal of Hazardous Materials 163 (2–3), 1008–1018.

Gubinelli, G., Cozzani, V., 2009b. The assessment of missile hazard: evaluation of fragment number and drag factors. Journal of Hazardous Materials 161, 439–449.

Hauptmanns, U., 2001a. A Monte Carlo-based procedure for treating the flight of missiles from tank explosions. Probabilistic Engineering Mechanics 16, 307–312.

Hauptmanns, U., 2001b. A procedure for analyzing the flight of missiles from explosions of cylindrical vessels. Journal of Loss Prevention in the Process Industries 14, 395–402.

Holden, P.L., Reeves, A.B., 1985. Fragment hazards from failures of pressurised liquefied gas vessels. IChemE Symposium Series 93, 205.

Holden, P.L., 1986. Assessment of Missile Hazards: Review of Incident Experience Relevant to Major Hazard Plant, Report No. SRD R 477. United Kingdom Atomic Energy Authority, Warrington, UK.

Health and Safety Executive (HSE), 1992. Analysis of Projectiles, HSE Report OTI 92 603. HSE, Ascot, UK.

Kar, A.K., 1979. Residual velocity for projectiles. Nuclear Engineering and Design 53, 87–95.

Khan, F.I., Abbasi, S.A., 1998. Models for domino effect analysis in chemical process industries. Process Safety Progress 17, 107.

Mannan, S., 2005. Lees' Loss Prevention in the Process Industries, third ed. Elsevier, Oxford, UK.

Mebarki, A., Nguyen, Q.B., Mercier, F., Saada, R.A., Reimeringer, M., 2008. Reliability analysis of metallic targets under metallic rods impact: towards a simplified probabilistic approach. Journal of Loss Prevention in the Process Industries 21, 518–527.

Mébarki, A., Mercier, F., Nguyen, Q.B., Saada, R.A., 2009a. Structural fragments and explosions in industrial facilities. Part I: probabilistic description of the source terms. Journal of Loss Prevention in the Process Industries 22, 408–416.

Mébarki, A., Nguyen, Q.B., Mercier, F., 2009b. Structural fragments and explosions in industrial facilities: Part II—projectile trajectory and probability of impact. Journal of Loss Prevention in the Process Industries 22, 417–425.

Moore, C.V., 1967. The design of barricades for hazardous pressure systems. Nuclear Engineering and Design 5, 81–97.

Neilson, A.J., 1985. Empirical equations for the perforation of mild steel plates. International Journal of Impact Engineering 3 (2), 137–142.

Neilson, A.J., Howe, W.D., Garton, G.P., 1987. Impact Resistance of Mild Steel Pipes: An Experimental Study. AEEW-R2125. UKAEA Atomic Energy Establishment, Winfrith, UK.

Ohte, S., Yoshzawa, H., Chiba, N., Shida, S., 1982. Impact strength of steel plates struck by projectiles (evaluation formula for critical fracture energy of steel plate). Bulletin of JSME 25 (206), 1226–1231.

Pettitt, G.N., Schumacher, R.R., Seeley, L.A., 1993. Evaluating the probability of major hazardous incidents as a result of escalation event. Journal of Loss Prevention in the Process Industries 6, 37.

Pula, R., Khan, F.I., Veitch, B., Amyotte, P.R., 2007. A model for estimating the probability of missile impact: missiles originating from bursting horizontal cylindrical vessels. Process Safety Progress 26 (2), 129–139.

Recht, R.F., Ipson, T.W., 1963. Ballistic perforation dynamics. Journal of Applied Mechanics Transactions of ASME 30 (3), 384.

Recht, R.F., 1974. Quasi-empirical models of the penetration programs. In: Pilkey, W., Saczalski, S., Schaeffer, H. (Eds.), Structural Mechanics Computer Programs, Surveys, Assessments, and Availability. Univ. of Virginia Press, Charlottesville, VA.

Scilly, N.F., Crowther, J.H., 1992. Methodology for Predicting Domino Effects from Pressure Vessel Fragmentation. In: Proc. CCPS Int. Conf. on Hazard Identification and Risk Analysis, Human Factors and Human Reliability in Process Safety. CCPS, New Orleans, LA, p. 1.

Stronge, W.J., 1985. Impact and perforation of cylindrical shells by blunt missiles. In: Reid, S.R. (Ed.), Metal Forming and Impact Mechanics. Pergamon Press, Oxford, UK, pp. 389–402.

US Naval Ordnance Laboratory (US-NOL), 1955. Explosion Effects Data Sheets. NAVORD Report 2986.

Van Den Bosh, C.J.H., Weterings, R.A.P.M., 1997. Methods for the Calculation of Physical Effects (Yellow Book), third ed. TNO, Committee for the Prevention of Disasters, The Hague, The Netherlands.

Westin, R.A., 1971. Summary of Ruptured Tank Cars Involved in Past Accidents, Report No. RA-01-2-7. Railroad Tank Car Safety Research and Test Project, Chicago, IL.

Westin, R.A., 1973. Summary of Ruptured Tank Cars Involved in Past Accidents, Report No. RA-01-2-7, ADDENDUM 5/1/73. Railroad Tank Car Safety Research and Test Project, Chicago, IL.

Westine, P.S., Vargas, L.M., 1978. Design Guide for Armoring Critical Aircraft Components to Protect from High Explosive Projectiles, Final Report, Contract No. F33615-77-C-3006. U. S. Air Force, San Antonio, TX.

7 Other Causes of Escalation

Valerio Cozzani, Elisabeth Krausmann†,
Genserik Reniers‡*

* LISES, Dipartimento di Ingegneria Civile, Chimica, Ambientale e dei
Materiali, Alma Mater Studiorum – Università di Bologna, Bologna, Italy,
† European Commission, Joint Research Centre, Institute for the Protection
and Security of the Citizen, Ispra, Varese, Italy, ‡ Centre for Economics and
Corporate Sustainability (CEDON), HUB, KULeuven, Brussels, Belgium

---

7.1   Introduction

In the previous chapters, the direct causes of domino effects were explored: blast
waves, fires or fragment projection causing damage to equipment items resulting in
one or more secondary accidents. Direct escalation is unambiguously recognized in
the literature as the main (or in some texts as the only) cause of domino scenarios.

However, some authors also classified as domino accidents events in which an
indirect escalation took place. An indirect escalation is defined as an escalation event
not resulting from the direct damage of a process item or a storage unit (see
Chapter 3). In indirect escalation events, escalation is usually the result of loss of
control of a process or storage unit due to indirect effects of the primary scenario: e.g.
control room damage due to a blast wave, escape of operators due to a toxic release,
and similar indirect events. Although unlikely in most industrial facilities, such events

Domino Effects in the Process Industries. http://dx.doi.org/10.1016/B978-0-444-54323-3.00007-5

are documented in past accident records (Khan and Abbasi, 1998) and may be a relevant hazard when poor information exchange occurs within different companies operating in a complex industrial cluster (Reniers et al., 2005).

Besides indirect causes of escalation, two more specific causes of domino scenarios will be discussed in the present chapter: domino effects triggered by natural events and escalation of intentional malicious acts of interference. While the conventional definition of domino accident provided in Chapter 3 does not include such scenarios, in recent years, several authors referred to NaTech (natural events triggering a technological disaster) events as a "domino effect" of the natural event that causes a technological accident with hazardous materials releases (e.g. see Krausmann and Cruz, 2008; Showalter and Myers, 1994). Furthermore, the analogies between conventional and security-related domino scenarios were pointed out in recent studies (Reniers et al., 2008; Pavlova and Reniers, 2011).

7.2   Indirect Causes of Escalation

7.2.1   Domino Accidents Caused by Indirect Escalation

In Chapter 3, escalation was described as the key element of a domino accident. As evident from Chapter 2, most domino accidents are caused by "direct" escalation: the physical effects of the primary event (blast wave, fire radiation, and fragment impact) damage one or more secondary units starting secondary scenarios that result in the escalation.

However, Khan and Abbasi (1998) discuss the possibility of a completely different escalation route. In 1995, a toxic cloud formed during a "runaway" accident in a rubber processing factory in Pilliyarkuppam, India, and spread to a nearby plant operated by a different company, producing perfumes. The operators affected by the cloud ran for their lives and left the plant unattended, without performing proper shutdown operations. The process ran without any control for several hours. Fortunately, after the toxic dispersion was stopped, it was possible to complete the shutdown without further damage a few hours later.

Clearly, in this case, a possible escalation hazard was caused by the operators who, in the absence of personal protection devices and without adequate training, were not able to provide a proper emergency response, leaving the plant unattended without attempting an emergency shutdown. Thus, in the domino scenario described above, the severe escalation hazard was not caused by direct damage of equipment due to the primary event.

7.2.2   Definition and Causes of Indirect Escalation

An indirect escalation may be defined as the escalation of a primary event in which the secondary scenarios are not caused by the direct damage of equipment items. Operator errors, loss of control due to loss of communication, and structural damage of civil structures (warehouses, support structures, etc.) can be identified as the most likely mechanisms.

The role of possible operator errors in triggering an escalation following a primary accident scenario is described in detail in the Pilliyarkuppam accident history reported above (Khan and Abbasi, 1998). It should be noted that inadequate response from the operators has been identified as a concurrent cause of most domino accidents (Abdolhamidzadeh et al., 2011; Darbra et al., 2010; Kourniotis et al., 2000). However, only errors directly resulting in the triggering of a secondary accident should be considered as causes of indirect escalation.

Loss of control and loss of communication is usually a consequence of the damage of the central control room, of local equipment rooms or of crucial wiring connections. When specific hazards are not recognized in the design phase, such elements may not be adequately protected. It is well-known that the heavy damage to the central control room following the first vapor cloud explosion was a key element of the escalation of events leading to the loss of the Piper Alpha rig (Cullen, 1990).

Structural damage of civil buildings was documented as an important element in the escalation of some accidents. In particular, an escalation of consequences was reported for several "runaway" explosions affecting indoor reactors, due to the collapse of the building hosting the reactors, starting secondary fires (Sales et al., 2007).

7.2.3 Prevention of Indirect Escalation

The majority of documented case histories involving indirect escalation date back to the early days of the process and chemical industry. Lessons learned since the 1960s caused a huge change in regulations, technical standards and safety attitude in the design and management of chemical and process plants. Nowadays, in most chemical and process plants, an adequate safety culture is present. Appropriate personal safety devices, remote controls, safety barriers, and automatic and manual shutdown systems are usually in place. Operators are trained to face emergency situations and to provide appropriate emergency response.

Thus, the possibility of indirect escalation due to unrecognized hazards may be reasonably considered as remote in most cases, in particular when appropriate safety barriers, risk management and emergency response procedures are adopted. However, the design of safety systems and operator training are mostly focused on the hazards and the accident scenarios identified for the specific site of concern. Recent studies highlighted the severe hazards related to "atypical scenarios", defined as scenarios not deemed as credible and thus not analyzed in the safety reports of the affected facility (Paltrinieri et al., 2012a). These may be the consequence either of inappropriate hazard identification or of unknown hazards related to new technologies (Paltrinieri et al., 2012b). Safety barriers and emergency procedures are prone to fail when facing unforeseen accident scenarios (Pidgeon and O'Leary, 2000), resulting in either direct or indirect escalation.

Indirect escalation is facilitated by inefficient knowledge management and ineffective or untimely information exchange. These elements were identified as one of the possible problems in the safety management of modern industrial clusters (Reniers et al., 2005). The evolution of the chemical and process industry toward integrated clusters where several companies operate different facilities side by side on the same site actually raises the issue of proper information exchange, in particular with respect to hazard scenarios. In this framework, indirect escalation may be considered an atypical scenario by the definition of Paltrinieri et al., (2012a). The few documented near misses should be interpreted as "early warning" providing a risk notion concerning the possibility of such events. Therefore, the possibility of atypical scenarios due to indirect escalation needs to be considered in hazard identification and may be ruled out only after a specific assessment of such a hazard.

The need to protect the facilities from accident scenarios taking place in nearby plants operated by other companies is a controversial point with respect to both economical aspects and regulation (Pavlova and Reniers, 2011), as discussed in detail in Chapter 1. In particular, information exchange concerning hazards and accident scenarios among different and competing companies poses several practical difficulties, with respect to information disclosure and responsibilities in the timely and effective management of the data exchange process.

Therefore, indirect escalation is mainly an issue that needs to be explored when the adequacy of the risk management process is verified and concerns mostly the assessment of the efficiency and adequacy of the hazard identification procedures, the knowledge management system, and the emergency response and operator training procedures adopted by the company. A detailed discussion of such issues is beyond the purview of the present book, which mainly discusses the analysis and prevention of direct escalation accidents. Resilience indicators were recently proposed as a valuable approach to the issue, although several alternative methods are present in the literature (Oien, 2001; Skodgdalen and Vinnem, 2011; Paltrinieri et al., 2012b).

7.2.4 Escalation Triggered by Toxic Releases

Toxic releases are a frequent final outcome of accident scenarios affecting the chemical and process industry and some authors indicated these events as possible causes of escalation. Thus, a short discussion of the topic is provided in order to frame the issue.

Toxic releases are not able to cause direct escalation. Both liquid and gaseous releases of a chemical acutely toxic to humans, with no other harmful property, will not cause damage to equipment and thus cannot trigger secondary scenarios directly.

With respect to indirect escalation, it should be remarked that only very few cases (mostly near misses) are documented in accident databases to be the result of a toxic release. Nevertheless, on the basis of the above discussion, there is evidence that toxic releases may actually cause an indirect escalation, if they affect facilities where unprotected and untrained operators are present (Khan and Abbasi, 1998).

However, such events are deemed highly unlikely or even not credible if

- Toxic hazards or the possibility of a toxic release in the area are known to the operators and an alert system is present
- Personal protective devices are available
- Emergency shut-down procedures are in place and operators are trained to apply them.

These conditions allow the successful prevention of indirect escalation of a toxic release.

7.3 Natural Events

7.3.1 Domino Scenarios Triggered by Natural Events

The damage of process equipment due to the impact of natural events is known to have triggered a number of severe accidents in the chemical and process industries (e.g. see Girgin, 2011; Krausmann et al., 2010; Krausmann and Cruz, 2013; Lindell and Perry, 1997). The impact of natural events on industrial facilities can cause structural damage starting loss of containment (LOC) of hazardous substances. LOCs can easily escalate to severe secondary scenarios like fires, explosions, water/soil contamination or toxic and/ or flammable cloud dispersions. The impact of a natural event on a chemical plant is usually characterized by the simultaneous failure of several equipment items at different locations. Thus, the overall scenario is somehow similar to that expected in a domino accident. Although recently the specificity of such accidents was recognized and a specific term ("NaTech") was introduced to classify such events (Showalter and Myers, 1994), in several technical and scientific documents, these scenarios are still indicated as "domino accidents caused by natural events". In the following discussion, the key elements concerning the specific issues and an approach to the assessment of such domino scenarios are described. The reader is referred to Krausmann et al. (2011a) for further details on this specific category of accidents.

7.3.2 The Characterization of NaTech Accidents

The increasing frequency of some severe natural events caused by climate change (Parry et al., 2007) is raising concern about the possible interference of these external hazards with potentially unprepared hazardous industrial activities (OECD, 2012; Krausmann and Baranzini, 2012; Cruz and Krausmann, 2009; Cruz et al., 2001). Industrial accidents triggered by natural events can be an important cause of direct damage to the population present in nearby residential areas, due to the accident scenarios triggered by equipment damage (blast waves, toxic releases, and heat radiation from fires). Analysis of the most important industrial accident databases, recently carried out in several studies by Cozzani, Krausmann and coworkers (Krausmann et al., 2011b; Cozzani et al., 2010; Renni et al., 2010), showed that between 2% and 5% of reported major industrial accidents were triggered by natural events. The analysis of past accidents also highlighted the specific features of NaTech scenarios deriving from the varied impact that different natural events can have on industrial sites. For instance, lightning clearly emerges as the most frequent cause of NaTech events (Renni et al., 2010). However, earthquakes resulted in the most severe scenarios, due to the contemporary damage of a high number of equipment items, as shown in Table 7.1. This is also confirmed by the detailed analysis of specific case studies (Krausmann and Cruz, 2013; Krausmann et al., 2010; Steinberg and Cruz, 2004).

The analysis of past accidents also showed that the final outcomes of release scenarios induced by natural events can have specific characteristics that would be impossible or unlikely in the case of conventional release scenarios. For instance, in the case of floods, the released chemicals can react with water, possibly resulting in

Table 7.1 Results Obtained from the Analysis of 29 Accidents Related to the Release of Flammable Substances Triggered by Earthquakes

Number of accidents analyzed	29
Number of damaged equipment items	254
Maximum number of equipment items damaged in a single event	97
Average number of equipment items damaged in a single event	9
Recorded number of LOCs following equipment damage	180
Recorded number of fires following LOC	137

explosions and in the formation of toxic clouds. In several other events, the flooding of catch basins caused extensive water and land contamination (Cozzani et al., 2010).

The failure of mitigation systems and safety barriers was also often observed in these accidents: flooding of catch basins in the case of floods and severe storms, damage of catch basins in earthquakes, and unavailability of fire-fighting water following earthquakes were reported in several NaTech events (Girgin, 2011; Krausmann and Mushtaq, 2008). It is obvious that the unavailability of safety barriers increases the possibility and probability of further escalation of the accident scenarios directly triggered by the natural event. A recent example is the fire in the Cosmo Oil refinery at Chiba, Japan, following the 11 March 2011, earthquake. The earthquake-induced collapse of a spherical tank resulted in a primary fire due to the rupture of liquefied petroleum gas (LPG) pipes. The accident was aggravated by the fact that a safety valve on an LPG pipe was manually switched to open in violation of safety regulations (Cosmo Oil, 2011). With the fire impinging on several nearby storage tanks, escalation due to heat radiation resulted in the boiling liquid evaporating vapor explosions (BLEVE) of four LPG spherical storage tanks. The blast waves and the fireballs following the BLEVEs caused further fires that resulted in the destruction of the other 12 storage tanks. In addition, the blast waves and thermal radiation effects triggered fires in two neighboring chemical facilities (Krausmann and Cruz, 2013).

NaTech events are characterized by the following main factors: (1) the cause of the event is external to the industrial site, thus the prevention and mitigation of such events may not be managed only at site level; (2) the spatial extent of the natural event triggering the technological accident is wide, thus multiple equipment items may be simultaneously affected and loss of utilities can take place, leading to common cause failures; and (3) emergency response may be hampered or delayed by the natural event. Factor (1) may find some correspondence with domino accidents in which the primary event takes place in a different facility than where the secondary targets are (see Chapter 3). Point (2) also may be a feature of conventional domino accidents, although Table 7.1 shows that the number of damaged items and that of simultaneous scenarios can be extremely high in comparison with a conventional domino accident. Factor (3) is specific for NaTech domino events.

Finally, as discussed above, factors (1) and (3), and specifically the possible unavailability of safety and mitigation barriers, as well as the delay of emergency response, cause a strong increase in the probability of further escalation events. NaTech events are thus likely to result in extremely severe scenarios, where the events directly triggered by the natural event can further escalate, causing an extremely complex and severe technological accident. The characteristics of this category of accidents clearly show the need for specific tools to address the analysis and the assessment of NaTech scenarios and of the further escalation events that can take place.

7.3.3 Approaches for the Quantitative Assessment of Domino Scenarios Triggered by Natural Events

As stated above, the specificity of this category of accidents was recognized only in recent years. Thus, consolidated assessment methodologies are not yet available. However, several methodologies provide an assessment of the NaTech hazard and/or risk at different levels of detail.

The first level in the assessment of NaTech hazards is the identification of the sites where such a hazard is relevant. The problem is usually of concern at the district, regional or national level, thus requiring the analysis of extended areas. Therefore, the assessment may be reasonably based on simplified screening methods. Cruz and Okada (2008) proposed a detailed screening methodology mostly useful at the district level.

Within the activities of the FP7 iNTeg-Risk project, a specific task was dedicated to the issue of NaTech risks. A methodology was developed to obtain a ranking of the NaTech hazard, based on four hazard classes (iNTeg-Risk, 2011). Cozzani and coworkers proposed an index method mainly aimed at ranking NaTech hazards at the regional or national level (Sabatini et al., 2008). More recently, Rota and coworkers proposed the application of the analytical hierarchy process to screening procedures for the ranking of NaTech hazards (Busini et al., 2011). This methodology is particularly useful since it allows the user to easily combine the assessment of conventional direct escalation with that triggered by natural events. The application of all these methods to case studies proved to yield effective results in the identification of "hot spots" and critical sites where the application of more detailed assessment techniques is recommended.

Methodologies allowing a more detailed NaTech risk assessment were recently developed. The rapid NaTech risk analysis and mapping software RAPID-N implements a probabilistic risk assessment methodology by a web-based software tool (Girgin and Krausmann, 2012). The tool provides the results of the risk analysis as interactive risk maps and summary reports (Figure 7.1) that can be used for land use and emergency planning. While the tool currently focuses on earthquakes, its framework is sufficiently flexible for expansion to other types of natural hazards.

Extending the bow-tie approach to NaTech hazards was proposed since the development of the MIMAH (Methodology for the Identification of Major Accident Hazards) technique within the ARAMIS (Accidental risk assessment methodology for industries) project (Delvosalle et al., 2006). Bow ties including natural events as failure causes were developed in the approach. More recently, specific efforts were made to

Risk Assessment Information

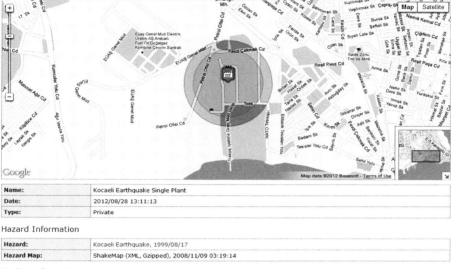

Name:	Kocaeli Earthquake Single Plant
Date:	2012/08/28 13:11:13
Type:	Private

Hazard Information

Hazard:	Kocaeli Earthquake, 1999/08/17
Hazard Map:	ShakeMap (XML, Gzipped), 2008/11/09 03:19:14

Facility Information

Facility:	▓▓▓▓ ▓▓▓ Power Plant, Turkey

Damage Estimation

Damage Classification:	Auto
Flexible fragility curve selection:	Yes

Facilities

1. ▓▓▓▓ ▓▓▓ Power Plant, Turkey

No	Process Unit	Hazard Parameters	Fragility Curve	Damage Estimate	Damage Parameters	End-point Distance
1.	Storage Tank (T-STR)[*] [Gasoline]	PGA: 18.777 %g; EMS: Slightly damaging; MM: Strong; MSK: Strong; MMI: 6.4866; d_e: 101.38 km; d_h: 102.79 km; PGA$_h$: 74.415 cm/s^2; PGV: 15.573 cm/s «	OS00-F50-G	≥ DS2: 4.0546%	Fire/Explosion Event: Vapor Cloud Explosion; $Q_{involved}$: 4250 kg; $f_{m, passive}$: 1; $P_{c, fire}$: 100%; $f_{v, involved}$: 10 %v; $V_{involved}$: 5.7432 m3; $P_{c, release}$: 30%; f_{yield}: 0.1; RMP Scenario: Worst-case; $t_{release}$: 10 min; $Q_{release}$: 425 kg/min; $Q_{released}$: 4250 kg; A_{pool}: 6146.1 ft2; h_{pool}: 1 cm; $q_{release, r}$: 425 kg/min; T_a: 1; R: 0.4; q_R: 5000 W/m2; t_{exp}: 40 s; D_T: 342 TDU; d_e: 270.58 m; Q_{fuel}: 4250 kg; P_{damage}: 4.0546%; P_{natech}: 4.0546% «	271 m: 4.0546%
				≥ DS3: 0.004631%	Fire/Explosion Event: Vapor Cloud Explosion; $Q_{involved}$: 8500 kg »	341 m: 0.004631%
				≥ DS4: Very low	-	-

Figure 7.1 Example of a Natech risk analysis report and associated risk map for earthquake impact on a gas power plant in Turkey (Girgin and Krausmann, 2012). (For color version of this figure, the reader is referred to the online version of this book.)

extend the conventional quantitative risk assessment (QRA) procedure to allow the quantitative assessment of domino and NaTech scenarios. A detailed procedure for the calculation of individual and societal risk due to NaTech scenarios was developed. Figure 7.2 shows the conceptual flowchart of the procedure. As shown in the figure,

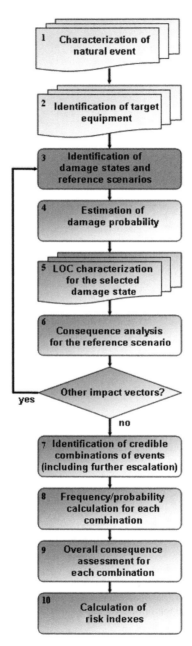

Figure 7.2 Conceptual flowchart for the quantitative assessment of NaTech scenarios. (For color version of this figure, the reader is referred to the online version of this book.) *Source*: Adapted from Antonioni et al. (2009).

several steps are very similar to those proposed for the quantitative analysis of domino scenarios (see Chapter 10). Specific steps include the identification of damage and release states and, most importantly, of equipment damage probability as a consequence of the impact of the natural event. Further escalation of scenarios triggered by the natural event can also be included in the assessment within step 7 of the procedure.

The procedure was obtained from the modification and extension of the well-known scheme used for conventional risk assessment (CCPS, 2000). As far as possible, the procedure required for NaTech risk assessment was integrated with the conventional QRA procedure used for the assessment of risk caused by internal failures, in order to limit the additional resources needed to perform the assessment of NaTech risk. However, specific steps and specific models and tools are required to extend the conventional procedure to NaTech risk assessment. In particular, specific damage models are required to understand the type and probability of equipment damage that may result in a LOC. Moreover, a specific procedure is required for the calculation of risk indices that should account for the possibility of simultaneous releases from more than a single process or storage unit (Antonioni et al., 2007). This procedure may be easily derived and combined with that required for the QRA of domino scenarios, discussed in Chapter 10 (Cozzani et al., 2005).

Pilot applications of the methodology led to the calculation of isorisk curves for chemical plants and refineries. An example of results obtained for NaTech scenarios triggered by earthquakes, and obtained with the Aripar-GIS 4.5 software (Campedel et al., 2008), is shown in Figure 7.3. The approach also allows the calculation of societal risk if data is available on the distribution of population. An example is reported in Figure 7.4.

7.4 Intentional Interferences

In the pre-9/11 era, management of industrial security focused on security measures aimed at preventing company personnel from filching company assets, at preventing theft (internal as well as external theft, e.g. by company employees or by drug dealers), or at preventing information losses and attacks on company software. Without any doubt, the September 11, 2001, attacks were the principal reason behind the corporate establishment's desire to focus substantially more attention on terrorism. Chemical companies thus became aware after 9/11 that should they not dedicate adequate resources on their security, the risk of becoming a target of terrorism was concrete.

It is not possible to quantitatively assess the level of "security efficiency" against terror, simply because empirical data is not available and cannot be simulated. Analogous to major safety-related accidents (e.g. indirect domino effects), the probability of terrorist attacks on-site of a chemical process plant is extremely low, which again leads to a very slow awareness of both politicians and captains of industry, hindering a proactive approach to such threats.

Moreover, terrorist attacks with the specific aim to induce one or several domino effects within a chemical industrial area never took place until present, at least in Europe. As a result, to date, chemical plant security technology and management for

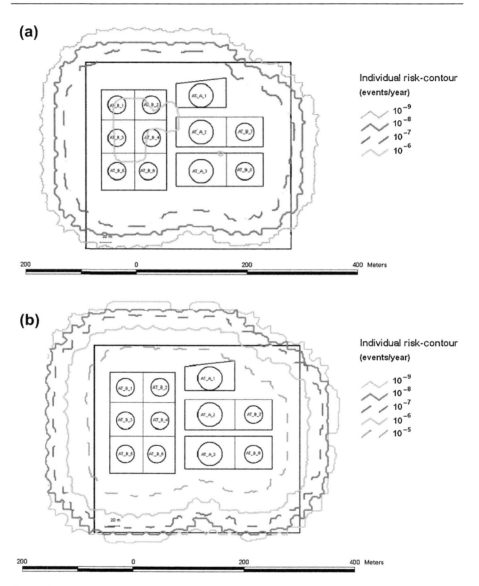

Figure 7.3 Individual risk curves for an atmospheric tank farm of an oil refinery: (a) no seismic events; (b) with accidental scenarios triggered by seismic events (Campedel et al., 2008). (For color version of this figure, the reader is referred to the online version of this book.)

a more effective and efficient prevention of induced escalation disasters is lagging behind and is largely insufficient to adequately deal with a terrorist attack.

However, the "intentional interference" threat does exist. Safety managers not always sufficiently recognize the threat. (Reniers 2012) indicates that safety managers

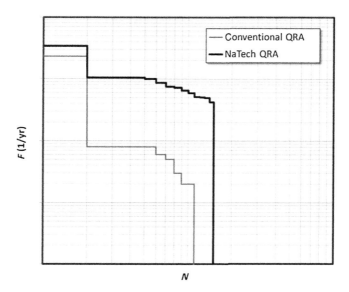

Figure 7.4 Example of societal risk calculations for NaTech scenarios: Overall expected frequency (*F*) of an accident with an expected number of fatalities equal or higher than *N*. (For color version of this figure, the reader is referred to the online version of this book.)

are divided on the importance of security management in general in the process industries: half of them consider security as an independent management domain and the others think of security as a subdomain of safety. Hence, there is still a long way to go in industrial practice before managers will be aware that an induced large-scale domino accident may become a reality. Hopefully, the industry does not have to wait for an intelligent attack before adequate security measures are taken against intentional domino effects.

In this chapter, an approach to possibly deal with the prevention of security-related domino effects is presented. Dealing with risks of intentional domino effects requires a chemical industrial area to be modeled as a networked system. The efficiency and effectiveness of the protection of such an area, represented as a network, can then be intelligently designed. The question thus is how a chemical industrial area can be modeled in a way that can be used to prevent terrorist attacks inducing domino effects within the area. The modeling approach suggested by Reniers and Dullaert (2008) is presented in the following discussion.

A chemical industrial area consists of a number of separate chemical installations. Each chemical installation has an inherent hazard depending on the amount of substances present, the physical and toxic properties of the substances and the specific process conditions. Hence, such installations present—to a greater or lesser extent— a danger to their surroundings. In this way, installations are mutually linked in terms of the level of danger they pose to each other. In this particular approach, installations are considered to be the nodes of a network and the amount of danger between the installations is represented by the edges linking the nodes together. Using this approach, a chemical industrial area can be modeled as a connected weighted network.

To this end, a variable that represents the weight of each link between two chemical installations is introduced. The weight of an arc (v_i,v_j) with $(v_i \neq v_j)$ represents the amount of danger (for initiating or continuing domino effects) outgoing from installation v_i onto installation v_j. Mathematically, the arc weight is simply a scalar (real number). Reniers (2010) refers to the latter as the "domino danger unit" (DDU). Let DDU_{ij} denote the weight of an arc (v_i,v_j) with $(v_i \neq v_j)$, such that $DDU_{ij} \in R^+$ if $v_i \neq v_j$ and $DDU_{ij} = 0$ if $v_i = v_j$. By calculating all (unidirectional) DDUs between all nodes in the entire network an installations danger matrix DDU of order $N \times N$ (with N the number of nodes in the network) is obtained. The factor DDU_{ij} can be seen as a measure of the danger that installation v_i represents for installation v_j in terms of domino effects. The DDU_{ij} can be calculated between every node v_i and the remaining nodes $v_j(v_j \in G; v_j \neq v_i)$ in the graph $G(N,A)$ representing the installation's network with N the number of nodes and A the number of arcs within the graph, implying that DDU_{ij} does not have to be equal to DDU_{ji}, although the possibility exists.

The effect-distance associated with a possible accident scenario from one installation to another is linked to the real distance between the two installations concerned. Depending on the difference in both distances (real distance and effect-distance), a standard *distance factor* (DF) is defined, using four possible categories. These numerical values represent the relative level of importance given to the pairs of installations with respect to their danger for inducing or continuing domino effects. Such a strategy allows for a relative ranking of the installation pairs in the network to identify the most dangerous ones.

The choice of the numerical DF values (giving rise to the four different categories) is actually a preference decision problem for security management. The set of possible (mutually exclusive) alternatives from which security management must choose balances the consequences of accident scenarios. For example, the outcome of a major accident scenario may be compared with the consequences of two accident scenarios ranked in one category below. In this illustrative example of the approach, we define the DF value as follows (see Reniers, 2010). For a specified scenario, if the real distance between both installation items does not exceed a quarter of the theoretical effect-distance, DF equals 100. On the other hand, if the real distance strictly exceeds the effect-distance, $DF = 0$. In the case where the real distance strictly exceeds one-quarter of the effect-distance and is lower than three-quarters of the theoretical effect-distance, $DF = 70$. In the final case where the real distance is restricted by the effect-distance and strictly exceeds three-quarters of the effect-distance, $DF = 40$. Figure 7.5 illustrates a hypothetical example of the calculated DFs between installation v_i and installation v_j.

A DDU_{ij} is defined to express the escalation dangerousness from one installation to another, by summing the DF values for the different possible scenarios $(1, ..., K)$ going out from installation v_i and coming into installation v_j:

$$DDU_{ij} = \sum_{\substack{\text{scenario } k = 1}}^{\text{scenario } K} (DF_{ij})_k \tag{7.1}$$

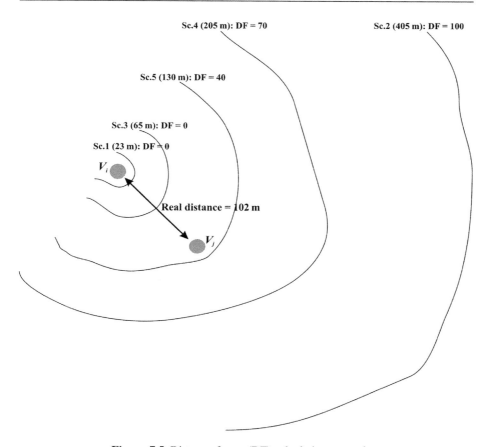

Figure 7.5 Distance factor (DF) calculation example.

Hence, for the example in Figure 7.5, $DDU_{ij} = 0 + 0 + 40 + 70 + 100 = 210$.

Equation (7.1) is used to calculate the weights of the directed arcs between every pair of installations in an industrial area consisting of a number of chemical installations. The formula allows us to obtain a matrix of DDUs, mathematically representing the network of the area.

Terrorist attacks have a very low probability combined with possibly very severe consequences, at least in Europe, Japan, or the United States. Hence, in the case of terrorist attacks, it is extremely difficult or even impossible to estimate the probability or the likelihood of an attack, simply because there is insufficient information available to make such an estimate. Therefore, designing a network of chemical installations solely based on the possible consequences of accident scenarios, and not considering the probabilities of terrorist attacks, seems to be the most justifiable option for further designing intelligent protection against such attacks. The threat assessment for domino effect terrorism prevention and protection thus needs to determine the installations to be protected effectively, in order to limit the consequences of an attack. To investigate where to take prevention and protection measures against intentional attacks, installations displaying very high dangerousness connectivity in the network need to be identified. Indeed, very few nodes may exhibit very high dangerousness connectivity (we call such nodes "danger hubs"), while others (the vast majority of nodes) have only relatively unimportant dangerousness links. By taking adequate precautions with those danger hubs, an industrial area would be much better secured against the propagation of domino events. A more mathematical approach to prevent and protect installations against deliberate actions is therefore required. To this end, the "out-strength" s_i^{out} of a node i is introduced:

$$s_i^{out} = \sum_{j(\neq i) \in G} DDU_{ij}$$

with DDU_{ij} being the weight on the arc (v_i, v_j). Hence, an installation out-strength distribution is obtained.

Let us now take the example of a real industrial area to verify the applicability of the methodology when viewed in function of security. To be able to understand the importance of the danger hubs in an industrial area, the installations showing the largest danger out-strength were removed from a network of a real industrial area situated in the Antwerp harbor area. Since the literature Barabasi, (2003) indicates a scale-free network to collapse by eliminating on an average 12% of its hubs, this percentage was used as an indicative value. In our case, the 27 most dangerous hubs were omitted. We notice that in an industrial area where the most dangerous installations are removed, "domino islands" emerge. Hence, in such a case, separate collections of installations with no domino danger connections between each other (we call them "domino islands") appear and the industrial area as a whole cannot be destroyed anymore. To illustrate the latter, maps reporting the networks with and without danger hubs have been developed (see Figures 7.6 and 7.7). The global Lambert coordinates allow positioning the installations in relation to each other. The x- and y-axes on the maps represent the x Lambert position coordinates and the y Lambert position coordinates of the installations, respectively. Installations are indicated on the maps by a triangle and an installation number. A purple-colored full line connection between a pair of installations indicates that the domino danger runs from the installation with the lowest installation number to the installation with the highest installation number. A green-colored dotted line connection indicates that the domino danger runs from the installation with the highest installation number to the one with the lowest installation number. If the connection is colored both purple and green, there is reciprocal domino danger between the installations.

The maps of the industrial area networks display very thorough danger interconnectedness between the installations. The maps also demonstrate that if the danger hubs are removed, (disjoined) domino islands emerge in the industrial area as a whole.

The impact of eliminating danger hubs in the network is further elaborated and studied more in depth by removing the hubs 1% at a time, starting with the hubs having the highest out-strength value. For each step, the network is recalculated

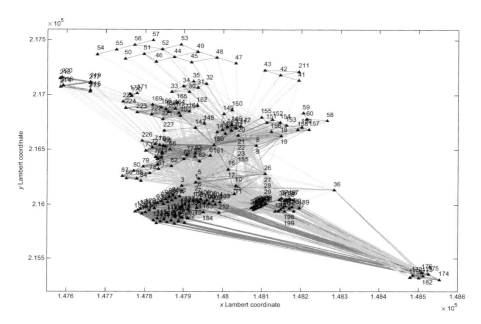

Figure 7.6 Cluster network, with danger hubs (number of installations equals 227). (For color version of this figure, the reader is referred to the online version of this book.)

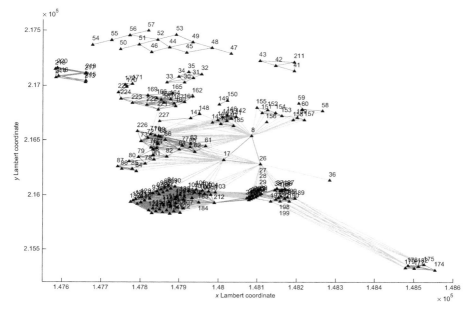

Figure 7.7 Cluster network, without danger hubs (number of installations equals 200). (For color version of this figure, the reader is referred to the online version of this book.)

(i.e. the number of edges as well as the total danger of the new network are determined and the structure of the new network is drafted). Using this simulation technique, the emerging domino islands (i.e. smaller areas in the network forming networks of their own with no links in between) can be observed and investigated. Table 7.2 illustrates the latter process, showing that by eliminating less than 1% of the domino danger hubs in the area, about 20% of the total domino danger and about 12% of the number of links are simultaneously eliminated. Moreover, the whole area is divided into three domino islands where no domino effects can occur in between. Table 7.2 further shows that by eliminating as less as 3% of the domino danger hubs in the area, about 30% of the total domino danger and about 20% of the number of links were simultaneously eliminated in this particular real-case industrial area. Moreover, the whole industrial area was divided into four domino islands where no domino effect can occur in between. As a result, the global danger for inducing domino effects in this particular industrial area decreased substantially. Furthermore, Table 7.2 clearly demonstrates that if more danger hubs are being removed, more disjoined domino islands emerge in the industrial area. Further removing domino danger hubs in the area thus reveals additional domino islands and further decreases the number of links and the total domino danger, compared with the original network. The higher the percentage of hubs removed, the lower the additional reduction of domino links and of the total domino danger. As a result, the global danger for inducing domino effects in the industrial area decreases even more.

One thus has to make a trade-off between security efforts (investments) and the preferred number of emerged domino islands.

A weakness of the presented approach concerns not including frequencies or probabilities of the initiating events into the methodology. However, as already mentioned, dealing with security risks indeed should be independent of frequencies and probabilities, taking only the potential consequences of designed domino effects into account, since terrorist acts are not concerned with coincidence at all. The proposed security part of the methodology is thus purely deterministic and takes merely the outcome of intentional escalation effects into account.

Furthermore, it should be noted that the suggested approach identifies those installations which should be better secured against potential threats in order to substantially decrease the possibility of a deliberate domino effect in the area, but it does not consider the question of *how to* eliminate those domino danger hubs. It is up to security management of the industrial complex to discuss the required countermeasures.

Using information provided by real-case chemical plants, it can thus be concluded that it only takes a relatively small number of domino effect hops to get from one installation to any other in a chemical industrial area. If the so-called danger hubs are made safe and secure in such an area, domino danger diminishes substantially. To mitigate the consequences of terrorist attacks in an industrial area, it is most efficient to consider the area as a whole (not per individual plant—although the method can of course be used for one single plant) and to determine the danger hubs of the area, to make sure that these high-danger installations are secured. Although there appear to be limitless possibilities as to what might go wrong and what industry might have to

Table 7.2 Overview of Characteristics of the Different Networks Emerged by Eliminating Domino Danger Hubs from the Original Real-Case Industrial Area Network

Reduction of Highly Connected Installations	Number of Installations	Number of Links between Installations	Additional Reduction of Links	Total Amount of Network Domino Danger	Additional Reduction of Domino Danger	Number of Domino Islands
0	227	5244	–	488,410	–	1
−1%	224	4625	−11.80%	394,170	−19.29%	3
−3%	220	4157	−8.92%	340,400	−11.01%	4
−5%	215	3675	−9.19%	292,330	−9.84%	4
−7%	211	3365	−5.91%	262,930	−6.02%	5
−9%	206	2978	−7.38%	236,310	−5.45%	5
−11%	202	2715	−5.01%	216,510	−4.05%	5
−13%	197	2358	−6.81%	189,670	−5.50%	6
−15%	192	2069	−5.51%	169,660	−4.10%	7

do to prevent or protect against (e.g. simultaneous) intentional threats, this research offers some exciting new insights into tackling this new security-driven risk paradigm.

7.5 Conclusions

Besides the "obvious" domino effect scenarios used to elaborate credible, worst-credible or worst-case scenarios involving propagation of accidents that are mostly related to single propagation or/and multilevel domino chains, there exist some "atypical" domino scenarios that might be very important to consider. The significance of these often overlooked scenarios may depend on the location of the chemical plant, e.g. in the case of NaTech accidents inside flood- or earthquake-prone areas, or other locations subject to natural disasters. In the case of intentional interferences, the site location may also play a role (e.g. the Middle East or Iraq vs. Europe or Japan), although one can never, at no location worldwide, be sure that a terrorist event will not take place. Regarding the third "atypical" type of scenarios, indirect causes of escalation, such scenarios may unfold anywhere at any time, and chemical plants or chemical clusters may often not be prepared for them. This chapter provides approaches to deal with all of these three often overlooked escalation scenarios.

References

Abdolhamidzadeh, B., Abbasi, T., Rashtchian, D., Abbasi, S.A., 2011. Domino effect in process-industry – an inventory of past events and identification of some patterns. Journal of Loss Prevention in the Process Industries 24, 575–593.

Antonioni, G., Spadoni, G., Cozzani, V., 2007. A methodology for the quantitative risk assessment of major accidents triggered by seismic events. Journal of Hazardous Materials 147, 48–59.

Antonioni, G., Bonvicini, S., Spadoni, G., Cozzani, V., 2009. Development of a general framework for the risk assessment of Na-Tech accidents. Reliability Engineering and Safety Systems 94, 1442–1450.

Barabasi, A., 2003. Linked. How everything is connected to everything else and what it means for business, science, and everyday life. Plume, New York.

Busini, V., Marzo, E., Callioni, A., Rota, R., 2011. Definition of a short-cut methodology for assessing earthquake-related Na-Tech risk. Journal of Hazardous Materials 192, 329–338.

Campedel, M., Cozzani, V., Garcia-Agreda, A., Salzano, E., 2008. Extending the quantitative assessment of industrial risks to earthquake effects. Risk Analysis 28, 1231–1246.

CCPS, 2000. Guidelines for Chemical Process Quantitative Risk Analysis, second ed. AIChE, New York (USA).

Cosmo Oil. Overview of the fire and explosion at Chiba refinery, the cause of the accident and the action plan to prevent recurrence, press release 2 August 2011. http://www.cosmo-oil. co.jp/eng/press/110802/index.html (accessed 01.07.12.).

Cozzani, V., Gubinelli, G., Antonioni, G., Spadoni, G., Zanelli, S., 2005. The assessment of risk caused by domino effect in quantitative area risk analysis. Journal of Hazardous Materials 127, 14–30.

Cozzani, V., Campedel, M., Renni, E., Krausmann, E., 2010. Industrial accidents triggered by flood events: analysis of past accidents. Journal of Hazardous Materials 175, 501–509.

Cruz, A.M., Steinberg, L.J., Luna, R., 2001. Identifying hurricane-induced hazardous material release scenarios in a petroleum refinery. Natural Hazards Review, 203–210.

Cruz, A.M., Okada, N., 2008. Consideration of natural hazards in the design and risk management of industrial facilities. Natural Hazards Review 46, 199–215.

Cruz, A.M., Krausmann, E., 2009. Hazardous-materials releases from offshore oil and gas facilities and emergency response following Hurricanes Katrina and Rita. Journal of Loss Prevention in the Process Industries 22, 59–65.

Cullen, D., 1990. The Public Inquiry into the Piper Alpha Disaster. Department of Energy, Cm 1310. HMSO, London (UK).

Darbra, R.M., Palacios, A., Casal, J., 2010. Domino effect in chemical accidents: main features and accident sequences. Journal of Hazardous Materials 183, 565–573.

Delvosalle, C., Fievez, C., Pipart, A., Debray, B., 2006. ARAMIS project: a comprehensive methodology for the identification of reference accident scenarios in process industries. Journal of Hazardous Materials 130, 200–214.

Girgin, S., 2011. The natech events during the 17 August 1999 Kocaeli earthquake: aftermath and lessons learned. Natural Hazards and Earth System Sciences 11, 1129–1140.

Girgin, S., Krausmann, E., 2012. RAPID-N: rapid Natech risk assessment and mapping tool. Submitted to Journal of Loss Prevention in the Process Industries.

iNTeg-Risk, 2011. Final deliverable of Task 1.5.3. FP7 EC INTeg-Risk Project (Early Recognition, Monitoring and Integrated Management of Emerging, New Technology Related Risks) CP-IP, 213345–2.

Khan, F.I., Abbasi, S.A., 1998. Models for domino effect analysis in chemical process industries. Process Safety Progress 17, 107.

Kourniotis, S.P., Kiranoudis, C.T., Markatos, N.C., 2000. Statistical analysis of domino chemical accidents. Journal of Hazardous Materials 71, 239–252.

Krausmann, E., Cruz, A.M., 2008. Natech disasters: when natural hazards trigger technological accidents. Natural Hazards 64 (2) (special issue).

Krausmann, E., Mushtaq, F., 2008. A qualitative Natech damage scale for the impact of floods on selected industrial facilities. Natural Hazards 46, 179–197.

Krausmann, E., Cruz, A.M., Affeltranger, B., 2010. The impact of the 12 May 2008 Wenchuan earthquake on industrial facilities. Journal of Loss Prevention in the Process Industries 23, 242–248.

Krausmann, E., Cozzani, V., Salzano, E., Renni, E., 2011a. Industrial accidents triggered by natural hazards: an emerging risk issue. Natural Hazards and Earth System Sciences 11, 921–929.

Krausmann, E., Renni, E., Campedel, M., Cozzani, V., 2011b. Industrial accidents triggered by earthquakes, floods and lightning: lessons learned from a database analysis. Natural Hazards 59, 285–300.

Krausmann, E., Baranzini, D., 2012. Natech risk reduction in the European Union. Journal of Risk Research 15, 1027–1047.

Krausmann, E., Cruz, A.M., 2013. Impact of the 11 March, 2011, Great East Japan earthquake and tsunami on the chemical industry. Natural Hazards, DOI 10.1007/s11069-013-0607-0.

Lindell, M.K., Perry, R.W., 1997. Hazardous materials releases in the Northridge earthquake: implications for seismic risk assessment. Risk Analysis 17, 147–156.

OECD, 2012. Draft Report of the Workshop on Natech Risk Management (23–25 May, 2012, Dresden, Germany), 22nd Meeting of the Working Group on Chemical Accidents, Paris, France, 17–19 October, Report ENV/JM/ACC(2012)2.

Øien, K., 2001. Risk indicators as a tool for risk control. Reliability Engineering and System Safety 74, 129–145.

Paltrinieri, N., Dechy, N., Salzano, E., Wardman, M., Cozzani, V., 2012a. Lessons learnt from Toulouse and Buncefield disasters: from risk analysis failures to the identification of atypical scenarios through a better knowledge management. Risk Analysis 32, 1404–1419.

Paltrinieri, N., Øien, K., Cozzani, V., 2012b. Assessment and comparison of two early warning indicator methods in the perspective of prevention of atypical accident scenarios. Journal of Reliability Engineering and System Safety 108, 21–31.

Parry, M.L., Canziani, O.F., Palutikof, J.P., Van der Linden, P.J., Hanson, C.E., 2007. Contribution of Working Group II to the 4th Assessment Report of the Intergovernmental Panel on Climate Change. Cambridge University Press, Cambridge (UK).

Pavlova, Y., Reniers, G.L.L., 2011. A sequential-move game for enhancing safety and security cooperation within chemical clusters. Journal of Hazardous Materials 186, 401–406.

Pidgeon, N., O'Leary, M., 2000. Man-made disasters: why technology and organizations (sometimes) fail. Safety Science 34, 15–30.

Reniers, G., 2012. Security within the chemical process industry: survey results from Flanders, Belgium. Chemical engineering transactions 26, 465–470.

Reniers, G., Dullaert, W., 2008. Knock-on accident prevention in a chemical cluster. Expert systems with applications 34 (1), 42–49.

Reniers, G.L.L., Dullaert, W., Ale, B.J.M., Soudan, K., 2005. The use of current risk analysis tools evaluated towards preventing external domino accidents. Journal of Loss Prevention in the Process Industries 18, 119–126.

Reniers, G.L.L., Dullaert, W., Audenaert, A., Ale, B.J.M., Soudan, K., 2008. Managing domino effect-related security of industrial areas. Journal of Loss Prevention in the Process Industries 21, 336–343.

Renni, E., Krausmann, E., Cozzani, V., 2010. Industrial accidents triggered by lightning. Journal of Hazardous Materials 184, 42–48.

Sabatini, M., Ganapini, S., Bonvicini, S., Cozzani, V., Zanelli, S., Spadoni, G., 2008. Ranking the Attractiveness of Industrial Plants to External Acts of Interference. Proc. Eur. Safety and Reliability Conf. Taylor & Francis, London (UK). pp. 1199–1205.

Sales, J., Mushtaq, F., Christou, M.D., Nomen, R., 2007. Study of major accidents involving chemical reactive substances analysis and lessons learned. Process Safety and Environmental Protection 85 (2B), 117–124.

Showalter, P.S., Myers, M.F., 1994. Natural disaster in the United States as release agents of oil, chemicals, or radiological materials between 1980–9: analysis and recommendations. Risk Analysis 14, 169–181.

Skogdalen, J.E., Vinnem, J.E., 2011. Quantitative risk analysis offshore—human and organizational factors. Journal of Reliability Engineering and System Safety 96, 468–479.

Steinberg, L.J., Cruz, A.M., 2004. When natural and technological disasters collide: lessons from the Turkey earthquake of August 17, 1999. Natural Hazards Review 5, 121–130.

Part II

Prevention of Domino Effects from a Technological Perspective

8 Approaches to Domino Effect Prevention and Mitigation

Valerio Cozzani*, Gigliola Spadoni*, Genserik Reniers†

* LISES, Dipartimento di Ingegneria Civile, Chimica, Ambientale e dei Materiali, Alma Mater Studiorum – Università di Bologna, Bologna, Italy, † Centre for Economics and Corporate Sustainability (CEDON), HUB, KULeuven, Brussels, Belgium

8.1 Approach to Domino Effect Assessment

It is clear from Part 1 of this book that escalation may lead to severe domino accidents that may be highly destructive. Fire, blast waves and fragment projection are the likely causes of such scenarios that may propagate the damage at very high distance from the position of the primary accident.

In recent years, several methodologies were proposed for the identification and assessment of domino scenarios and for the analysis of hazard, damage and risk that may derive from this category of accidents.

As stated in Chapter 3, a domino accident is actually the result of a complex propagation and escalation process of a primary event. Thus, the identification and assessment of domino scenarios actually requires a detailed analysis of the consequences of the primary scenario and of the potential structural damage caused to secondary targets and the evaluation of probability and intensity of secondary scenarios triggered by the primary event. Only a sufficiently detailed assessment of such steps may provide a comprehensive and complete information on the hazards related to domino events and on the likely scenarios involving escalation.

However, due to the complexity of the problem and due to the rather few years having passed since the specificity of domino scenarios was recognized, there is

Domino Effects in the Process Industries. http://dx.doi.org/10.1016/B978-0-444-54323-3.00008-7

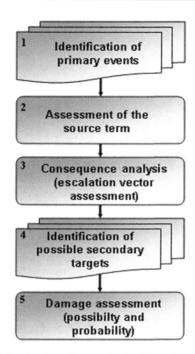

Figure 8.1 Steps required for domino effect analysis. (For color version of this figure, the reader is referred to the online version of this book.)

not yet a widely accepted methodology for the assessment of escalation scenarios and domino accidents. This is also due to the different levels of detail that may be required in such studies. Actually, when carrying out an assessment of domino accident scenarios, the required level of detail may be extremely different depending on the aim of the study. In several contexts, a preliminary assessment is sufficient, while in specific cases, a detailed analysis of worst-case scenarios would be required. Thus, a number of different methodologies are available in the literature in order to carry out the different steps of the analysis at different levels of detail. Resources and time required by the application of the different techniques may be extremely different. Attention should be given to correctly understand the level of detail required by a domino assessment and to select the correct methodology. As a matter of fact, the detailed study provided as an example in Chapter 4 may have costs as high as several hundreds of k€ for a single process unit, while simplified assessments may require only few man-hours to be completed for an entire plant.

If the assessment procedure shown in Figure 8.1 is taken as a reference, the analysis of domino effect may be divided into two main stages: the assessment of the primary scenario and the escalation vector (Steps 1–3, Figure 8.1) and the assessment of possible escalation effects (Steps 4–5, Figure 8.1). The first stage of the procedure requires the definition and assessment of the source term (Step 2) and the modeling of the consequences of the final outcomes of the primary scenario (Step 3). The second phase requires the identification of possible target equipment (Step 4) and the assessment of probability and consequences of the potential failure of the identified secondary targets (Step 5).

A correspondence may be identified among three widely used conventional approaches to hazard and risk assessment (Lees, 1996; CCPS, 2008) and the corresponding level of detail required for the assessment of domino scenarios.

1. Preliminary hazard analysis (PHA) is widely used in early design phases and may be applied to understand the possibility of escalation events
2. Quantitative risk assessment (QRA) may be suitable to understand the risk due to domino scenarios, to identify "hot spots" or critical scenarios and regulatory purposes
3. Distributed parameters modeling aimed at worst-case accident or maximum credible accident analysis may be carried out to assess the design basis or to obtain detailed data on a specific scenario of interest

The above techniques are well known and widely used in different stages of design and lifetime of a plant, as well as for different purposes. Considering domino effects in each of the above contexts is possible (although not usually done).

8.2 Preliminary Analysis of Domino Hazard

In the PHA phase of safety assessment or when a qualitative assessment of domino hazards is required, the analysis of escalation may be carried out on the basis of a simplified assessment of primary scenarios, using threshold values for the identification of escalation targets.

The approach requires that the primary events of interest are identified and that the consequences of the final outcomes of these primary scenarios are calculated. Conventional integral models for consequence assessment as those described, e.g. in the "yellow book" (Van den Berg and Weterings, 1997) or by Lees (1996) may be used at this stage to calculate the intensity of the escalation vectors selected for the primary scenarios.

A preliminary assessment concerning the possibility of escalation may be provided comparing the intensity of the escalation vector at the position of a potential target vessel with the appropriate "escalation threshold". Escalation thresholds consist in values of physical effects (e.g. a maximum peak overpressure for blast waves or a radiation intensity for a fire) below which no damage to the target item is expected. An example of escalation thresholds is provided in Table 8.1.

When conventional scenarios in the process industry are of concern, the use of well-recognized models for consequence assessment combined with escalation thresholds allows a further simplification in the analysis. Plots correlating the vessel inventory or the release equivalent diameter to a safety distance, as those shown in Figures 8.2, 8.3 and 8.4, may be used to obtain a preliminary indication concerning the possible exceedance of escalation thresholds at the location of a secondary target (Cozzani et al., 2007; Tugnoli et al., 2008a,b; Cozzani et al., 2009). A very

Table 8.1 Example of Suggested Threshold Values and Safety Distances for Escalation (Cozzani et al., 2007)

Primary Scenario	Escalation Vector	Equipment Category	Threshold Value	Safety Distance
Fireball	Heat radiation	Atmospheric	15 kW/m^2	Fireball radius
		Pressurized	45 kW/m^2	0
Jet fire	Heat radiation	Atmospheric	15 kW/m^2	Flame length + 50 m
		Pressurized	45 kW/m^2	Flame length + 25 m
Pool fire	Heat radiation	Atmospheric	15 kW/m^2	Pool border + 50 m
		Pressurized	45 kW/m^2	Pool border + 15 m
Vapor cloud explosion (VCE)	Overpressure ($F \geq 5$; $M_f \geq 0.35$)	Atmospheric	22 kPa	$R = 1.75$
		Pressurized	20 kPa	$R = 2.10$
BLEVE	Overpressure	Atmospheric	22 kPa	$R = 1.80$
		Pressurized	20 kPa	$R = 2.00$
	Fragment projection	Any	Undefined	Undefined
Mechanical explosion	Overpressure	Atmospheric	22 kPa	$R = 1.80$
		Pressurized	20 kPa	$R = 2.00$
	Fragment projection	Any	Undefined	Undefined

R, F and M_f are respectively the Sachs energy-scaled distance, the strength factor as in the Multi-Energy Method (Van den Berg, 1985) and the flame Mach number in the Baker-Sthrelow-Tang methodology (Tang and Baker, 1999).

preliminary analysis may be carried out by the use of such plots without even the need to run consequence analysis models. A detailed discussion of these shortcut methods is provided elsewhere (Cozzani et al., 2007).

A further easy-to-use tool that was developed in 2003 to conduct a preliminary analysis of domino hazard is the so-called Instrument for Domino Effects (IDE) (RIVM, 2003). The IDE has been developed to identify "domino establishments" in the Netherlands. The instrument is based on connecting "causing establishments" and "vulnerable establishments" with respect to domino effects. Establishments are coupled determining typical escalation distances. Distances are related to possible physical effects that cause the propagation and escalation of the primary event in the other installations considered. Accident scenarios included in the IDE are the bursting of pressure vessels, boiling liquid evaporating vapor explosions giving rise to fragment projection, vapor cloud explosions, pool fires, and jet fires. The definitions of these accident scenarios are based on the definitions given in the Purple Book (Uijt de Haag and Ale, 1999). In order to define a domino distance, the IDE only considers containment systems and installations with explosive, flammable, highly flammable or very highly flammable substances. The IDE offers tabulations of calculated domino distances for standard couples of installations. The figures obtained are based on various parameters such as scenarios considered, quantities of hazardous substances, substance categories, generic substances and vulnerability levels of the exposed

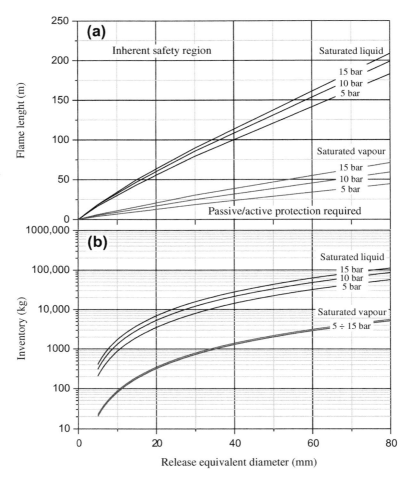

Figure 8.2 Average length of a jet fire flame (a) and minimum vessel inventory for a 15-min release (b) with respect to the equivalent release diameter for hot saturated hydrocarbons at high pressure. Safety distances for escalation are obtained summing to flame length 50 m for unprotected atmospheric vessels and 25 m for unprotected pressurized vessels (Cozzani et al., 2007).

installations. For cases deviating from the standard procedure (e.g. mixtures of substances), a method described in the IDE can be used to manually calculate the domino distances. Table 8.2 shows an excerpt of table from the IDE reporting safety distances for the case of two quantities (i.e. smaller than 5 t or between 5 and 10 t) of flammable gas under pressure stored in bulk or processed. For vulnerable installations situated at a distance higher than the "domino distance" linked to a causing installation for a certain scenario, the escalation of the domino event with cardinality 0 (i.e. the initiating accident) is not possible. In other words, a secondary event is considered not credible in such case.

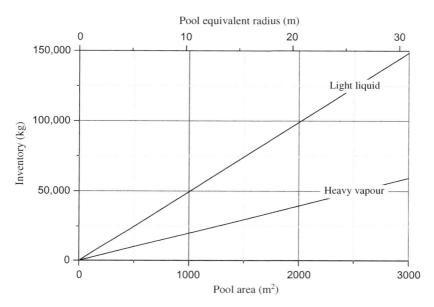

Figure 8.3 Critical mass released in pool fire scenarios with respect to catch basin surface area and equivalent radius. Safety distances for escalation are obtained summing to catch basin equivalent radius 50 m for atmospheric vessels and 15 m for pressurized vessels (Cozzani et al., 2007).

The IDE documentation is user friendly (although only available in Dutch to date) and the data needed to implement it are already included in safety reports required for sites falling under the obligations of the "Seveso" directives. This makes it very useful for regulators as well as for company safety managers or consultants to quickly determine domino distances. The use of the IDE is intended for the evaluation of as many pairs of installations as possible as regards domino effects without explicitly having to perform the rather cumbersome calculation steps to determine the domino distances.

The use of these "shortcut" methods is attractive and useful at a preliminary stage of domino hazard assessment. However, it is very important to understand and always keep in mind that the escalation thresholds represent an oversimplification of the problem and should be used with extreme caution. Escalation thresholds are inherently empirical and are heavily influenced both by the features of the primary scenario and by the geometrical and mechanical characteristics of the secondary target (Cozzani et al., 2006a). Moreover, a number of different threshold values are provided in the literature (Cozzani and Salzano, 2004; Cozzani et al., 2006a). Thus, appropriate values need to be selected in order to carry out a consistent analysis. This issue will be discussed in detail in Chapter 9.

Therefore, it may be concluded that a simplified analysis based on escalation thresholds (e.g. by the methods discussed above) may provide useful indications on the presence of an escalation hazard, which may be used to understand if a more

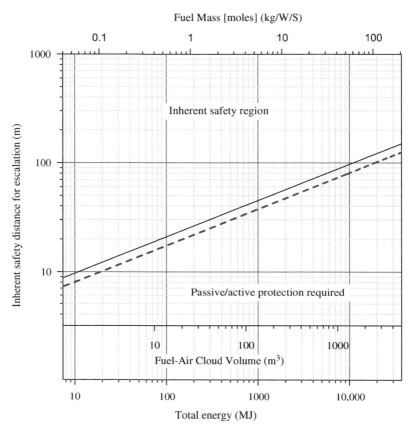

Figure 8.4 Safety distance for escalation calculated for equipment loaded by vapor cloud explosions generated by the more common hydrocarbons. (W, molecular weight; S, stoichiometric concentration expressed in volume percent of fuel–air mixture). Distances are intended from the flammable cloud border. For more reactive fuels, a factor 1.2 should be used to correct energy values based on cloud volume. Solid line, pressurized target vessel; dashed line, atmospheric target vessel (Cozzani et al., 2007).

detailed analysis is needed and/or if protection and mitigation measures with respect to escalation of the primary scenarios selected should be considered. However, consistent, sound and coherent values need to be selected for the escalation thresholds and the set of primary scenarios analyzed needs to be correctly selected and representative, in order to obtain significant data.

Clearly enough, this method only provides a yes/no answer concerning the likelihood of the target being affected by the primary scenario of concern. Results of application provide a screening that may useful to identify the possible relevant targets of escalation caused by the set of primary scenarios considered. More detailed approaches are needed to assess the likelyhood of the possible escalation events identified by this procedure.

Table 8.2 Safety Distances (m) of Vulnerable Installations from Pressurized Bulk Storage/Pressurized Processes Involving Flammable Gases

M_{sys} (t)	Bursting/Peak Overpressure (bar)				Missiles	BLEVE ($R_{fireball}$)	LFL-Distance	VCE/Peak Overpressure (bar)				Pool Fire/Heat Radiation (kW/m²)		
	0.1	0.2	0.3	0.45				0.1	0.2	0.3	0.45	D_{max} (m)	8	37.5
≤5														
GF1	30	21	16	11		52	40	141	90	72	58	24.8	67	39
GF2	35	23	18	16			120	204	168	157	150	23.9	97	54
GF3	61	37	27	21	296		115	187	146	138	135	–	–	–
>5–10														
GF1	37	26	20	14		65	55	177	113	90	73	35.0	89	53
GF2	43	29	23	16			157	266	220	205	195	33.4	130	74
GF3	77	47	34	27			134	237	172	161	155	–	–	–

GF1, GF2, or GF3 are hazard categories attributed to the flammable gas based on physical and chemical properties. M_{sys}: overall quantity of subtance in the system considered; D_{max}: maximum diameter of the pool fire; $R_{fireball}$: radius of impact of the expected fireball; LFL: lower flammability limit.
Source: Adapted from Instrument Domino Effects (RIVM, 2003).

The threshold approach does not need a high level of detail on the site and equipment, and only requires a limited additional work with respect to that needed to issue an ordinary safety report (e.g. as that required for sites falling under the obligations of Seveso Directives: Directive 96/82/EC and Directive 2012/18/EU) is required. Thus the assessment may also be carried out in early design stages, when limited details concerning the plant equipment and layout are available. If a correct selection of threshold values is carried out, results are likely to be overconservative.

8.3 Quantitative Risk Assessment of Domino Scenarios

If a relevant escalation hazard is identified, a more detailed assessment than that based on damage threshold values may be necessary. Quantitative Risk Assessment (QRA) is nowadays used as a standard tool to analyze and compare the risk due to industrial installations (Lees, 1996; CCPS, 2000).

A framework for the quantitative risk assessment of domino scenarios was developed in several previous studies (Delvosalle, 1998; Khan and Abbasi, 1998; Abdolhamidzadeh et al., 2010; Reniers et al., 2005; Cozzani et al., 2005) and is summarized in Figure 8.5. Specific methods were proposed for the calculation of individual (Abdolhamidzadeh et al., 2010; Reniers and Dullaert, 2007; Cozzani et al., 2005) and societal risk (Cozzani et al., 2006b) caused by domino scenarios.

Two main difficulties arise when affording the inclusion of domino scenarios in a QRA study.

On one hand, if the analysis of a complex plant is undertaken, a huge number of possible scenarios need to be considered. It has been demonstrated that if escalation is considered for n equipment items, up to 2^n different first-level domino scenarios are possible (see Chapter 10). Thus, a suitable computational approach needs to be applied to carry out steps 6–9 of Figure 8.5. Several examples of prototypes and research software tools are provided in the literature (Reniers and Dullaert, 2007; Cozzani et al., 2006b), but a standard commercial software able to support such analysis is not yet available. This issue is further discussed in Chapter 14.

On the other hand, the quantitative assessment of risk due to domino scenarios requires reliable and sufficiently simple models for equipment vulnerability (Step 4 of Figure 8.5). Only recently, simplified damage probability models were developed and became available for escalation assessment of different categories of process equipment. In Part 1 of this book several of such models were revised and discussed. These tools allow the estimation of the equipment damage probability as a function of severity parameters of the escalation vector (maximum peak overpressure, radiation intensity, etc.). Clearly enough, the reliability and the accuracy of the results are heavily influenced by the quality of these models.

Nevertheless, QRA for domino scenarios has proved to be feasible and to provide valuable results. Quantitative risk figures allow the ranking of domino scenarios and the identification of criticalities. In particular, the relevance of domino scenarios, the presence of critical equipment or primary events, and land use planning issues may be obtained by its application. Several examples are provided in the literature (Cozzani

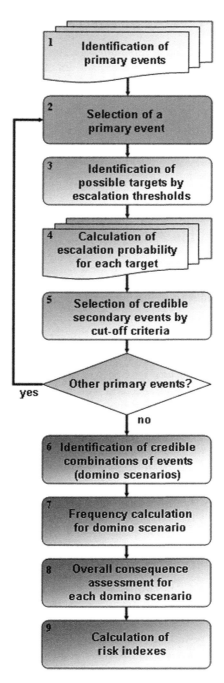

Figure 8.5 Steps required in the quantitative risk assessment of domino scenarios. (For color version of this figure, the reader is referred to the online version of this book.)
Source: Adapted from Cozzani et al. (2005).

et al., 2005; Antonioni et al., 2009). The issue will be further discussed in Chapters 10 and 14.

Quantitative risk assessment of domino scenarios requires the availability of detailed data on the layout of the site of concern. Limited data are needed on equipment items (mainly concerning the type, the volume and the design pressure range) and relatively low additional work with respect to that needed to issue an ordinary QRA is required. Thus, data input and analysis of results are likely to be definitely more time requiring than in the case of the threshold-based approach, although the significance and accuracy of results is expected to be much higher. Thus, as in the case of an ordinary QRA, quantitative assessment of domino effects may be carried out during basic design or for existing installations.

8.4   Distributed Parameter Models for Worst-Case or Worst-Credible Accident Analysis

Distributed parameter models, which require a detailed description of equipment design and layout involved in the primary scenario, may be used when a high detail is required in the analysis, in particular when consequences of the primary scenarios in the near field are addressed. These requirements are usually derived from offshore application, or, more generally, from critical installations whose structural integrity should be assessed with a high level of confidence, in particular in the design phase in order to address the design procedure. Usually, single worst-case or worst-credible scenarios are identified and selected as design basis and are analyzed in detail to assess the required structural resistance and protection measures.

In recent years, computational fluid dynamics (CFD) codes based on Reynolds-averaged Navier–Stokes equations have been specifically developed for the consequence assessment of gas explosions in medium- and large-scale scenarios. Relevant progress was also made in the development of CFD tools able to simulate jet flames and pool fires. The detailed 3D pressure–time profile or radiation–time profile obtained by the CFD codes may be used as an input to analyze the structural resistance of process equipment to blast wave or radiation. Also in this case, distributed parameter models may be applied. In particular, the use of finite element codes may allow the analysis of the local stresses induced by the dynamic impulse caused by the blast wave or the effect of thermal stresses induced by flame radiation, impingement or engulfment. Although finite element simulations by standard codes are usually limited to the elastic region, in the framework of safety assessment, the possible plastic deformation of an equipment item may be assumed as a condition sufficient to cause a loss of containment. Thus, finite element codes may be applied to obtain a detailed simulation of the effect of fires and explosions. An example of such studies in the case of blast waves was provided in Chapter 4. The issue will be further discussed in Chapter 11.

Despite the progress made in recent years, it is quite clear that this approach has important limitations due to the computer time and man-hours required to build the numerical domain and to carry out the simulations. In particular, it is not practical to use CFD in order to reproduce the behavior of blast wave propagation or radiation intensity in the far-field, due to the huge amount of computer resources and time usually required to increase the numerical domain. On the other hand, it is not possible to perform a specific finite element simulation for all the equipment items present in a plant. Moreover, the finite element simulation of an equipment item requires high amount of design details, which is usually not available in early design phases. Thus, while near-field fire and explosion effects may be addressed by these tools, in particular for design cases, simplified methods seem more practical to apply when escalation assessment and domino effects are of concern.

It should be recalled that in the assessment of domino accidents the main issue is the identification of the possible escalation events that may involve nearby units. Thus, as far-field escalation is usually of concern, the use of the available distributed parameter modeling tools is actually extremely time consuming and scarcely practicable at the time. Distributed parameter approaches, due to time and costs, presently should reasonably be limited to the analysis of a single or very limited set of explosion or fire scenario, while in far-field domino effect assessment, a huge number of different scenarios are usually possible.

In any case, when the above approach is applied to domino effect assessment, it is strongly suggested to integrate it with a preliminary threshold analysis and a QRA. Actually, these less-detailed approaches can effectively provide clear and sound criteria to identify the more relevant scenario to be selected as a design case or a study case for the application of distributed parameter modeling.

8.5 Conclusions

Three levels having increasing complexity may be identified for domino effect hazard assessment: (1) threshold-based approach, (2) quantitative risk assessment, and (3) distributed parameters modeling. The framework in which the use of these methods is acceptable, the robustness and the detail of the results obtained differ widely. Thus, the level of the analysis should be carefully selected, on the basis of the aims of the specific study undertaken. Further details on each of these levels of approach are provided in Chapters 9 to 11 of the present book.

References

Abdolhamidzadeh, B., Abbasi, T., Rashtchian, D., Abbasi, S.A., 2010. A new method for assessing domino effect in chemical process industry. Journal of Hazardous Materials 182, 416–426.

Antonioni, G., Spadoni, G., Cozzani, V., 2009. Application of domino effect quantitative risk assessment to an extended industrial area. Journal of Loss Prevention in the Process Industries 19, 463–477.

CCPS, 2000. Guidelines for Chemical Process Quantitative Risk Analysis, second ed. Center for the Chemical Process Safety, AIChE, New York (USA).

CCPS, 2008. Guidelines for Hazard Evaluation Procedures, third ed. Center for the Chemical Process Safety, AIChE, New York (USA).

Cozzani, V., Salzano, E., 2004. Threshold values for domino effects caused by blast wave interaction with process equipment. Journal of Loss Prevention in the Process Industries 17, 437–447.

Cozzani, V., Gubinelli, G., Antonioni, G., Spadoni, G., Zanelli, S., 2005. The assessment of risk caused by domino effect in quantitative area risk analysis. Journal of Hazardous Materials 127, 14–30.

Cozzani, V., Gubinelli, G., Salzano, E., 2006a. Escalation thresholds in the assessment of domino accidental events. Journal of Hazardous Materials 129, 1–21.

Cozzani, V., Antonioni, G., Spadoni, G., 2006b. Quantitative assessment of domino scenarios by a GIS-based software tool. Journal of Loss Prevention in the Process Industries 19, 463–477.

Cozzani, V., Tugnoli, A., Salzano, E., 2007. Prevention of domino effect: from active and passive strategies to inherently safe design. Journal of Hazardous Materials 139, 209–219.

Cozzani, V., Tugnoli, A., Salzano, E., 2009. The development of an inherent safety approach to the prevention of domino accidents. Accident Analysis and Prevention 41, 1216–1227.

Delvosalle, C., 1998. A Methodology for the Identification and Evaluation of Domino Effects, Rep. CRC/MT/003, Belgian Ministry of Employment and Labour, Bruxelles (B).

Directive 96/82/EC, Council Directive 96/82/EC of 9 December 1996 on the Control of Major-Accident Hazards Involving Dangerous Substances. Official Journal of the European Communities, L 10/13, Brussels, 14.1.97.

Directive 2012/18/EU, European Parliament and Council Directive 2012/18/EU of 4 July 2012 on Control of Major-Accident Hazards Involving Dangerous Substances, Amending and Subsequently Repealing Council Directive 96/82/EC. Official Journal of the European Communities, L 197/1, Brussels, 24.7.2012.

Khan, F.I., Abbasi, S.A., 1998. Models for domino effect analysis in chemical process industries. Process Safety Progress 17, 107.

Lees, F.P., 1996. Loss Prevention in the Process Industries, second ed. Butterworth-Heinemann, Oxford (UK).

Reniers, G.L.L., Dullaert, W., Ale, B.J.M., Soudan, K., 2005. The use of current risk analysis tools evaluated towards preventing external domino accidents. Journal of Loss Prevention in the Process Industries 18, 119.

Reniers, G.L.L., Dullaert, W., 2007. DomPrevPlanning: user-friendly software for planning domino effects prevention. Safety Science 45, 1060.

RIVM, 2003. Instrument Domino Effecten. Rijksinstituut voor Volksgezondheid en Milieu, Bilthoven (NL).

Tang, M.J., Baker, Q.A., 1999. A new set of blast curves from vapour cloud explosion. Process Safety Progress 18, 235.

Tugnoli, A., Khan, F.I., Amyotte, P., Cozzani, V., 2008a. Safety assessment in plant layout design using indexing approach: implementing inherent safety perspective. Part 1 – guideword applicability and method description. Journal of Hazardous Materials 160, 100–109.

Tugnoli, A., Khan, F.I., Amyotte, P., Cozzani, V., 2008b. Safety assessment in plant layout design using indexing approach: implementing inherent safety perspective. Part 2 – domino hazard index and case study. Journal of Hazardous Materials 160, 110–121.

Uijt de Haag, P.A.M., Ale, B.J.M., 1999. Guidelines for Quantitative Risk Assessment (Purple Book). Committee for the Prevention of Disasters, The Hague (NL).

Van den Berg, A.C., 1985. The Multi-energy method – a framework for vapor cloud explosion blast prediction. Journal of Hazardous Materials 12, 1.

Van Den Bosh, C.J.H., Weterings, R.A.P.M., 1997. Methods for the Calculation of Physical Effects (Yellow Book). Committee for the Prevention of Disasters, The Hague (NL).

9 Threshold-Based Approach

Valerio Cozzani*, Alessandro Tugnoli*,
Sarah Bonvicini*, Ernesto Salzano†

* LISES, Dipartimento di Ingegneria Civile, Chimica, Ambientale e dei
Materiali, Alma Mater Studiorum - Università di Bologna, Bologna, Italy,
† Istituto di Ricerche sulla Combustione, Consiglio Nazionale delle Ricerche
(CNR), Napoli, Italy

9.1   Introduction

In the assessment of domino accident hazard, the starting point is the identification of equipment that may be "damaged" by the physical effects (escalation vectors) generated by the primary accidental scenarios. In this framework, the straightforward identification of possible damage following a primary event is greatly enhanced by the availability of so-called damage threshold values, i.e. minimum values of physical effects that may cause the damage of target equipment. As discussed in Chapter 8, escalation thresholds are a valuable tool to develop "rules of thumb" or screening

Domino Effects in the Process Industries. http://dx.doi.org/10.1016/B978-0-444-54323-3.00009-9

approached for the preliminary assessment of possible escalation scenarios resulting in domino accidents. In this context, the results of consequence analysis models applied to the simulation of primary scenarios may be easily compared to threshold values, identifying a maximum credible escalation radius (Cozzani et al., 2007). This straightforward procedure is attractive due to its simplicity and transparency. The approach has the merit of reducing escalation assessment to a monodimensional problem, thus limiting the computational resources needed to carry out the analysis even on extended plant layouts. Several simplified models for the assessment of damage caused by blast waves in quantitative risk analysis (Lees, 1996; Bagster and Pitblado, 1991), as well as many "safety distance" criteria adopted in technical standards (Gledhill and Lines, 1998; Van Den Bosh and Weterings, 1997) and even in the legislation (DM 151/2001), are based on threshold values (Cozzani et al., 2006).

However, a wide uncertainty exists in threshold values for domino escalation. As shown in Table 9.1, proposed values span over an order of magnitude. This may result

Table 9.1 Escalation Thresholds Reported in the Literature

Escalation Vector	Threshold	Equipment Category	Reference
Radiation (kW/m^2)	9.5	All	Tan (1967)
	12.5	All	DM 151/2001
	15.6	All	API RP 510 (1990)
	24.0	All	Bagster and Pitblado (1991)
	25.0	All	Van den Bosh et al. (1989)
	37.0	All	Khan and Abbasi (1998)
	37.5	All	HSE (1978)
	37.5	All	BS 5908 (1990)
	37.5	All	Mecklenburgh (1985)
	38.0	All	Kletz (1980)
Overpressure (kPa)	7.0	Atmospheric	Gledhill and Lines (1998)
	10.0	Atmospheric	Barton (1995)
	10.0	Atmospheric	Bottelberghs and Ale (1996)
	10.0	Atmospheric	Kletz (1980)
	14.0	Atmospheric	Gugan (1979)
	20.3	Atmospheric	Brasie and Simpson (1968)
	20.7	Atmospheric	Clancey (1972)
	23.8	Atmospheric	Glasstone (1980)
	30.0	All	DM 151/2001
	30.0	Pressurized	Bottelberghs and Ale (1996)
	35.0	All	Wells (1980)
	35.0	All	Bagster and Pitblado (1991)
	38.0	Pressurized	Gledhill and Lines (1998)
	42.0	Pressurized	Cozzani and Salzano (2004c)
	55.0	Pressurized	Brasie and Simpson (1968)
	65.0	Pressurized	Glasstone (1980)
	70.0	All	Khan and Abbasi (1998)
Fragments (m)	800.0	All	DM 151/2001
	1150.0	All	HSE (1978)

in safety distances ranging from few tens of meters to several hundreds of meters. The wide uncertainty in threshold values derives from the complexity of the escalation phenomenon. Several problems need to be defined in order to obtain a robust set of escalation threshold:

- the damage mechanism of the primary scenario (e.g. impingement or engulfment vs. distant radiation, blast wave impact vs. fragment impact, etc.);
- the geometry and the characteristics of the target equipment (e.g. shape, wall thickness, connections, protections as pressure relief valves (PRVs) or fireproofing);
- the critical extension of damage needed for a relevant escalation (e.g. minor damage, full bore rupture of connections, collapse, etc.).

Among the factors influencing the possibility of propagation, the specific features of the escalation vectors in the scenario considered may play an important role (e.g. the duration of the scenario may influence the possibility of escalation due to radiation). Furthermore, the design features of the possible target equipment may also result in a quite different resistance to damages caused by the escalation vectors. The last point in particular is often overlooked in escalation analysis. Actually, damage of secondary equipment is not sufficient to trigger a domino accident as defined in Chapter 3, since an important part of the definition is that the overall consequences of the domino scenario should be higher than those of the primary event. Thus, a minor pool fire triggered as a consequence of a devastating explosion may not be considered a relevant escalation, since the explosion consequences may be by fare more relevant. If this definition is not applied, almost all severe industrial accidents should be considered as domino effects.

In this chapter, threshold criteria for different categories of process equipment were revised with respect to the three escalation vectors that more frequently lead to domino accidents: radiation/fire impingement, overpressure, and fragment projection. The concepts of "damage state" (DS) and "loss intensity" (LI) are introduced and related to the escalation potential of the scenario. Finally, the specific features of different primary scenarios are discussed, proposing detailed criteria for the definition of escalation thresholds. These were revised also in the perspective of recent research results obtained in the modeling and in the assessment of escalation events. Table 9.2 summarizes the different categories of primary accidental scenarios considered, derived from definitions widely used in the current practice (CCPS, 2000; Uijt de Haag and Ale, 1999).

9.2 Escalation Vectors and Equipment Damage Mechanisms

9.2.1 Radiation and Fire Impingement from Stationary Fires

As shown in Table 9.2, several primary scenarios may result in an escalation due to radiation and/or fire impingement. Besides, Table 9.1 evidences that escalation hazard is generally addressed considering radiation intensity. However, three factors should actually be taken into account: (1) the intensity and mode of radiation; (2) the time evolution of the accidental event; and (3) the characteristics of the secondary target.

Table 9.2 Escalation Vectors and Expected Secondary Scenarios for the
Different Primary Scenarios Considered in the Present Analysis

Primary Scenario	Escalation Vector	Expected Secondary Scenarios*
Pool fire	Radiation and fire impingement	Jet fire, pool fire, BLEVE, and toxic release
Jet fire	Radiation and fire impingement	Jet fire, pool fire, BLEVE, and toxic release
Fireball	Radiation and fire impingement	Tank fire
Flash fire	Fire impingement	Tank fire
Mechanical explosion[†]	Fragments and overpressure	All[‡]
Confined explosion[†]	Overpressure	All[‡]
BLEVE[†]	Fragments and overpressure	All[‡]
VCE	Overpressure and fire impingement	All[‡]
Toxic release	–	–

BLEVE, boiling liquid expanding vapor explosion; VCE, vapor cloud explosion.
*Expected scenarios also depend on the hazards of target vessel inventory.
[†]Following primary vessel failure, further scenarios may take place (e.g. pool fires, fireballs, and toxic releases).
[‡]All: any of the scenarios listed in column 1 may be triggered by the escalation vector.
Source: Cozzani et al. (2006).

The radiation mode is influenced by the accidental scenario and by the relative position of the secondary target vessel: the vessel may be fully or partially engulfed by a fire or may receive heat radiation from the fire with no engulfment or impingement.

When time evolution is taken into account, the main element to consider is that the duration of the primary scenario should be at least comparable with the characteristic "time to failure" (ttf) of the target equipment involved in the fire. Pool fires and jet fires are the final scenarios that more likely generate long-lasting stationary flames. However, the duration of the fire is influenced by a number of factors as inventory, ignition delay, presence and effectiveness of mitigation systems as emergency shutdown valves able to cut the fuel feed to the fire, etc. A specific assessment is needed to understand the credible duration of the primary fire considered.

The vessel ttf also depends on the equipment design (e.g. pressurized vessels usually have a higher ttf than atmospheric storage tanks due to higher shell thickness), as well as on the presence of active and passive protections (e.g. water deluges, thermal insulation, etc.) that may delay the vessel shell heat-up.

Further considerations can be drawn considering the characteristics of specific accident scenarios. A jet fire is a turbulent flame that may have a relevant length in the direction of the release, due to the high kinetic energy of the jet (Van Den Bosh and Weterings, 1997). A jet fire may cause an escalation by two different mechanisms: direct flame impingement on a target vessel, and far-field radiation from the flame zone. Jet fire impingement is a well-known cause of escalation, as shown by the analysis of past accidents where domino effects took place (Mannan, 2005). Damage due to heat transfer caused by distant stationary radiation may, as well, cause target vessel failure, although higher values of ttf are expected and more time is available for active mitigation measures (Cozzani et al., 2006).

Also in the case of pool fires two different scenarios may be identified with respect to the possibility of escalation. A target vessel may be fully engulfed in the flames, or may be distant from the pool, thus receiving a stationary heat radiation from the flames. As in the case of jet fire impingement, it is well known that the engulfment in a pool fire may cause the failure of the target vessel, resulting in an escalation. Thus, the escalation should be considered possible for any target vessel located inside the pool area. In the case of a target vessel receiving a stationary heat radiation from the flame, without flame impingement or engulfment, the possibility of escalation should be addressed taking into account the intensity of heat radiation and the characteristics of the target vessel (Cozzani et al., 2006). Passive and active protection barriers may be effective in preventing escalation also in this case (see Chapter 5).

9.2.2   Radiation and Fire Impingement from Transient Flames

Transient flames are typical of scenarios having a quick time evolution, such as flash fires, thermal effects from vapor cloud explosions (VCEs) and fireballs. These events have a characteristic duration that in general is lower than the ttf due to heat radiation of any type of process or storage vessel.

As discussed in detail in Chapter 5, flash fires are not likely to result in the damage of a secondary vessel due to heat radiation. Nevertheless, an escalation may be caused by secondary flames due to the direct ignition of flammable of material. A single case is usually of relevance within a process plant: the ignition of vapors above the roof of a floating roof tank, starting a tank fire. Similar considerations apply limitedly to the thermal effects from the explosive combustion of a cloud of flammable material (VCE).

The fireball scenario is originated by the diffusive combustion of the flammable vapors released in the catastrophic failure of a process unit. Fireballs have typically a limited duration, usually much lower than 1960s (CCPS, 2000; Van Den Bosh and Weterings, 1997). As discussed in detail in Chapter 5, the possibility of escalation following the damage of equipment items caused by fireball heat radiation, in the case of both full engulfment and distant source radiation, needs a careful assessment and the issue may found different approaches in the Quantitative Risk Assessment (QRA) practice. As a matter of fact, the relatively short duration of the fireball makes questionable the possibility of radiation damage to process vessels. The study by Cozzani et al. (2006) suggests that, even in the case of full engulfment in flames, the duration of the fireball event is lower of about an order of magnitude than the ttf of any typical unprotected pressurized vessel, even neglecting any protection provided (e.g. thermal insulation). Thus, it may be concluded that an escalation caused by fireball radiation seems unlikely for target pressurized equipment. On the other hand, in the case of atmospheric vessels, the ttf may be of the same order of magnitude of the duration of the fireball. Moreover, as discussed above, in the case of flame engulfment of floating roof storage tanks, the escalation may be caused by the ignition of flammable vapors above the roof sealing or by the failure of the roof sealing. Thus, even if the escalation due to fireball radiation seems to be credible only for atmospheric vessels involved in very severe fireball scenarios, a specific assessment may be necessary if no thermal insulation is present.

9.2.3   Blast Waves

Accidental scenarios in which escalation effects may be caused by pressure can be summarized as (1) unconfined and partially confined gas and vapor gas explosions; (2) confined explosions (including gas, vapor and dust explosions inside vented or unvented equipment; runaway reactions, boiling liquid expanding vapor explosion (BLEVE)); (3) mechanical explosions (caused by vessel failure following the gas or liquid mechanical compression to pressures above the vessel design pressure); and (4) point-source explosion of explosives or reactive solids. For each of these categories of explosion, the produced blast wave is characterized by different shape and characteristic time duration, depending on the geometrical scenario and the total available energy.

As discussed in detail in Chapter 4, the expected damage due to overpressure is usually assessed considering only the peak static overpressure on the target item, even if it is widely recognized that many other factors may influence the damage due to blast waves. In particular, the dynamic pressure (drag forces), the rise time of the positive phase of the wave and the total impulse, as well as other complex phenomena such as the reflection of pressure wave either on the ground or on the loaded equipment, flow separation, effects due to the geometry and the relative position of the loaded equipment and blast wave may influence the damage caused by the blast wave (Baker et al., 1983). Besides, the geometric characteristics of the target equipment, the design pressure, and the natural period of the structure also greatly influence the damage experienced.

The single degree of freedom method is widely used in the process industry for predicting structural response of object (equipment, buildings, and structures) to blast loads. It is a simple approach which idealizes the actual structure into a spring mass model and is very useful in routine design procedures in order to obtain accurate results for relatively simple structures subjected to limited ductility. A more refined alternative is the multiple degree of freedom system, which may have a large number of modal periods. Details can be found in classical textbook as Biggs (1964).

Complex structural models based on finite element analysis are used to simulate the behavior of structures under blast loads, as discussed in Chapter 11. However, such detailed approaches may be hardly justified in a QRA framework, given the uncertainty in the feature of the primary scenarios. Moreover, even the assessment of the resistance of a simple planar target facing an idealized triangular blast waves is a complex task (HSE, 2004; UKOOA, 2002). On the other hand, quasi-static regime can be assumed as conservative for escalation assessment (UKOOA, 2002; Schneider, 1998; Whitney et al., 1992). Thus, when far-field interactions between the explosion source and the target equipment are of concern, or when relatively low-pressure explosion are considered (maximum peak static overpressure lower than 50 kPa, as in most industrial explosions), the damage caused by a blast wave may be effectively correlated to the peak static overpressure only.

9.2.4 Fragment Impact

As discussed in Chapter 6, the primary scenarios that are likely to result in fragment projection include all types of explosions. The fragment number, shapes and weights are mainly dependent on the characteristics of the vessel that undergoes the fragmentation. On the other hand, it is well known that the distance of fragment projection is mainly dependent on the initial fragment velocity, on the initial direction of projection and on the drag factor of the fragment (Baker et al., 1983; Holden and Reeves, 1985; Mannan, 2005). The analysis of a wide number of past accidents involving the projection of missiles from the fragmentation of different equipment items carried out in several studies allowed the identification of the mass range and of the fragment shapes more frequently experienced in accidental events (Holden and Reeves, 1985; Gubinelli and Cozzani, 2009a,b). In Chapter 6, a detailed discussion of the issue is provided. Both experience from past accidents and models for fragment projection evidence that maximum fragment projection distances that are far too high to allow the definition of any useful safety distance criterion for escalation (Mannan, 2005; Hauptmanns, 2001; Gubinelli et al., 2004). Therefore, escalation criteria for fragment projection may only be derived taking into account fragment impact probability.

9.2.5 Definition and Calculation of Escalation Vectors

Table 9.2 shows the primary scenarios that are likely to trigger escalation effects, and the escalation vectors identified for each scenario on the basis of the above discussion. The intensity of the escalation vector generated by a given primary scenario depends on the total amount of energy (or substance) which is possibly released from the primary system of containment (reactor, storage tank, etc.), and may be defined as the maximum distance at which escalation effects may be considered credible. The latter may be determined on the basis of threshold values for escalation that will be discussed in the following. However, it should be remarked that the above discussion evidences that the escalation thresholds depend on the structural features of the target equipment. Thus, although conservative values may be obtained considering the more vulnerable equipment categories, more accurate escalation thresholds and safety distances may be obtained considering the structural characteristics of specific equipment categories.

Furthermore, the determination of the escalation vector or of safety distances for escalation may also depend on the secondary scenario generated by the failure of the secondary equipment of concern. Actually, as discussed in Chapter 3, a domino accident requires an escalation beside the damage of secondary equipment: the overall consequences of a relevant domino event should be more severe than those of the primary scenario taken alone (Section 3.2). Thus, besides equipment damage thresholds, the assessment of escalation scenarios may suggest to consider as well thresholds based on the expected LI following equipment damage. Such thresholds may account for the expected severity of the secondary scenario. This issue is addressed in the following section.

9.3 Damage State and Loss Intensity Criteria

9.3.1 The Concept of Equipment Damage State

Table 9.1 evidences that threshold values for equipment damage reported in the literature range over an order of magnitude (Cozzani and Salzano, 2004a; Cozzani et al., 2006). These strong uncertainties widely affected technical standards and regulations concerning the assessment of domino events. Actually, damage thresholds reported in the literature are mainly derived from the analysis of past accident data. The first cause of data dispersion is thus possibly caused by the different definitions used for structural damage (buckling, complete collapse, rupture of connected pipes, etc.). The second origin of uncertainty is that the design features of target equipment are usually not considered by the original references reporting the damage thresholds listed in Table 9.1. Thus, in the definition of damage threshold values, two important points should be considered: threshold values (1) should refer to a specific damage definition and (2) should take into account the specific structural resistance (i.e. the design) of the equipment.

Nevertheless, the framework for the application of damage thresholds only allows considering limited details on equipment features and expected damage. Broad equipment categories having similar structural characteristics may be considered. Salzano and Cozzani ((2000c) introduced four equipment categories (atmospheric vessels, pressurized vessels, elongated equipment, and auxiliary equipment) for the assessment of damage thresholds and equipment vulnerability models.

The concept of DSs was then introduced to obtain a less ambiguous definition of equipment damage for the categories of equipment considered in the analysis. A DS is a well-defined type of damage experienced by an equipment item. DSs are a widely applied concept for damage classification following, e.g. internal failures (e.g. see the "Purple Book" (Uijt de Haag and Ale, 1999)), blast wave damage (Tam and Corr, 2000; Schneider, 1998; Whitney et al., 1992), natural events (e.g. see the HAZUS guidelines (HAZUS, 1997)) and damages to vessels in the transportation of hazardous substances (ACDS, 1991 and CCPS, 1995 guidelines). In several studies referred to the quantitative assessment of domino effect as well as to equipment damage in accidents caused by natural events, Salzano and Cozzani have introduced the following three DSs (Cozzani et al., 2006, 2007; Fabbrocino et al., 2005):

- DS1: minor damage to the structure, to the piping connections or to the auxiliary equipment;
- DS2: relevant structural damage, associated to an intense loss of containment;
- DS3: intense or catastrophic damage, or even the total collapse of structure.

The definition of a limited number of DSs and of categories suitable to classify equipment items having a structural similarity allowed a significant progress in the analysis of the data on equipment damage due to blast waves and heat radiation.

Several studies (Cozzani and Salzano, 2004a, 2004b; Cozzani et al., 2006, 2007) evidenced that this approach results in a significant correlation of available literature data on equipment damage, that allowed the development of the vulnerability models for equipment damage described in Chapters 4 and 5.

9.3.2 Loss Intensity and Damage State

As discussed above, in the analysis of the escalation process it is convenient to assess the extent of the structural (mechanical) damage of the target equipment items defining a set of DSs. However, it is evident from the above definitions that different intensities of loss of containment may be associated to the different DS categories. Linking a well-defined loss of containment intensity (LI state) to a DS category may allow the quantitative assessment of the source term in the consequence analysis of the secondary scenario. Thus, the definition of LI classes and their association to DS categories is a crucial element in the framework of a quantitative risk assessment of domino scenarios. It should be remarked that the definition of a set of LI classes is a common practice in quantitative risk assessment. The "Purple Book" (Uijt de Haag and Ale, 1999) suggests considering at least three different LI classes in the QRA of a storage or process vessel: a release from a small diameter pipe (10 mm equivalent diameter), a release of the entire vessel inventory in 10 min, and the "instantaneous" release of the entire vessel content. A similar approach is adopted to define the source term in the analysis of release scenarios due to accidents in the transportation of hazardous substances (ACDS, 1991; CCPS, 1995).

This approach was introduced in the analysis of domino effect by Salzano and Cozzani (2005, 2006) that defined three LI classes based on the suggestions of the "Purple Book" (Uijt de Haag and Ale, 1999) and associated them to three DSs:

- LI1: release from a 10 mm average release diameter, associated to DS1 (minor damage to the structure, to piping connections or to the auxiliary equipment);
- LI2: release of the entire vessel inventory in 10 min, associated to DS2 (relevant structural damage);
- LI3: instantaneous release of the entire vessel content, associated to DS3 (intense or catastrophic damage, or even the total collapse of structure).

Clearly enough, the definition of the above LI classes is functional to the quantitative assessment of the consequences of the secondary scenario. Thus, different definition of LI classes may be introduced (e.g. LI2 and LI3 may be merged in a single LI class if nonevaporating flammable liquids are considered, since limited differences are present in the final outcomes of the releases). However, caution should be used in particular when a higher number of LI classes are proposed, since a more detailed analysis may be justified only if actual damage data or results of structural modeling became available.

9.3.3 The Concept of Damage Threshold

In the scientific and technical literature, a damage threshold is interpreted as a given intensity of a physical effect (blast wave intensity, radiation intensity, etc.) below which the damage of any equipment item is considered not possible.

As discussed in Section 9.1, damage thresholds are widely used in the framework of domino effect analysis, often not considering the specific features of the primary scenario that may trigger the escalation process, as well as the features of the target equipment.

From the above discussion it is clear that different damage thresholds should be defined for different equipment categories and different primary scenarios. Furthermore, specific damage thresholds may be introduced for each of the damage categories considered in the analysis.

Clearly enough, if different DSs are considered, the damage threshold for a specific equipment item or an equipment category may be considered as the minimum value of the intensity of physical effects needed to obtain any of the DSs considered. The values of the damage thresholds suggested in the literature for different types of primary scenarios will be discussed in Section 9.4.

However, it may be worth to remark that although the concept of damage threshold is widely used and understood, it may not be the more suitable for the assessment of escalation hazard. Actually, the same structural damage may result in secondary scenarios having a widely different severity, depending on the characteristics of the substance released. As discussed in Chapter 3, structural damage in the absence of a relevant secondary scenario may not produce an escalation of the primary event that is the key feature of a domino accident. Thus, a less conservative approach aimed at the specific assessment of the escalation potential of a primary scenario may be based on a different concept than damage thresholds.

9.3.4 The Concept of Escalation Threshold

In the framework of domino effect assessment, thresholds for structural damage may not be correspondent to the threshold values related to a "relevant escalation" of accident scenarios. Indeed, the possibility of escalation following the damage is dependent also on the intensity of loss of containment. Increasing LIs usually result in an increase of the severity of the secondary scenario and in a decrease of the time available for successful mitigation. However, beside the LI, the hazard posed by the substance released from the damaged equipment item also plays an important role. In particular, if the same LI is considered, toxic substances may cause more severe scenarios than flammable substances in the case of volatile releases. On the other hand, in the case of nonvolatile releases, flammable substances may cause more severe hazards than toxic substances.

Thus, an escalation threshold may be defined as a given intensity of a physical effect (blast wave intensity, radiation intensity, etc.) below which, although structural damage of the secondary equipment of concern is possible, a relevant escalation is deemed not credible. As stated in Section 9.2, a "relevant escalation" should be interpreted as a secondary scenario having the potential to provide a significant contribution to the overall consequences of the domino event.

As an example, Table 9.3 shows the expected secondary scenarios and the estimated escalation potential for different LIs and DSs. It is clear from the table that in the case of flammable materials the possibility of escalation following

Table 9.3 Analysis of Escalation Potential Following Equipment Damage Classified by Equipment Category and Damage State: Expected Secondary Scenarios and Severity Ranking

Damage State	Loss Intensity	Substance Hazard (Target Unit)	Equipment Category (Target unit)			
			Atmospheric	Pressurized	Elongated	Auxiliary
DS1	LI1	Flammable	*Pool fire (minor)*	**Jet fire**	*Pool fire (minor); flash fire (minor)*	*Pool fire (minor); flash fire (minor)*
		Toxic	*Toxic dispersion (minor)*	**Toxic dispersion**	**Toxic dispersion**	*Toxic dispersion (minor)*
DS2	LI2	Flammable	**Pool fire; flash fire; VCE**	**Jet fire; flash fire; VCE**	**Pool fire; flash fire; VCE**	*Pool fire (minor); flash fire (minor)*
		Toxic	**Toxic dispersion**	**Toxic dispersion**	**Toxic dispersion**	*Toxic dispersion (minor)*
DS3	LI3	Flammable	**Pool fire; flash fire; VCE**	**BLEVE/fireball; flash fire; VCE**	**Pool fire; flash fire; VCE**	*Pool fire (minor); flash fire (minor)*
		Toxic	**Toxic dispersion**	**Toxic dispersion**	**Toxic dispersion**	**Toxic dispersion**

VCE, vapor cloud explosion; BLEVE, boiling liquid evaporating vapor explosion; high severity, bold; low severity, italic.

a blast wave is credible in the case of LI1 state only for pressurized equipment, while an escalation involving an atmospheric or elongated vessel requires at least an LI2 loss. On the other hand, when toxic materials are concerned, LI1 seems a credible cause of escalation also for elongated vessels (due to the possible higher temperatures of the release, e.g. in distillation processes). This approach was used to estimate more detailed threshold values for escalation that will be discussed in Section 9.4.

9.4   Damage and Escalation Thresholds

9.4.1   Values for Damage and Escalation Threshold

Table 9.1 reports a list of damage thresholds collected in the technical and scientific literature. The values in the table derive from a number of different studies, and are based on different and nonhomogeneous approaches. Thus, the reader is addressed to the original reference in order to understand the origin and the context of application of the proposed threshold value.

A more systematic approach to the issue, based on the above defined concepts of damage and escalation thresholds, was undertaken by Cozzani et al. (2006, 2007). A wide number of representative case studies were defined, in order to assess the possibility of escalation of different scenarios. A sensitivity analysis of all factors affecting the escalation possibility was also performed, in order to assess critical values for the different parameters. The results obtained by this approach are discussed in detail in the following.

9.4.2   Damage and Escalation Thresholds for Fire Scenarios

As mentioned earlier, most of the available criteria for accident escalation due to radiation only define a threshold for radiation intensity, neglecting or making implicit assumptions for other parameters (Tan, 1967; Bagster and Pitblado, 1991; Van Den Bosh et al., 1989; Khan and Abbasi, 1998; Salzano et al., 2003). A conventional lumped-parameters model was used by Cozzani et al. (2006) to simulate the behavior of a number of reference vessels exposed to different fire scenarios. Different equipment designs and sizes were considered in the simulations, while the effect of thermal protections or active mitigation systems was neglected. They found a much lower resistance of atmospheric vessels if compared to pressurized vessels. In particular, the ttf of any atmospheric vessel considered was higher than 10 min for radiation intensities lower than 15 kW/m^2 and was higher than 30 min for radiation intensities lower than 10 kW/m^2. In the case of pressurized vessels, the ttf resulted slightly dependent on the design pressure (less than a factor 2 in the reference vessels considered in the study). Nevertheless, in the range of the design pressures considered (1.5–2.5 MPa), the ttf resulted higher than 10 min for a radiation intensity of 60 kW/m^2, and higher than 30 min for a radiation intensity lower than 45 kW/m^2. These results confirmed that escalation caused by vessel wall heating due to radiation from stationary fires is possible even in the absence of flame impingement or engulfment. Since no thermal protection was considered, the results reported above should be considered as rather conservative, in particular for pressurized vessels, where usually both thermal protections as fireproofing materials and active mitigation systems as water deluges are present.

Further considerations can be drawn considering the characteristics of specific accident scenarios. In the case of jet fire impingement or pool fire engulfment, radiation intensities on exposed vessels usually exceed by far the threshold values given above, thus escalation should always be considered possible. In the case of transient flames, as discussed in Section 9.2, the time evolution of the scenario becomes extremely relevant. Flash fires are not likely to result in the damage of a secondary vessel due to heat radiation. Nevertheless, an escalation may be caused by secondary flames due to the direct ignition of flammable material in the flame envelope (e.g. ignition of vapors above the roof of a floating roof tank). Similar considerations apply limitedly to the thermal effects from the explosive combustion of a cloud of flammable material (VCE). In the case of fireball scenarios, more controversial results are obtained. While an escalation caused by fireball radiation seems unlikely for pressurized equipment, in the case of atmospheric vessels, the ttf resulted of the same order of magnitude of the duration of the fireball (Cozzani et al., 2006). This, together with the potential for ignition of flammable vapors, indicates that an escalation potential may be considered for atmospheric equipment involved in the flame zone of a fireball.

Therefore, on the basis of the discussion in Section 9.3, the results obtained allowed the definition of the damage and escalation thresholds summarized in Table 9.4.

9.4.3 Damage and Escalation Thresholds for Blast Wave Scenarios

The study by Cozzani et al. (2006) proposed threshold values for equipment damage and escalation based on past accident data. The threshold values were obtained according to the DS and LI approach described in Section 9.3 (Cozzani and Salzano, 2004c; Salzano and Cozzani, 2006).

When damage is considered, the analysis of available data for the four categories of equipment considered provided the following damage thresholds (Cozzani and Salzano, 2004c): (1) 7 kPa for atmospheric equipment; (2) 20 kPa for pressurized equipment; (3) 14 kPa for elongated equipment; and (4) 12 kPa for auxiliary equipment.

If escalation is considered, the analysis of potential secondary scenarios is also required (Section 9.3.4). In this case, threshold values of 22 and 20 kPa were obtained, respectively, for atmospheric and pressurized equipment. The values are a consequence of the higher LI required to have relevant escalation scenarios when atmospheric vessels are damaged. For elongated vessels, two different thresholds were identified for flammable and toxic hazards, again on the basis of the different relevance of secondary scenarios. In the case of a flammable release, a threshold of 31 kPa may be assumed, while for a toxic release a threshold of 20 kPa was obtained.

Table 9.4 Damage and Escalation Thresholds

Scenario	Escalation Vector	Modality	Target Category	Damage Threshold	Escalation Threshold
Flash fire	Heat radiation	Fire impingement	Floating roof tanks	Damage unlikely	Flame envelope
			All other units	Damage unlikely	Escalation unlikely
Fireball	Heat radiation	Flame engulfment	Atmospheric	$I > 100\ \text{kW/m}^2$	$I > 100\ \text{kW/m}^2$
			Pressurized	Damage unlikely	Escalation unlikely
		Distant radiation	Atmospheric	$I > 100\ \text{kW/m}^2$	$I > 100\ \text{kW/m}^2$
			Pressurized	Damage unlikely	Escalation unlikely
Jet fire	Heat radiation	Fire impingement	All	Flame envelope	Flame envelope
		Stationary radiation	Atmospheric	$I > 15\ \text{kW/m}^2$	$I > 15\ \text{kW/m}^2$
			Pressurized	$I > 45\ \text{kW/m}^2$	$I > 45\ \text{kW/m}^2$
Pool fire	Heat radiation	Flame engulfment	All	Flame envelope	Flame envelope
		Stationary radiation	Atmospheric	$I > 15\ \text{kW/m}^2$	$I > 15\ \text{kW/m}^2$
			Pressurized	$I > 45\ \text{kW/m}^2$	$I > 45\ \text{kW/m}^2$
VCE	Overpressure	Blast wave interaction	Atmospheric	$P > 7\ \text{kPa}$	$P > 22\ \text{kPa}$
			Pressurized	$P > 20\ \text{kPa}$	$P > 20\ \text{kPa}$
			Elongated (toxic)	$P > 14\ \text{kPa}$	$P > 20\ \text{kPa}$
			Elongated (flammable)	$P > 14\ \text{kPa}$	$P > 31\ \text{kPa}$
Confined explosion	Heat radiation	Fire impingement	See flash fire	See flash fire	See flash fire
	Overpressure	Blast wave interaction	Atmospheric	$P > 7\ \text{kPa}$	$P > 22\ \text{kPa}$
			Pressurized	$P > 20\ \text{kPa}$	$P > 20\ \text{kPa}$
			Elongated (toxic)	$P > 14\ \text{kPa}$	$P > 20\ \text{kPa}$
			Elongated (flammable)	$P > 14\ \text{kPa}$	$P > 31\ \text{kPa}$

Mechanical explosion	Overpressure	Blast wave interaction	Atmospheric	$P > 7$ kPa	$P > 22$ kPa
			Pressurized	$P > 20$ kPa	$P > 20$ kPa
			Elongated (toxic)	$P > 14$ kPa	$P > 20$ kPa
			Elongated (flammable)	$P > 14$ kPa	$P > 31$ kPa
	Missile projection		All	Fragment impact	Fragment impact
BLEVE	Overpressure	Blast wave interaction	Atmospheric	$P > 7$ kPa	$P > 22$ kPa
			Pressurized	$P > 20$ kPa	$P > 20$ kPa
			Elongated (toxic)	$P > 14$ kPa	$P > 20$ kPa
			Elongated (flammable)	$P > 14$ kPa	$P > 31$ kPa
	Missile projection		All	Fragment impact	Fragment impact
Point-source explosion	Overpressure	Blast wave interaction	Atmospheric	$P > 7$ kPa	$P > 22$ kPa
			Pressurized	$P > 20$ kPa	$P > 20$ kPa
			Elongated (toxic)	$P > 14$ kPa	$P > 20$ kPa
			Elongated (flammable)	$P > 14$ kPa	$P > 31$ kPa

I, heat radiation intensity; P, maximum peak static overpressure.

Further details on the derivation of the threshold values quoted above are reported elsewhere (Cozzani et al., 2006; Salzano and Cozzani, 2006). Table 9.4 summarizes the damage and escalation thresholds obtained and reports further details concerning the specific escalation features of the different scenarios that may generate a blast wave.

Nevertheless, it should be remarked that the damage in the near field is affected by specific and complex phenomena and a generalization based on simple overpressure thresholds is not possible (Cozzani et al., 2006). In the far field, the identified escalation threshold may be obtained only for relatively intense VCEs (explosion strength factor $F \geq 5$) in the Multi-Energy method (Van den Berg, 1985) or flame Mach number $M_f \geq 0.29$ in the Baker–Strehlow method (Strehlow et al., 1979; Tang and Baker, 1999). Hence, any slow, subsonic deflagration, independently on the total energy of explosion and on the target equipment, may be excluded as a credible cause of escalation in the far field.

Other types of explosion phenomena, as mechanical explosions, BLEVEs, and confined/vented explosions (Ferrara et al., 2008), are typically characterized by the production of shock waves, i.e. a pressure history with an initial, abrupt rise of pressure. These waves can cause structural damage only in the very near field but are generally negligible in the far field.

9.4.4 Fragment Projection

Damage and escalation caused by fragment impact require two conditions to take place: the distance of the target vessel must be lower than the maximum credible projection distance and the impact must be followed by a loss of containment at the target vessel. The latter requirement is usually assumed to be verified in a conservative approach to the assessment of missile damage (Gubinelli et al., 2004; Mannan, 2005). Thus, if the target vessel is within the circle with a radius equal to the maximum fragment projection distance, the escalation should be considered possible. For most accident scenarios resulting in fragment projection in the process industry, the maximum calculated projection distance of fragments is usually higher than 1000 m (Gledhill and Lines, 1998; Gubinelli et al., 2004). Even if this theoretical value might be overestimated, projection distances up to 900 m were observed in past accidents involving storage vessel commonly used in the process industry (Holden and Reeves, 1985). Therefore, a probabilistic approach might be introduced to assess the credibility of escalation events as a function of distance from the primary vessel that undergoes the fragmentation. If credible ranges of initial fragment velocities and uniform probability distributions are assumed for horizontal projection angles and fragment drag factors, the average probability to hit a secondary target may be calculated (Gubinelli et al., 2004). In several literature studies the estimated impact probabilities are shown to be generally lower than 3×10^{-1} at 100 m and 5×10^{-2} at 300 m (Cozzani et al., 2006). More specific estimates should take into account the target geometry and the specific range of explosion energy. Thus, a probability cutoff value may be considered to define a damage and escalation threshold. This issue is discussed in detail in Chapter 6.

9.5   Conclusions

In the assessment of domino hazard it is crucial to define thresholds for the physical effects of the primary scenarios above which the escalation should be considered credible. Although a strong uncertainty exists in the literature when domino thresholds are considered, the introduction of DS and LI classes for a number of broad equipment categories allows obtaining more significant results.

A further distinction may be significant, defining and assessing separately threshold values for damage and escalation. Suggested threshold values derived from recent studies are reported in this chapter.

The threshold values reported in this chapter may be used to define safety distances for escalation. This issue will be discussed in detail in Chapter 1.

References

ACDS, 1991. Major Hazard Aspect of the Transport of Dangerous Substances. Advisory Committee on Dangerous Substances. Health and Safety Commission, London (UK).

API RP 510, 1990. Pressure Vessel Inspection Code: Maintenance, Inspection, Rating, Repair, Alteration. American Petroleum Institute, Washington D.C. (USA).

Bagster, D.F., Pitblado, R.M., 1991. The estimation of domino incident frequencies: an approach. Process Safety and Environmental Protection 69, 196.

Baker, W.E., Cox, P.A., Westine, P.S., Kulesz, J.J., Strehlow, R.A., 1983. Explosion Hazards and Evaluation. Elsevier, Amsterdam (NL).

Barton, R.F., 1995. Fuel gas explosion guidelines – practical application. In: Proc. IChemE Symposium Series No.139. IChemE, Rugby, UK, pp. 285–286.

Biggs, J.M., 1964. Introduction to Structural Dynamics. McGraw Hill, London (UK).

Bottelberghs, P.H., Ale, B.J.M., 1996. Consideration of domino effects in the implementation of the Seveso II directive in the Netherlands. In: Proc. European Seminar on Domino Effects, Leuven (B), pp. 45–58.

Brasie, W.C., Simpson, D.W., 1968. Guidelines for estimating damage from chemical explosions. In: Proc. 2nd Loss Prevention Symposium, p. 91.

BS 5908, 1990. Code of Practice for Fire Precautions in Chemical Plant. British Standards Institution, London, (UK).

CCPS, Center for Chemical Process Safety, 1995. Guidelines for Chemical Transportation Risk Analysis. American Institute of Chemical Engineers, New York (USA).

CCPS, Center for Chemical Process Safety, 2000. Evaluating Process Safety in the Chemical Industry: A User's Guide to Quantitative Risk Analysis. American Institute of Chemical Engineers, New York (USA).

Clancey, V.J., 1972. Diagnostic features of explosion damage. In: Proc. 6th Int. Meeting of Forensic Sciences, Edinburgh.

Cozzani, V., Salzano, E., 2004a. The quantitative assessment of domino effects caused by overpressure. Part I: Probit Models. Journal of Hazardous Materials 107, 67–80.

Cozzani, V., Salzano, E., 2004b. The quantitative assessment of domino effects caused by overpressure. Part II: Case studies. Journal of Hazardous Materials 107, 81–94.

Cozzani, V., Salzano, E., 2004c. Threshold values for domino effects caused by blast wave interaction with process equipment. Journal of Loss Prevention in the Process Industries 17, 437–447.

Cozzani, V., Gubinelli, G., Salzano, E., 2006. Escalation thresholds in the assessment of domino accidental events. Journal of Hazardous Materials 129, 1–21.

Cozzani, V., Tugnoli, A., Salzano, E., 2007. Prevention of domino effect: from active and passive strategies to inherently safe design. Journal of Hazardous Materials 139, 209–219.

DM 151/2001, Decreto Ministeriale 9/5/2001 n.151, Official Journal of Italian Republic (Gazzetta Ufficiale), Rome (I).

Fabbrocino, G., Iervolino, I., Orlando, F., Salzano, E., 2005. Quantitative risk analysis of oil storage facilities in seismic areas. Journal of Hazardous Materials 123, 61–69.

Ferrara, F., Willacy, S.K., Phylaktou, H.N., Andrews, G.E., Di Benedetto, A., Salzano, E., Russo, G., 2008. Venting of gas explosion through relief ducts: interaction between internal and external explosions. Journal of Hazardous Materials 155, 358–368.

Glasstone, S., 1980. The Effects of the Nuclear Weapons. Atom. Energy Comm., Washington, DC (USA).

Gledhill, J., Lines, I., 1998. Development of Methods to Assess the Significance of Domino Effects from Major Hazard Sites. CR Report 183. Health and Safety Executive, London, UK.

Gugan, K., 1979. Unconfined Vapour Cloud Explosions. IChemE, Rugby (UK).

Gubinelli, G., Zanelli, S., Cozzani, V., 2004. A simplified model for the assessment of the impact probability of fragments. Journal of Hazardous Materials 116, 175–187.

Gubinelli, G., Cozzani, V., 2009a. Assessment of missile hazards: identification of reference fragmentation patterns. Journal of Hazardous Materials 163, 1008–1018.

Gubinelli, G., Cozzani, V., 2009b. The assessment of missile hazard: evaluation of fragment number and drag factors. Journal of Hazardous Materials 161, 439–449.

Hauptmanns, U., 2001. A procedure for analyzing the flight of missiles from explosions of cylindrical vessels. Journal of Loss Prevention Process Industries 14, 395–402.

HAZUS, 1997. Earthquake Loss Estimation Methodology. National Institute of Building Sciences, Risk Management Solutions, Menlo Park, California, USA.

Holden, P.L., Reeves, A.B., 1985. Fragment Hazards from Failures of Pressurised Liquefied Gas Vessels. IChemE Symposium Series, 93, p. 205.

HSE, Health and Safety Executive, 1978. Canvey: An Investigation of Potential Hazards from Operations in the Canvey Island/Thurrock Area. HM Stationary Office, London, (UK).

HSE, Health and Safety Executive, 2004. Analysis and Design of Profiled Blast Walls. Research Report 146, HM Stationary Office, London, UK.

Khan, F.I., Abbasi, S.A., 1998. Models for domino effect analysis in chemical process industries. Process Safety Progress 17, 107.

Kletz, T.A., 1980. Plant layout and location: method for taking hazardous occurrences into account. Journal of Loss Prevention in the Process Industries 13, 147.

Lees, F.P., 1996. Loss Prevention in the Process Industries, second ed. Butterworth-Heinemann, Oxford (UK).

Mannan, S., 2005. Lees' Loss Prevention in the Process Industries, third ed. Elsevier, Amsterdam (NL).

Mecklenburgh, J.C., 1985. Process Plant Layout. George Goodwin, London (UK).

Salzano, E., Picozzi, B., Vaccaro, S., Ciambelli, P., 2003. Hazard of pressurized tanks involved in fires. Industrial Engineering Chemistry Research 42, 1804.

Salzano, E., Cozzani, V., 2006. A fuzzy set analysis to estimate loss of containment relevance following blast wave interaction with process equipment. Journal of Loss Prevention in the Process Industries 19, 343–352.

Salzano, E., Cozzani, V., 2005. The analysis of domino accidents triggered by vapor cloud explosions. Reliability Engineering and System Safety 90, 271–284.

Schneider, P., 1998. Predicting damage of slender cylindrical steel shells under pressure wave load. Journal of Loss Prevention in the Process Industries 11, 223–228.

Strehlow, R.A., Luckritz, R.T., Adamczyk, A.A., Shimp, S.A., 1979. The blast wave generated by spherical flames. Combustion and Flame 35, 297.

Tam, V.H.Y., Corr, B., 2000. Development of a limit state approach for design against gas explosions. Journal of Loss Prevention in the Process Industries 13, 443.

Tan, S.H., 1967. Flare system design simplified. Hydrocarbon Processing 46, 172.

Tang, M.J., Baker, Q.A., 1999. A new set of blast curves from vapour cloud explosion. Process Safety Progress 18, 235.

Uijt de Haag, P.A.M., Ale, B.J.M., 1999. Guidelines for Quantitative Risk Assessment (Purple Book). Committee for the Prevention of Disasters, The Hague (NL).

UKOOA, 2002. Fire and Explosion Guidance, Part 1: Avoidance and Mitigation of Explosions. UK Offshore Operators Association Limited, London (UK).

Van den Berg, A.C., 1985. The multi-energy method – a framework for vapor cloud explosion blast prediction. Journal of Hazardous Materials 12, 1.

Van Den Bosh, C.J.H., Merx, W.P.M., Jansen, C.M.A., De Weger, D., Reuzel, P.G.J., Leeuwen, D.V., Blom-Bruggerman, J.M., 1989. Methods for the Calculation of Possible Damage (Green Book). Committee for the Prevention of Disasters, The Hague (NL).

Van Den Bosh, C.J.H., Weterings, R.A.P.M., 1997. Methods for the Calculation of Physical Effects (Yellow Book). Committee for the Prevention of Disasters, The Hague (NL).

Wells, G.L., 1980. Safety in Process Plant Design. Wiley, Chichester (UK).

Whitney, M.G., Barker, D.D., Spivey, K.H., 1992. Ultimate capacity of blast loaded structures common to chemical plants. Plant/Operations Progress 11, 205–212.

10 Quantitative Assessment of Risk Caused by Domino Accidents

Valerio Cozzani*, Giacomo Antonioni*, Nima Khakzad†, Faisal Khan†, Jerome Taveau‡, Genserik Reniers§

* LISES, Dipartimento di Ingegneria Civile, Chimica, Ambientale e dei Materiali, Alma Mater Studiorum – Università di Bologna, Bologna, Italy, † Safety and Risk Engineering Research Group, Faculty of Engineering and Applied Science, Memorial University of Newfoundland, St. John's, Newfoundland, Canada, ‡ Scientific advisor at Fike Corporation, Industrial Explosion Protection Group, Blue Springs, USA, § Centre for Economics and Corporate Sustainability (CEDON), HUB, KULeuven, Brussels, Belgium

Chapter Outline

10.1 Introduction

The relevance of domino hazard was recognized since the early times of process safety. In the 1980s, three milestone projects were promoted in Europe to answer public concerns about the risks posed by large concentrations of hazardous facilities: the Canvey Island (United Kingdom) (HSE, 1978, 1981), Rijnmond (The Netherlands) (COVO, 1982), and Ravenna (Italy) (Egidi et al., 1995) quantitative area

Domino Effects in the Process Industries. http://dx.doi.org/10.1016/B978-0-444-54323-3.00010-5

risk analysis studies. The results of these studies explicitly recognized the hazard due to domino effects, in particular in highly congested areas where several industrial facilities are present.

In 1982, the first European Directive aimed at the control of major accident hazard already required the identification and assessment of domino hazards (Directive 82/501/EEC). Efforts to develop quantitative methodologies for the assessment of domino accidents are documented since the early 1990s. Bagster and Pitblado (1991) described an approach for the inclusion of domino events in risk assessment, by considering it as an external event in a fault tree and increasing the frequencies of corresponding incidents. Several other studies followed after 1995, mainly concerned with methodologies for domino assessment (Contini et al., 1996; Gledhill and Lines, 1998) and focusing on specific issues such as escalation frequency assessment (Pettitt et al., 1993) and escalation triggered by fires (Latha et al., 1992; Morris et al., 1994).

A relevant step forward toward the assessment of domino accident scenarios was made in the late 1990s, with the studies of Khan and Abbasi (Khan and Abbasi, 1998a,b) and that of Delvosalle and coworkers (Delvosalle, 1998). However, the limited computational resources available at the time still made complex the application of the methodology to real industrial facilities. More recently, a relevant - research effort was devoted to develop methodologies and tools aimed at encompassing the quantitative analysis of domino scenarios within quantitative risk assessment (QRA) methodologies (Antonioni et al., 2009; Cozzani et al., 2005, 2006).

QRA is nowadays the tool more widely applied to provide quantitative information on the risk caused by conventional accidents in chemical and process plants. Despite the obvious fact that QRA is not an exact description of reality, it is the best available, analytic predictive tool to date to assess the risks of complex process and storage facilities. QRA consists of a set of methodologies for estimating the risk posed by a given system in terms of human loss or, in some cases, economic loss (CCPS, 2000; Mannan, 2005; Uijt de Haag and Ale, 1999). In the risk assessment of industrial facilities, QRA is aimed at the calculation of risk expressed by risk indexes such as (local specific) individual risk (LSIR), individual risk per annum, and societal risk (expressed by Frequency-Fatality (F-N) plots, individual risk-number of persons exposed (I-N) hystograms, potential life loss (PLL) and/or expectation value indicators) (Carter and Hirst, 2000; Egidi et al., 1995; Uijt de Haag and Ale, 1999).

The general approach of QRAs is unchanged since its origin in the early 1980s. QRAs have been refined, are nowadays more sophisticated and more detailed, and are based on better and more accurate models. The continuous optimization of the software programs and models (e.g. dispersion calculations, consequence analysis, etc.) goes hand in hand with the advances in computer technology and the continuous improvements of calculation power with computers. The use of QRA to study the risks posed by hazardous facilities has been facilitated by the increase of computer power in a way that now thousands of scenarios can be treated in a reasonable time. QRA techniques are now widely used to assess risks posed by installations extracting, processing and/or storing hazardous substances, and QRA has become an essential tool for the development, continued operation and expansion of process installations that meet society's growing safety expectations.

Although QRA is a mature and consolidated tool, its application to domino effect to date has been limited. Despite their importance, domino effects are usually excluded from QRAs. Actually, important computational resources are required for the quantitative assessment of domino effects that were not available or not easy to obtain until recent years. Moreover, layout information and meteorological conditions (e.g. wind direction) play an extremely important role in determining the escalation probability. Thus, a much more extended use of geographical information is required than in conventional QRA applications.

In recent years, methods and models become available to allow the quantitative assessment of domino accidents in a QRA framework (Cozzani et al., 2005), supported by specific software tools based on geographic information systems (Cozzani et al., 2006). Further progress in the field also explored alternative routes, based, e.g. on Bayesian analysis and Monte Carlo methods, to assess the frequency of escalation scenarios. In the following discussion, these up-to-date approaches to the quantitative assessment of domino effect will be presented, starting from the tools dedicated to allow the quantitative assessment of domino effects within conventional QRA procedures.

10.2   Quantitative Risk Assessment of Domino Accidents

10.2.1   Procedure for the Quantitative Assessment of Domino Effect within a QRA

The quantitative assessment of industrial risk caused by domino scenarios may be effectively carried out starting from a conventional QRA of the facilities considered. The assessment of the contribution of domino scenarios may be interpreted as an additional activity to complete the risk frame obtained for the site of interest. Table 10.1 shows the steps needed to carry out the quantitative assessment of domino scenarios (Antonioni et al., 2009; Cozzani et al., 2005), referencing the four classical steps of QRA (identification, frequency assessment, consequence assessment and risk calculation/recomposition) (CCPS, 2000; Uijt de Haag and Ale, 1999). The table also shows the tools needed to carry out each step of the assessment. The tools underlined in the table are specific for the assessment of domino scenarios. As evident from the table, three main categories of tools are required for the quantitative assessment of domino scenarios:

1. Threshold values for the identification of potential targets of escalation
2. Equipment damage models
3. Specific tools and procedures for the assessment of frequency and consequences of the overall domino scenarios

Threshold values were discussed in Chapter 9. Probabilistic models for equipment damage due to escalation vectors were discussed in Chapters 4, 5 and 6. The theory and the tools for the overall assessment of domino scenarios are discussed in the following.

Figure 10.1 shows a flowchart of the procedure required for the quantitative assessment of risk due to domino scenarios. Although the procedure was originally developed by Cozzani et al. (2005), it has a general value since it provides the basis for the quantitative assessment of domino effects independent of the specific tools

Table 10.1 Steps and Tools Required for the Quantitative Assessment of Risk Caused by Domino Scenarios with Reference to the Conventional Steps of Quantitative Risk Assessment.

Stage	Step	Tools Needed
1. Risk identification	**1.1** Escalation assessment of the primary event considered	Consequence analysis models
	1.2 Identification of possible target units	Threshold values
	1.3 Identification of each secondary scenario that may follow the damage of each target unit	
2. Frequency calculation	**2.1** Estimation of the damage probability for each target unit	Equipment damage probability models
	2.2 Identification of each possible combination of secondary scenarios	Specific software tools
	2.3 Conditional probability calculation for each possible combination of secondary scenarios	Specific software tools
3. Consequence assessment	**3.1** Consequence assessment of the primary scenario	Consequence analysis models
	3.2 Consequence assessment of each secondary scenario considered	Consequence analysis models
4. Risk recomposition	**4.1** Calculation of vulnerability maps for each possible combination of scenarios	Human vulnerability models
	4.2 Calculation of individual risk for each possible combination of scenarios and for each primary event considered	Risk recomposition software
	4.3 Calculation of overall risk indexes	

Underlined: models or tools specific for domino effect quantitative assessment.
Source: Adapted from Antonioni et al. (2009).

used to carry out any step of the assessment. Furthermore, depending on the aim of the study, specific steps of the procedure may be skipped or carried out with a different level of detail. Thus, the flowchart in Figure 10.1 provides the basis for the quantitative assessment of domino effects.

As shown in the figure, the procedure may be iterated (Gate G3 after Step 9) to allow the assessment of secondary or even higher level domino effects. However, as discussed in the following, the complexity and computational resources required for the assessment increase exponentially with the level of the assessment, and usually "first level" scenarios are those providing the main contribution to risk indexes.

10.2.2 Data Required for the Quantitative Assessment of Domino Effect

The data required to apply the procedure defined in Figure 10.1 are discussed in detail in the following. However, it is useful to summarize the main information that is necessary for the quantitative assessment of domino effect.

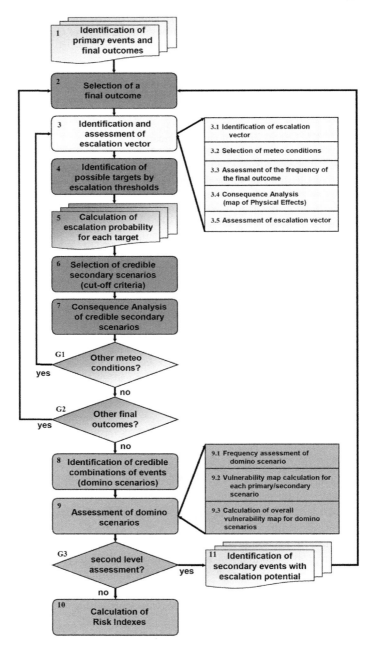

Figure 10.1 Detailed flowchart for the quantitative assessment of risk due to domino scenarios (G: gates). (For color version of this figure, the reader is referred to the online version of this book.)

- A layout of the site examined
- The position on the layout of the risk sources that may generate the primary events of concern
- The full characterization of all the primary events of concern (expected frequencies and consequence analysis)
- The position of all the possible escalation targets of concern (equipment with relevant inventories of hazardous substances, etc.)
- The consequence analysis of the secondary event that is supposed to take place following the damage of the target equipment

The list of top events identified by HazOp analysis may also be required to identify low-severity initiating events (see Chapter 3). It is quite evident that most of the data required are available from a conventional QRA, thus the assessment procedure does not require relevant additional work for data collection. On the other hand, HazOp may be adapted to be used explicitly for investigating the possibility of domino effects, as further discussed in Chapter 13.

10.2.3   Identification of Relevant Primary Events

The starting point of the assessment is the identification of primary events that have a nonnegligible escalation potential (step 1 in Figure 10.1). The event tree technique may be applied to identify the possible final outcomes (e.g. pool fires, vapor cloud explosions, and jet fires) of relevant "top events" identified by conventional HazOp (CCPS, 2008). As an alternative, a specific domino-HazOp technique (see Chapter 13), the "Instrument Domino Effecten" (IDE) methodology (RIVM, 2003), approaches based on reference scenarios (Uijt de Haag and Ale, 1999), or bow–tie diagrams available for the "critical events" identified (Delvosalle et al., 2006) may be applied. These methods also allow the assessment of the expected frequency of each final outcome (Step 3.3 in Figure 10.1).

In a QRA framework, all final outcomes that generate a credible escalation vector should be considered for the assessment. Table 3.3 reports a list of final outcomes and of the escalation vectors generated.

Thus, at the end of Step 1 a list of escalation vectors and final outcomes associated to all relevant primary events should be obtained. The actual relevance of each of the final outcomes identified in Step 1 is assessed in Steps 3 and 4 of the procedure. Conventional models for consequence analysis may be used to calculate a map of the physical effects generated by the final outcome (e.g. radiation intensity due to a jet fire and maximum overpressure due to a blast wave), thus obtaining the intensity of the escalation vector (Step 3 of procedure in Figure 10.1). The comparison between escalation thresholds (see Chapter 9) and the value of the escalation vector at the position of all relevant secondary targets allows the identification of equipment items that may be damaged by the escalation vector, thus starting a secondary scenario. In carrying out this procedure, specific attention should be devoted to scenarios influenced by meteorological conditions (wind direction and speed and category of atmospheric stability), such as pool fires or dispersions leading to a vapor cloud explosion; when such scenarios are considered, the assessment should be repeated considering all different meteorological conditions relevant in the area (Step 3.2 and gate G1), as usually done in a QRA study (Egidi et al., 1995; Spadoni et al., 2000).

Clearly enough, if no secondary target is identified in Step 4 for a given final outcome, this may be disregarded in the following steps of the analysis. Thus, Step 4 is crucial to limit the complexity of the analysis.

10.2.4 Assessment of the Expected Frequencies of Domino Scenarios

Steps 1–4 allow the identification of the relevant final outcomes and the assessment of their expected frequency, defined as f_{pe} in the following discussion (frequency of the primary event in the domino sequence). The expected frequency of a single escalation event (that is, a primary event triggering a secondary accidental scenario) may be calculated as:

$$f_{de} = f_{pe} \cdot P_d \tag{10.1}$$

where f_{de} is the expected frequency of the domino event (events/year), f_{pe} is the expected frequency (events/year) of the primary event (PE) of the domino sequence (obtained considering the frequency of the top event, the probability of the final outcome and the probability of meteorological conditions if relevant) and P_d is the probability of escalation (E) given the primary event:

$$P_d = P(E|PE) \tag{10.2}$$

The expected frequency of the primary event is obtained in Step 3.1 of the procedure (see Figure 10.1) and may be calculated considering the frequency of the critical event, the quantified event trees and the meteorological data usually available in a QRA study or in the safety report of the facility of concern (Egidi et al., 1995; Uijt de Haag and Ale, 1999; Spadoni et al., 2000). The escalation probability should be assessed using specific probabilistic equipment damage models, as those discussed in Chapters 4, 5 and 6, or alternative models proposed in the literature.

The above relations are only valid if it is reasonable to assume that the primary and the secondary event may take place at the same time only due to escalation. This means that the primary and the secondary event should be considered "mutually exclusive" from a probabilistic point of view, unless escalation effects take place. This assumption is justified if the expected frequencies of the primary event and the secondary event not triggered by escalation have sufficiently low values. It must be remarked that in the following, the secondary scenarios will be conservatively defined as simultaneous or "contemporary" to the primary event, even if they will actually always take place in sequence (few seconds to few minutes after the primary event, depending on the primary escalation vector and on the loss intensity at the secondary unit damaged by the primary event).

In a complex layout, usually a single primary event may be able to trigger simultaneously more than one secondary event. This has been documented in several past accidents (see Chapter 2). In this framework, Eqn (10.1) is still valid, yielding the overall probability that a given secondary event is initiated by the primary event considered. However, the frequencies of domino scenarios should be calculated taking into account the possibility of having more than one secondary scenario triggered by the same primary event (Steps 8 and 9 in Figure 10.1).

If the possible further simultaneous escalation of secondary events is neglected, the escalation events may be reasonably considered as independent from a probabilistic point of view. Therefore, if N secondary events are possible, the probability of a secondary scenario given by a generic combination m of k secondary events ($k \leq N$) is the following:

$$P_d^{(k,m)} = \prod_{i=1}^{N}[1 - P_{d,i} + \delta(i, \mathbf{J}_m^k)(2 \cdot P_{d,i} - 1)] \tag{10.3}$$

where $P_{d,i}$ is the probability of escalation for the ith secondary event defined by Eqn (10.2), $\mathbf{J}_m^k = [\gamma_1, ..., \gamma_k]$ is a vector whose elements are the indexes of the mth combination of k secondary events, and the function $\delta(i, \mathbf{J}_m^k)$ is defined as follows:

$$\delta(i, \mathbf{J}_m^k) = \begin{cases} 1 & i \in \mathbf{J}_m^k \\ 0 & i \notin \mathbf{J}_m^k \end{cases} \tag{10.4}$$

The total number of domino scenarios in which the primary event triggers k contemporary secondary events is:

$$\nu_k = \frac{N!}{(N-k)! \cdot k!} \tag{10.5}$$

Thus, the total number of different domino scenarios that may be generated by the primary event is:

$$\nu = \sum_{k=1}^{N} \nu_k = 2^N - 1 \tag{10.6}$$

where ν is the total number of domino scenarios that need to be assessed in the quantitative analysis of domino effects, unless cutoff criteria based on frequency values are applied. Clearly enough, Eqn (10.6) evidences that a very high number of possible combinations are to be generated. In the present framework, where computational resources are constantly increasing, this is no more a problem if a dedicated software tool is available (Antonioni et al., 2009; Cozzani et al., 2006).

The expected frequency of a generic combination m of k events is thus

$$f_{de}^{(k,m)} = f_{pe} \cdot P_d^{(k,m)} \tag{10.7}$$

In the application of the procedure, the (k,m) combination may be neglected if the frequency value falls below a given threshold. This should be decided on the basis of the values of risk that are considered of interest in the analysis.

The total probability that an escalation will take place thus becomes

$$P_e = \sum_{k=1}^{N} \sum_{m=1}^{\nu_k} P_d^{(k,m)} \tag{10.8}$$

The expected frequency of the primary event in the absence of escalation may be calculated as follows:

$$f_{pe,n} = f_{pe} \cdot (1 - P_e) \tag{10.9}$$

Figure 10.2 shows an example of the results obtained in the assessment of domino scenarios for a primary event and five possible secondary targets (Cozzani et al., 2006). As shown in the figure, a clear ranking emerges for the importance of domino sequences from a probabilistic point of view.

The approach presented above may be extended to assess higher level domino events (G3 in Figure 10.1). If a second-level domino effect is considered, the escalation probability, P_d, needs to be assessed for each secondary combination of events (primary scenario + simultaneous secondary scenarios) selected (Step 10 in Figure 10.1). This requires to consider a sum of physical effects (e.g. the overall radiation at a given location from simultaneous fires) or the sum of escalation probabilities (e.g. in the case of explosion followed by fire). Equation (10.7) may be used to assess the frequency of the upper-level domino scenario, where f_{pe} actually becomes the frequency of the domino combination of interest. It may be easily understood that extending this approach to higher level domino effects greatly increases the computational resources required to carry out the calculations.

10.2.5   Assessment of the Consequences of Domino Scenarios

Extremely complex scenarios with multiple simultaneous events may be triggered by the escalation of a primary event leading to a domino sequence. A complete assessment of the consequences of such complicated scenarios is a difficult aim even using advanced tools such as computational fluid dynamics codes, as discussed in Chapter 11. The empirical or integral models used for consequence assessment in a QRA framework are not conceived to model the effects of multiple scenarios (e.g. the overall radiation caused by more than one pool fire). Moreover, synergetic effects are not taken into account in the available models for consequence analysis.

A detailed consequence assessment would require that each scenario should be analyzed with specific tools, taking into account the layout and introducing in the analysis a full geometrical characterization of the problem. However, in a QRA framework, the necessity to limit the computational effort requires to introduce simplifying assumptions in order to carry out the consequence assessment. Thus, accident consequences may be analyzed superimposing the physical effects (radiation, overpressure, toxic gas concentration) separately calculated for each of the primary and secondary final outcomes considered, neglecting the assessment of possible synergetic effects. This approach obviously results in an oversimplification of the problem, allowing only a rough estimate of the actual potential consequences of domino scenarios. Nevertheless, this assumption seems acceptable in a QRA framework.

Therefore, the consequence assessment of the possible domino scenarios may be carried out in three steps: (1) the assessment of the consequences of the primary scenario and of each of the secondary events by conventional models used for consequence

Domino effect

Primary risk source

Name	Coordinates	Top	Scenario
AT2_F_	757167 926718	02 LIQUID FUELS	05 Pool fire

Secondary risk source

	Name	Coordinates	Top	Scenario	Prob.
1	PV1___	757156 926715	01 LPG	03 Fireball	6.06E-02
2	PV3___	757165 926712	01 LPG	03 Fireball	5.64E-01
3	AT1_F_	757159 926720	02 LIQUID FUELS	05 Pool fire	2.27E-01
4	AT3_T_	757131 926728	12 FLUORIDRIC ACID	09 Gaussian dispersion	7.08E-02
5	AT4_T_	757139 926725	12 FLUORIDRIC ACID	09 Gaussian dispersion	3.82E-01

Combinations

1	2	3	4	5	6	7	8	9	10	11	12	13	14	15	16	Prob.
	X															2.35E-01
	X			X												1.45E-01
				X												1.13E-01
	X	X														6.89E-02
		X														5.34E-02
	X	X		X												4.26E-02
		X		X												3.30E-02
	X		X													1.79E-02
X	X															1.52E-02
			X													1.39E-02
X																1.17E-02
	X		X	X												1.11E-02
X	X			X												9.37E-03
			X	X												8.58E-03
X				X												7.26E-03
	X	X	X													5.25E-03
X	X	X														4.44E-03
		X	X													4.06E-03
X		X														3.44E-03
	X	X	X	X												3.24E-03
X	X	X		X												2.75E-03
		X	X	X												2.51E-03
X		X		X												2.13E-03
X	X		X													1.15E-03
X			X													8.94E-04
X	X		X	X												7.14E-04
X			X	X												5.53E-04
X	X	X	X													3.38E-04
X		X	X													2.62E-04
X	X	X	X	X												2.09E-04
X		X	X	X												1.62E-04

Progetto: domino ex-enichem 2 Page 4 of 4

Figure 10.2 Example of probabilities of domino sequences calculated by the procedure presented in Section 10.2.4 (Cozzani et al., 2006).

assessment (CCPS, 2000; Van Den Bosh and Weterings, 1997), (2) the calculation of a "vulnerability map" (Leonelli et al., 1999) for each of the scenarios of concern, and (3) the combination of the "vulnerability maps" of the primary and secondary events involved to yield the overall consequences of the domino scenario of interest.

"Vulnerability maps" (a matrix yielding the death probability due to the event as a function of the position with respect to the source of the event (Leonelli et al., 1999)) are introduced to allow a homogenous assessment of consequences. Vulnerability maps may be obtained for each event from the maps of physical effects of the single final outcomes (calculated in Step 3.1) by the application of probit models. The "probit analysis" (Finney, 1971) is a reference procedure used in QRA to evaluate the dose–effect relation for human responses to toxic substances, thermal radiation and over-pressure (Van Den Bosh et al., 1989). An extended review of probit models for human vulnerability is reported elsewhere (Mannan, 2005; Van Den Bosh et al., 1989).

The consequences of a domino scenario involving multiple simultaneous events may thus be calculated by a combination of vulnerability maps. Considering an individual in a generic position with respect to a domino scenario involving a primary event and n secondary scenarios, the vulnerability (death probability) due to the domino event has the following general expression:

$$V_{de} = \varphi(D_{pe}, D_{d,1}, \ldots D_{d,n}) \tag{10.10}$$

where φ is a function that needs to be defined, D_{pe} is the "dose" due to the phys-ical effects caused by the primary event that triggers the domino scenario and $D_{d,i}$ (i=1 to n) is the dose due to the i-th secondary scenario. A proper definition of function φ should take into account both the effects due to the combination of the physical effects of the contemporary scenarios and the synergetic effects arising from the exposure to physical effects due to multiple scenarios.

If synergetic effects are neglected, V_{de} may be calculated as a combination of the vulnerabilities due to the single scenarios that take place in the domino event. Nevertheless, vulnerabilities are actually probability values, thus requiring the application of probabilistic rules for their combination. The results of an extended study aimed at the analysis of possible alternative combination strategies suggested the adoption of the following relation for the assessment of the combined vulnera-bility map of a complex domino scenario (Antonioni et al., 2004; Cozzani et al., 2005):

$$V_d^{(k,m)} = \min\left[\left(V_p + \sum_{i=1}^{n} \delta(i, \mathbf{J}_m^k) V_{d,i}\right), 1\right] \tag{10.11}$$

This corresponds to the assessment of the sum of the death probabilities due to all the scenarios involved in the domino sequence, with an upper limit of 1.

10.2.6 Risk Calculation (Risk Recomposition)

The final step of the analysis is the recomposition of risk, which is usually aimed at the calculation of risk indexes. The two more significant indexes used to assess

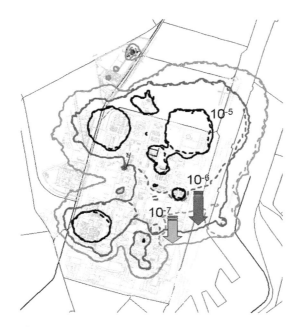

Figure 10.3 Example of individual risk (events/year) maps obtained from the quantitative assessment of domino scenarios carried out by the methodology presented in Section 10.2 (Antonioni et al., 2009).

industrial risk are LSIR and societal risk (expressed by $F–N$ curves) (CCPS, 2000; Uijt de Haag and Ale, 1999). Simplified numerical risk indexes, such as the PLLs, may be obtained from $F–N$ curves (Carter and Hirst, 2000; Cozzani et al., 2005).

Individual risk corresponding to each domino scenario may be calculated as follows:

$$R_{i,\text{de}}^{(k,m)} = f_{\text{de}}^{(k,m)} \cdot V_{\text{de}}^{(k,m)} \tag{10.12}$$

Overall individual risk at a given location may thus be calculated by summing the risk obtained from Eqn (10.12) for each primary event, every relevant outcome, every relevant meteorological condition and every relevant domino sequence.

An example of a risk map obtained from the above procedure using the Aripar-GIS software is provided in Figure 10.3 (Antonioni et al., 2009). If data are also available for the distribution of population in the area affected by the accident, the procedure allows the calculation of societal risk curves. An example is reported in Figure 10.4 (Cozzani et al., 2005). The approach outlined in this section and based on the flow-chart in Figure 10.1 was applied in several pilot studies for the quantitative assessment of domino effect, demonstrating its validity in real life applications. The more significant results were obtained in the analysis of the industrial area of Ravenna (Antonioni et al., 2009).

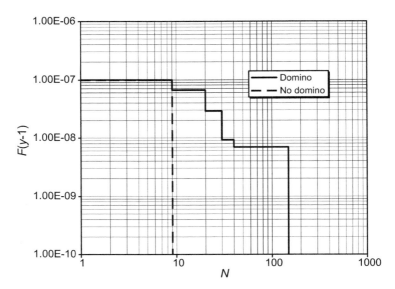

Figure 10.4 Example of a F-N societal risk plot obtained from the quantitative assessment of domino scenarios carried out by the methodology presented in Section 10.2 (Cozzani et al., 2005).

10.3 Domino Scenarios and Escalation Frequencies by Bayesian Analysis

10.3.1 Use of Bayesian Networks for the Identification and Assessment of Domino Scenarios

As an alternative to the frequency assessment procedure presented in Section 10.2.4, Bayesian networks may be applied, taking advantage of their flexible graphical structure to show the sequential order of scenarios. The layout of a chemical or process plant may be considered as a set of variables (i.e. each variable stands for a unit or an equipment item, depending on the level of detail of the analysis). A Bayesian network can thus be used to qualitatively and quantitatively model the influence of units on one another via causal arcs and probability tables, to estimate the probability of domino effect at different levels, and to update initial beliefs as new information becomes available (Khakzad et al., 2013).

A Bayesian network is a directed acyclic graph for reasoning under uncertainty in which the nodes represent variables and are connected by means of directed (causal) arcs. The arcs denote dependencies or causal relationships between the linked nodes, while the conditional probability tables assigned to the nodes determine the type and strength of such dependencies. In a Bayesian network, nodes from which arcs are directed are called parent nodes, whereas nodes to which arcs are directed are called child nodes. In fact, a node can simultaneously be the child of a node and the parent of another node. Nodes with no parent and nodes with no children are called root nodes and leaf nodes, respectively (Jensen and Nielsen, 2007). Using the chain rule and the d-separation criterion, a Bayesian network expands the joint probability distribution of a set of linked nodes, e.g. $U = \{X_1, X_2, \ldots, X_n\}$. In other words, by considering only local dependencies, a Bayesian network factorizes the joint probability distribution as the multiplication of the probabilities of the nodes given their immediate parents. The main application of Bayesian networks is in probability updating. A Bayesian network may take advantage of Bayes theorem to update the probability of variables given new observations. In the following section, the results obtained by Khan and coworkers (Khakzad et al., 2013) concerning the identification and frequency assessment of domino scenarios using Bayesian network is presented.

10.3.2 Identification of Domino Sequences

In order to model the likely propagation pattern of a domino effect, the following steps may be carried out, as schematized in the flowchart reported in Figure 10.5:

Step 1: According to the layout of the process plant of interest, a node is assigned to each process unit. These units (e.g. distillation columns, atmospheric storage tanks, pressurized storage tanks, etc.) are either susceptible to the accident or capable of escalating the accident. In order to discuss an example, a process plant with six units may be assumed (X_i with i ranging from 1 to 6, as shown in Figure 10.6).

Step 2: Using safety reports usually available for process plants or through risk assessment methods, the primary unit where the domino accident is likely to start is determined (e.g. in Figure 10.6). This step is conceptually equivalent to Step 1 in Figure 10.1, and also in this case the IDE methodology may also be applied (RIVM, 2003).

Step 3: According to the final outcomes that may follow the primary scenario (e.g. radiation, blast wave, or fragment projection), the escalation vectors should be assessed. This step is conceptually equivalent to Steps 2 and 3 in Figure 10.1.

Step 4: In the first phase (4.1), the potential secondary targets are identified comparing the value of the escalation vector to escalation thresholds (e.g. X_2, X_3 and X_4 in Figure 10.6). This step is conceptually equivalent to Step 4 in Figure 10.1. In the following phase (4.2), the probabilistic damage models are selected and, when necessary, the probit values are calculated. In phase (4.3) the escalation probability of potential secondary units given the primary event, i.e. $P(X_2|X_1)$, $P(X_3|X_1)$ and $P(X_4|X_1)$, are calculated, and in phase (4.4) the units with the highest escalation probability are chosen as the secondary unit(s) (for example, X_3 in Figure 10.6). Since the secondary events are caused by the primary event, a causal arc must be directed from X_1 to X_3, showing that the occurrence of X_3 is conditional on the occurrence of X_1.

Step 5: Given that the secondary units have been damaged, potential accident scenarios and occurrence probabilities for the secondary units may be specified. The final outcomes of these scenarios can effectively be identified using advanced hazard identification methods such as the Methodology for the Identification of Major Accident Hazards (MIMAH) considering the type of equipment, type of substance released, type of damage (e.g. catastrophic rupture, vessel collapse, large breach on the shell, and pipe leakage), and the vicinity of ignition sources (Delvosalle et al., 2006; Paltrinieri et al., 2012; Tugnoli et al., 2013).

Step 6: Substituting the secondary units for the primary unit, Steps 3 through 5 are repeated to determine potential tertiary units (e.g. X_2 and X_4), potential quaternary

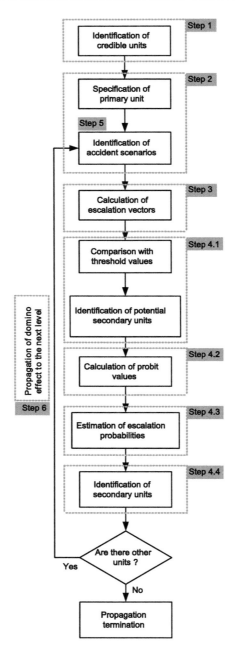

Figure 10.5 Flow diagram for Bayesian analysis of domino scenarios. (For color version of this figure, the reader is referred to the online version of this book.)

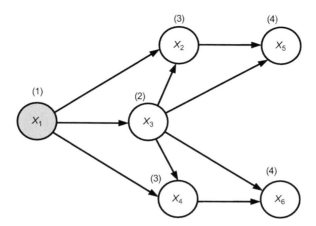

Figure 10.6 A likely propagation pattern of a domino effect in a process plant comprising six units (X_1 through X_6). The numbers in parentheses show the possible sequential order of events (Step 4 of the methodology). (For color version of this figure, the reader is referred to the online version of this book.)

units (e.g. X_5 and X_6) and so forth. In this case, it has been assumed since X_2 and X_4 (X_5 and X_6) have the same escalation probabilities, and are both selected as tertiary (quaternary) units.

It is worth noting that when repeating the same procedure (i.e. Steps 3–5) for either the secondary units or higher order units, synergistic effects should be considered. Generally speaking, through synergistic effects, the escalation vectors of newly engaged units (of order i) cooperate with those of already engaged units (of order $i-1$) to impact the units of higher order (of order $i+1$) which had not passed the threshold criteria in previous levels. For example, in Figure 10.6, X_2 and X_3 cooperate with each other (i.e. their escalation vectors are superimposed) to trigger an accident in X_5. Thus, causal arcs have to be directed from X_2 and X_3 to X_5, showing the conditional dependency of the latter on the former units. Accordingly, when assigning the conditional probability table of X_5, the escalation probability of X_5 due to the synergistic effect is also considered as $P(X_5|X_2,X_3)$.

After the likely propagation pattern of the domino effect is developed as a Bayesian network, and the probability of the primary event and the conditional probabilities of other events are calculated, the joint probability distribution of the events contributing to the domino effect can be derived. For instance, in Figure 10.6, the joint probability distribution of the events contributing to the domino effect $U = \{X_1,\ldots, X_6\}$ is calculated as

$$P(U) = P(X_1)P(X_3|X_1)P(X_2|X_1,X_3)P(X_4|X_1,X_3)P(X_5|X_2,X_3)P(X_6|X_3,X_4)$$

$$(10.13)$$

It should be noted that choosing another starting point rather than X_1 would result in a Bayesian network different from that developed in Figure 10.6 and consequently a joint probability distribution different from that shown in Eqn (10.13). However, assuming X_1 as the primary unit and according to Figure 10.6 and Eqn (10.13), the likely timeline or sequential order of the events would be $X_1 \rightarrow X_3 \rightarrow X_2$ or $X_4 \rightarrow X_5$ or X_6.

Knowing the propagation pattern of a domino effect, its occurrence probability at different levels can be estimated. Generally, the probability of the domino effect is calculated as the multiplication of the probability of the primary event and the escalation probability. For a domino effect to be in the first level, it is necessary that the accident in the primary unit propagates into at least one of the nearby secondary units. For example, in Figure 10.6, considering X_3 as the secondary unit, the probability of the first-level domino effect may be calculated as

$$P_{\text{First level}} = P(X_1)P(X_3|X_1) \tag{10.14}$$

Similarly, the domino effect could proceed to the second level only if at least one of the tertiary units X_2 and X_4 is impacted by the first-level domino accident (i.e. by a combination of X_1 and X_2). Accordingly, the probability of the second-level domino effect is calculated as

$$P_{\text{Second level}} = P(X_1)P(X_3|X_1)P(X_2 \cup X_4|X_1, X_3) \tag{10.15}$$

To account for the union of X_2 and X_4 represented in Eqn (10.15), Figure 10.6 can be modified by adding the auxiliary node L_1 such that $L_1 = X_2 \cup X_4$ (see Figure 10.7). As such, X_2 and X_4 are connected to L_1 using OR gate causal arcs (Bobbio et al., 2001; Khakzad et al., 2011) resulting in the conditional probability table shown in Table 10.2 for the node L_1. It should be noted that the probability of L_1 equals the propagation probability of the domino effect to the second level, i.e. the probability that at least one of the tertiary units X_2 or X_4 is involved in the accident. Likewise, for

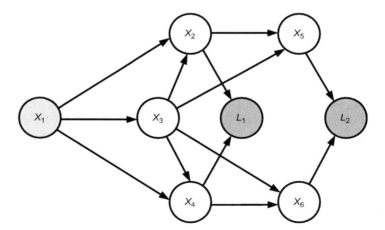

Figure 10.7 Modified Bayesian Network to incorporate the union of tertiary and quaternary events using auxiliary nodes L_1 and L_2, respectively. (For color version of this figure, the reader is referred to the online version of this book.)

Table 10.2 Conditional Probability Table Assigned to the Auxiliary Node L_1 Showing Its Conditional Dependence on Its Parents via An OR Gate

		$P(L_1|X_2, X_4)$	
X_2	X_4	**Accident**	**Safe**
Accident	Accident	1	0
Accident	Safe	1	0
Safe	Accident	1	0
Safe	Safe	0	1

the domino effect to proceed to the third level, it is necessary that the accident in the tertiary units propagate into at least one of the quaternary units.

The probability of the first-level domino effect can be estimated as the product of $P(X_1)$ and $P(X_3|X_1)$. Thus, if DL_1 is connected to X_1 and X_3 by AND gate causal arcs, $P(DL_1)$ would be equal to the probability of the first-level domino effect. This implies that for the first-level domino effect to occur, not only the primary event X_1 but also the secondary event X_3 are needed.

Likewise, if DL_2 is connected to nodes DL_1 and L_1 by AND gate causal arcs, $P(DL_2)$ would be equal to the probability of the second-level domino effect. This indicates that for the second-level domino effect to occur, the first-level domino effect (i.e. DL_1) and at least one of the tertiary events are needed, i.e. L_1. Table 10.3 shows the conditional probability table which has to be assigned to DL_1 (and also DL_2) to model intersection dependencies.

In addition to the domino effect probability, the probability of each event due to the domino effect can be calculated by marginalizing the joint probability distribution of the domino effect propagation network. For example, according to Figure 10.7, the probability of accident in X_3 caused exclusively by the domino effect is calculated as

$$P(X_3) = \sum_{U/X_3} P(U) = \sum_{X_1} P(X_1)P(X_3|X_1) \tag{10.16}$$

where U/X_3 denotes that the marginalization should be implemented over all variables except X_3. Equation (10.16) can be used either to estimate the domino-derived probability of X_3 or to estimate the domino-affected probability of X_3.

Table 10.3 AND Gate Conditional Probability Table for Node DL_1

		$P(DL_1|X_1, X_3)$	
X_1	X_3	**Accident**	**Safe**
Accident	Accident	1	0
Accident	Safe	0	1
Safe	Accident	0	1
Safe	Safe	0	1

10.4   Other Approaches to the Quantitative Assessment of Domino Effect

As discussed in the introduction to the present chapter, in the 1980s and 1990s, a number of different approaches were proposed for the quantitative assessment of domino effects (e.g. Bagster and Pitblado, 1991; Gledhill and Lines, 1998; Latha et al., 1992; Morris et al., 1994; Pettitt et al., 1993). Most of these methods, which had extreme importance at the time they were developed to allow at least a preliminary quantitative assessment of domino effects in risk assessment and safety studies, are now outdated. As a matter of fact, the methods were based on oversimplifying assumptions that were introduced at the time due to the limited knowledge in equipment damage mechanisms and/or to the limited computational resources available. Nowadays, the use of such approaches should be considered with attention and may hardly be justified in the light of results obtained in consequence analysis, equipment damage models and software for quantitative risk calculation.

Nevertheless, besides QRA and Bayesian estimators, other approaches were developed for the assessment of domino effect. Some approaches go toward the assessment of domino effect by the use of simplified risk indexes (Cozzani et al., 2009; Zhang and Chen, 2011). A detailed original approach to quantitative assessment of domino frequencies was recently proposed by Abdolhamidzadeh et al. (2010). These authors present a new methodology based on Monte Carlo simulation with which the expected frequencies of domino effects can be assessed. Monte Carlo simulation enables iterative evaluation of a system using sets of random numbers as inputs. This technique of simulation is particularly suitable when the underlying probabilities of a process are known, but their interaction is difficult to determine. The work of Abdolhamidzadeh et al. (2010) allowed the development of a Monte Carlo algorithm for the assessment of domino effects named FREEDOM (FREquency Estimation of DOMino accidents). The model was applied to the simulation of a multiunit system which may experience domino effects. The results of FREEDOM were compared to those of the QRA method presented in Section 10.2, showing a sufficient agreement even for a rather low number of simulations ($n = 1000$).

10.5   Conclusions

The relevant and considerable progress in equipment damage models and the increasing availability of computational resources make the quantitative assessment of risk due to domino scenarios now possible. In this chapter, a well-assessed methodology was presented for the QRA of domino scenarios, based on the equipment damage models described in Chapters 4, 5 and 6. The methodology, supported by software tools based on geographical information systems, allows the assessment of individual and societal risk, taking into account escalation of primary scenarios. A further progress in the quantitative assessment of domino scenarios is expected integrating the QRA approach to the development of innovative techniques based on Bayesian analysis and Monte Carlo methods.

References

Abdolhamidzadeh, B., Abbasi, T., Rashtchian, D., Abbasi, S.A., 2010. A new method for assessing domino effect in chemical process industry. Journal of Hazardous Materials 182, 416–426.

Antonioni, G., Spadoni, G., Cozzani, V., 2009. Application of domino effect quantitative risk assessment to an extended industrial area. Journal of Loss Prevention in the Process Industries 22, 614–624.

Antonioni, G., Cozzani, V., Gubinelli, G., Spadoni, G., Zanelli, S., 2004. The estimation of vulnerability in domino accidental events. In: Proc. Eur. Safety and Reliability Conf. Springer, London, UK, pp. 3212–3217.

Bagster, D.F., Pitblado, R.M., 1991. The estimation of domino incident frequencies: an approach. Process Safety Environmental Protection 69, 196.

Bobbio, A., Portinale, L., Minichino, M., Ciancamerla, E., 2001. Improving the analysis of dependable systems by mapping fault trees into Bayesian networks. Journal of Reliability Engineering and System Safety 71, 249–260.

Carter, D.A., Hirst, I.L., 2000. "Worst case" methodology for the initial assessment of societal risk from proposed major accident installations. Journal of Hazardous Materials 71, 117–128.

CCPS, Center for Chemical Process Safety, 2000. Guidelines for Chemical Process Quantitative Risk Analysis. American Institute of Chemical Engineers, New York, USA.

CCPS (Center for Chemical Process Safety), 2008. Guidelines for Hazard Evaluation Procedures, third ed. AIChE, New York.

Contini, S., Boy, S., Atkinson, M., Labath, N., Banca, M., Nordvik, J.P., 1996. Domino effect evaluation of major industrial installations: a computer aided methodological approach. In: Proc. European Seminar on Domino Effects. Leuven (B), pp. 21–34.

Cozzani, V., Gubinelli, G., Antonioni, G., Spadoni, G., Zanelli, S., 2005. The assessment of risk caused by domino effect in quantitative area risk analysis. Journal of Hazardous Materials 127, 14–30.

Cozzani, V., Antonioni, G., Spadoni, G., 2006. Quantitative assessment of domino scenarios by a GIS-based software tool. Journal of Loss Prevention in the Process Industries 19, 463.

Cozzani, V., Tugnoli, A., Salzano, E., 2009. The development of an inherent safety approach to the prevention of domino accidents. Accident Analysis and Prevention 41, 1216–1227.

COVO, 1982. Risk Analysis of Six Potentially Hazardous Industrial Objects in the Rijnmond Area, A Pilot Study. A Report to the Rijnmond Public Authority. Central Environmental Control Agency, Schiedam, NL.

Delvosalle, C., 1998. A Methodology for the Identification and Evaluation of Domino Effects, Rep. CRC/MT/003, Belgian Ministry of Employment and Labour, Bruxelles, B.

Delvosalle, C., Fievez, C., Pipart, A., Debray, B., 2006. ARAMIS project: a comprehensive methodology for the identification of reference accident scenarios in process industries. Journal of Hazardous Materials 130, 200–219.

Directive 82/501/EEC, Council Directive 82/501/EEC of 24 June 1982 on the major accident hazards of certain industrial activities. Official Journal of the European Communities L230/25, Brussels, 5.8.82.

Egidi, D., Foraboschi, F.P., Spadoni, G., Amendola, A., 1995. The ARIPAR project: an analysis of the major accident risks connected with industrial and transportation activities in the Ravenna area. Journal of Reliability Engineering and System Safety 49, 75.

Finney, D.J., 1971. Probit Analysis. Cambridge University Press, Cambridge, UK.

Gledhill, J., Lines, I., 1998. Development of Methods to Assess the Significance of Domino Effects from Major Hazard Sites, CR Report 183. Health and Safety Executive, London, UK.

HSE, Health and Safety Executive, 1978. Canvey: An Investigation of Potential Hazards from Operations in the Canvey Island/Thurrock Area. HM Stationary Office, London, UK.

HSE, Health and Safety Executive, 1981. Canvey: A Second Report. A Review of the Potential Hazards from Operations in the Canvey Island/Thurrock Area Three Years after Publication of the Canvey Report. HM Stationery Office, London, UK.

Jensen, F.V., Nielsen, T.D., 2007. Bayesian Networks and Decision Graphs, second ed. Springer, New York, USA.

Khakzad, N., Khan, F., Amyotte, P., 2011. Safety analysis in process facilities: comparison of fault tree and Bayesian network approaches. Journal of Reliability Engineering and System Safety 96, 925–932.

Khakzad, N., Khan, F.I., Amyotte, P., Cozzani, V., 2013. Domino effect analysis using Bayesian networks. Risk Analysis 33, 292–303.

Khan, F.I., Abbasi, S.A., 1998a. Models for domino effect analysis in chemical process industries. Process Safety Progress 17, 107.

Khan, F.I., Abbasi, S.A., 1998b. DOMIFFECT (DOMIno eFFECT): user-friendly software for domino effect analysis. Environmental Modelling and Software 13, 163–177.

Latha, P., Gautam, G., Raghavan, K.V., 1992. Strategies for the quantification of thermally initiated cascade effects. Journal of Loss Prevention in the Process Industries 5, 18.

Leonelli, P., Bonvicini, S., Spadoni, G., 1999. New detailed numerical procedures for calculating risk measures in hazardous materials transportation. Journal of Loss Prevention in the Process Industries 12, 507–515.

Mannan, S., 2005. Lees' Loss Prevention in the Process Industries, third ed. Elsevier, Oxford, UK.

Morris, M., Miles, A., Copper, J., 1994. Quantification of escalation effects in offshore quantitative risk assessment. Journal of Loss Prevention in the Process Industries 7, 337.

Paltrinieri, N., Dechy, N., Salzano, E., Wardman, M., Cozzani, V., 2012. Lessons learned from Toulouse and Buncefield disasters: from risk analysis failures to the identification of atypical scenarios through a better knowledge management. Risk Analysis 32, 1404–1419.

Pettitt, G.N., Schumacher, R.R., Seeley, L.A., 1993. Evaluating the probability of major hazardous incidents as a result of escalation events. Journal of Loss Prevention in the Process Industries 6, 37.

RIVM, 2003. Instrument Domino Effecten. Rijksinstituut voor Volksgezondheid en Milieu, Bilthoven, NL.

Spadoni, G., Egidi, D., Contini, S., 2000. Through ARIPAR-GIS the quantified area risk analysis supports land-use planning activities. Journal of Hazardous Materials 71, 423–437.

Tugnoli, A., Gyenes, Z., Van Wijk, L., Christou, M., Spadoni, G., Cozzani, V., 2013. Reference criteria for the identification of accident scenarios in the framework of land use planning. Journal of Loss Prevention in the Process Industries 26, 614–627.

Uijt de Haag, P.A.M., Ale, B.J.M., 1999. Guidelines for Quantitative Risk Assessment (Purple Book). Committee for the Prevention of Disasters, The Hague, NL.

Van Den Bosh, C.J.H., Merx, W.P.M., Jansen, C.M.A., De Weger, D., Reuzel, P.G.J., Leeuwen, D.V., Blom-Bruggerman, J.M., 1989. Methods for the Calculation of Possible Damage (Green Book). Committee for the Prevention of Disasters, The Hague, NL.

Van Den Bosh, C.J.H., Weterings, R.A.P.M., 1997. Methods for the Calculation of Physical Effects (Yellow Book). Committee for the Prevention of Disasters, The Hague, NL.

Zhang, X.M., Chen, G.H., 2011. Modeling and algorithm of domino effect in chemical industrial parks using discrete isolated island method. Safety Science 49, 463–467.

11 Detailed Studies of Domino Scenarios

Gabriele Landucci*, Ernesto Salzano†,
Jerome Taveau‡, Gigliola Spadoni§, Valerio Cozzani§

* Dipartimento di Ingegneria Civile e Industriale, Università di Pisa, Pisa, Italy, † Istituto di Ricerche sulla Combustione, Consiglio Nazionale delle Ricerche (CNR), Napoli, Italy, ‡ Scientific advisor at Fike Corporation, Industrial Explosion Protection Group, Blue Springs, USA, § LISES, Dipartimento di Ingegneria Civile, Chimica, Ambientale e dei Materiali, Alma Mater Studiorum – Università di Bologna, Bologna, Italy

---

11.1   Introduction

The assessment of domino effect may be supported by the use of advanced tools for the analysis of complex or critical escalation scenarios. As discussed in Chapter 8, distributed parameters models may be used when a high detail level is required for the analysis of domino accidents impact or to evaluate the resistance of target equipment to the primary scenario. Advanced models may provide a detailed evaluation of near-field consequences that have particular relevance when congested layouts are analyzed (e.g. offshore platforms, reactors systems, pipe racks, etc.) (Morris et al., 1994; I et al., 2009a,b; Salzano et al., 2002; Lea and Ledin, 2002; Paris et al., 2010). Moreover, the structural integrity assessment of target equipment may support the design phase of protection barriers (Lea and Ledin, 2002; Schleyer and Langdon, 2006; Health Safety Executive (HSE), 2012; I et al., 2009a; Landucci et al., 2009a). Nevertheless, the use of

Domino Effects in the Process Industries. http://dx.doi.org/10.1016/B978-0-444-54323-3.00011-7

these tools requires a high level of detail in input data, which might not be available, in particular during the early stages of project development. Furthermore, in the context of risk assessment, key input data as the precise location and features of the primary scenario that may trigger the escalation process are affected by an inherent uncertainty. Moreover, despite the progress made in recent years, it is quite clear that this approach has important limitations due to the computer time and man hours required to build the numerical domain and to carry out the simulations that results in high costs and in a relevant effort. Therefore, the use of advanced tools should be limited to the more critical cases, following a preliminary screening phase (Chapters 8–10), in which simplified correlations or lumped parameters models are instead applied.

The first part of this chapter deals with the potentialities of distributed parameter models for the simulation of primary scenarios. The state of the art of computational fluid dynamics (CFD) codes for the simulation of radiation and blast wave effect aimed at escalation assessment is briefly revised. The second part of this chapter shows some examples of detailed analysis on target structural elements, evidencing the potentialities of the distributed parameters codes. The attention will be focusing on the application of finite element models (FEM) for the analysis of process and storage equipment exposed to accidental fires. In particular, integrated thermal and mechanical FEM analyses will be presented, showing the simulation of the dynamic temperature and stress behavior of target equipment given different primary fire scenarios. The analysis is finalized at the prediction of equipment failure leading to domino escalation. Both for fire and blast wave escalation analyses, some case studies will be presented providing elements on the necessary input data for the setup of simulations and showing the possible results obtained by the application of detailed models.

11.2   CFD Codes for the Simulation of Accident Scenarios

11.2.1   CFD Codes for the Simulation of Gas Dispersions

CFD models solve Navier–Stokes equations of fluid flow (conservation of mass, momentum and scalar quantities) in a three-dimensional (3D) space (Ferziger and Peric, 2002; Yeoh and Yuen, 2008; Tu et al., 2013). The problem is reduced to the solution of a system of partial differential equations that needs to specify appropriate boundary conditions for their numerical solution. In the field of consequence analysis of accident scenarios, CFD models are mainly used to assess the dispersion of gaseous toxic and/or flammable materials. A CFD model can handle obstacles in a 3D environment, therefore representing an ideal candidate for consequence analysis in complex scenarios. CFD simulations are capable of assessing the effectiveness of passive barriers such as dikes, as shown in Figure 11.1 (Pontiggia et al., 2009, 2010, 2011; Landucci et al., 2011). CFD consequence modeling is a unique tool in providing results for gas concentration in the near field, allowing a detailed assessment of specific dispersion scenarios.

However, one must keep in mind that CFD models require a rigorous verification and validation process, as well as a comprehensive documentation, in order to allow

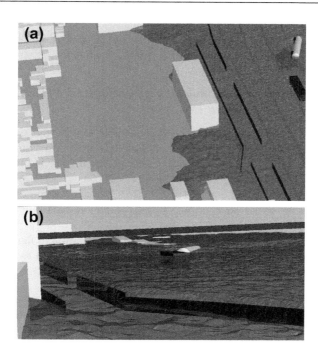

Figure 11.1 Example of a CFD simulation for the dispersion of a heavy cloud of flammable gas. Cloud boundary overcoming the wall with a weir effect: (a) side view; (b) front view (Landucci et al., 2011). (For color version of this figure, the reader is referred to the online version of this book.)

the user to apply them with good confidence. In CFD computations the setup of initial and boundary conditions is critical for the accuracy of prediction. Even though the open literature has been growing on the use of CFD as consequence modeling, there are still little details on how to address complex scenarios and how to verify the calculation involved. Moreover, the demand of large amounts of memory and computing time still limits the use of these models in everyday job to the analysis of specific scenarios for which a high level of detail is required. In the open literature, for several CFD codes reasonable details are available to set up the simulation of dispersion modeling (Yeoh and Yuen, 2008; Tu et al., 2013): e.g. FEM3 CFD (Luketa-Hanlin et al., 2007), FLACS (Dharmavaram et al., 2005; Hanna et al., 2004), FLUENT (Gavelli et al., 2008), CFX (Sklavounos and Rigas, 2004), and fire dynamics simulator (FDS) (Chan, 1992).

CFD models have become more and more convenient to use thanks to the improvement of computer power in last twenty years. Beside the simulation of the dispersion, submodels were included in CFD codes to account for the physical and chemical processes occurring in a fire or an explosion.

When dealing with domino effects in the near field, the capability to provide detailed results in the near field is an important factor, allowing a more precise assessment of the escalation vector and of escalation probability. Thus, CFD application to the simulation of vapor cloud explosions (VCEs) and of fires is the key factor of interest for the detailed assessment of domino scenarios.

11.2.2 CFD Codes for the Simulation of Unconfined or Partially Confined Gas and Vapor Cloud Explosion

In the last twenty years, CFD codes became a routine tool for the simulation of the consequences of unconfined gas or VCEs or partially confined explosion as in the case of vented equipment or offshore system characterized by high level of congestion and confinement (Lea and Ledin, 2002; Salzano et al., 2002).

Due to the large scale and computational effort, the CFD commonly adopted for such simulation are based on distributed parameter models as Reynolds averaged Navier–Stokes (RANS) code.

The effect of turbulence, which is essential for the definition of flame acceleration and for the production of shock waves, is generally implemented by empirical submodels, typically the k–ε model (Launder and Spalding, 1972) or similar, more updated models.

The combustion reaction rate is evaluated in the laminar and turbulent regime. In the second case, the turbulent burning velocity as calculated by empirical correlations reported in the literature (e.g. Gulder equation) is adopted for the definition of source model. A "subgrid" or "porosity" formulation has been often considered to keep into account of the turbulence induced by obstacles when the volume of the same obstacle is smaller than the volume of the single computational cell.

The use of RANS for the reproduction of large scale, compressible, reactive system is questionable and very large uncertainties are intrinsically produced by the several assumptions on the complex combustion phenomena involved in VCEs or partially confined explosions. Results can be only adopted by assuming a number of conservative options. On the other hand, this technique is the only available option for engineering analysis of large-scale industrial explosion, even if well-established skill is needed. Hence, several commercial codes are on the market with full success as, e.g. FLACS (Gexcon) or FLUENT/CFX (ANSYS).

Recently, the use of large eddy simulation (LES) has been proposed. This technique is still under development and is limited by the available computational power (Di Sarli et al., 2010).

11.2.3 CFD Codes for the Simulation of Fires

Many studies have been developed (Smith et al., 2003; Malalasekara et al., 1996; Hostikka et al., 2003) to investigate the possibility to model the fire characteristics by the use of the CFD codes which would be able to simulate the net effect of many different parameters as turbulence, reactions, soot formation and the effect of thermal radiation on the substrate.

Among several proposed CFD, the public code is named FDS, published by the American National Institute of Standards and Technology (NIST) as part of their Fire and Building research division. Several versions of the model have been developed along the years, thus making FDS and associated 3D visual application (Smokeview) a tool largely used by the international fire engineering community for the assessment of fire scenarios and design purposes (McGrattan et al., 2010). The computer software FDS solves a particular form of Navier–Stokes and it is integrated by submodels necessary to describe the combustion process and the interaction with the environment. The software is also equipped with many other submodels able to simulate pyrolysis, presence of different type of detectors, sprinklers, etc. making it a more complete fire engineering tool than other "general purpose" CFD codes.

It is worth saying that recent versions of FDS adopt successfully the LES methodology under the assumption of low Mach number. Indeed, with specific reference to industrial fires, the evolution of large eddy structures, which is the characteristic of most fire plumes, is lost with RANS. The application of LES techniques to fire is aimed at extracting greater temporal and spatial fidelity from simulations of fire performed on the more finely meshed grids allowed by ever faster computers (McGrattan et al., 1998).

11.3 Application of Finite Element Models to Assess Escalation Triggered by Fire

A distributed parameters model is aimed at the detailed evaluation of a physical phenomenon evolution in a domain of interest (Cook, 1995; Spyrakos, 1996; Van de Wouwer, 2013). After the definition of the domain, the model is chosen to represent the physical phenomena related to the quantities of interest. Hence a set of differential equations is obtained, which analytical solution might not be possible in the case of complex phenomena, such as the analysis of structures and equipment exposed to accidental fires (Najjar and Burgess, 1996; Landesmann et al., 2005; Wang and Li, 2009; Maljaars et al., 2009) or blast load (Yang, 1997; Wang et al., 2005; Liew, 2008). Finite Elements Models (FEM) are mathematical techniques aimed at providing approximate solutions to this problem (Roberts et al., 2000; Landucci et al., 2009a,b,c; Birk, 2006; Manu et al., 2009; Paltrinieri et al., 2009; Tan et al., 2003; Bi et al., 2011). FEM is based on the discretization of the domain of interest in small portions (elements) which form a grid on the domain itself (mesh). Among the elements, approximate functions of the quantities of interest are defined (shape functions) and evaluated in specific points (nodes), typically on the edges of each element, but also inside the element itself. Hence, the problem is discretized and locally solved in each node, deriving the behavior of the entire domain.

Several commercial codes allow for this type of evaluation, being ABAQUS (http://www.3ds.com/products/simulia/portfolio/abaqus/overview/), ANSYS (http://www.ansys.com/), MARC (http://www.mscsoftware.com/Products/CAE-Tools/Marc.aspx), COMSOL (http://www.comsol.com/), SAMCEF (http://www.lmsintl.com/samcef-solver-suite), NASTRAN (http://www.mscsoftware.com/products/cae-tools/msc-nastran.aspx), and LS DYNA (http://www.ls-dyna.com/) some of the more commonly used for industrial applications (Noor, 1986; Cross and Slone, 2005). The mentioned codes allow for the definition of a domain of complex geometry and the "meshing" operation for determining the elements, on which the different types of loads (forces, heat fluxes, and imposed temperatures) can be applied.

In the case of fire triggered domino scenarios, the key aspect of FEM consists in the combined thermal and stress analysis (Birk, 2006; Manu et al., 2009; Landucci et al., 2009a; Tan et al., 2003). As discussed in Chapter 5, the failure of fired structural elements and, in particular, process equipment and pipes, is related to the thermal weakening of the construction material caused by the heat up. Besides, higher temperatures lead to the increase of mechanical stress due to thermal dilatations and, in the case of pressurized equipment, due to the inner pressure rise. The advantage of using FEM codes is the possibility of evaluating in each point of the target (e.g. in each defined node) the stress and temperature conditions, obtaining a more precise prediction on potential failure conditions. Therefore, critical points of the structure such as connection with frameworks, supports, saddles, or strong discontinuities (e.g. interface between stored liquid and vapor space of a pressurized vessel, nonuniform fire exposure, etc.), which can be affected by stress intensification (Birk, 2006; Manu et al., 2009; Landucci et al., 2009a; Tan et al., 2003), may be taken into account with high level of detail.

Specific approaches based on FEM technique were proposed in the literature to analyze the problem (Roberts et al., 2000; Landucci et al., 2009a,b,c; Birk, 2006; Manu et al., 2009; Paltrinieri et al., 2009; Tan et al., 2003; Bi et al., 2011). In order to evidence the potentialities of the model and the setup requirements, the model developed by Landucci et al. (2009a,b,c) and implemented on the ANSYS™ code (ANSYS™, 2007) will be presented. The model is specifically set up for the simulation of different fire exposure modes, deriving the wall temperature and stress evolution for fired process and storage equipment. Hence, the model can predict the target time to failure, e.g. the time that occurs between the fire starting and the eventual equipment failure bringing a loss of containment.

In the following, brief model details and an example of FEM application will be presented analyzing large scale and pressurized vessels exposed to fires. More details on the model setup are reported elsewhere (Landucci et al., 2009a,b; Paltrinieri et al., 2009).

The first step in the FEM simulations is the detailed calculation of the temperatures on the vessel shell as a function of time and radiation mode. The vessels are modeled as a cylindrical body with different types of ends (conic roof and a flat base for atmospheric tanks and hemispherical heads for pressurized vessels). A proper mesh needs to be defined for each geometry considered. An example is provided in Figure 11.2 in which the calculation mesh is set up for the modeling of the large-scale atmospheric vertical tanks (Figure 11.2(a)) and pressurized horizontal vessels (Figure 11.2(b)). The features of the vessels considered in the case study are reported in Table 11.1.

Thermal loads are applied to simulate the shell temperature distribution. Two different types of thermal loads were applied:

- Constant heat loads: radiation from external fire, convection and for the surface emission.
- Time-dependent heat loads: heat flux from the inner steel wall to the fluid (gas or liquid phase).

A nonuniform thermal load distribution can be obtained by the FEM discretization, thus the method can be interfaced with complex CFD tools able to simulate

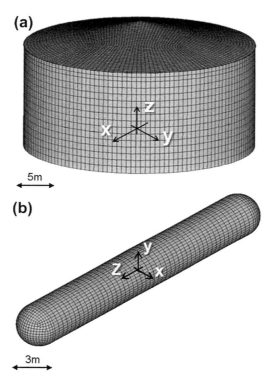

Figure 11.2 Example of geometry and mesh implemented in the FEM for the analysis of the case studies: (a) atmospheric vessel; (b) pressurized vessel. See Table 11.1 for the vessels' geometrical features.

large-scale jet flames and pool fires. In the case study presented, the atmospheric vessel is exposed to a distant pool fire radiation, while the pressurized vessel is fully engulfed by the fire. The values of thermal loads (Table 11.1) were obtained from the analysis of relevant fire scenarios for escalation (Section 5.3) while the heat transfer coefficients were derived from a literature data analysis (Knudsen et al., 1999).

Figure 11.3 shows two examples of the results of the detailed temperature simulations, respectively, for the atmospheric (Figure 11.3(a)) and the pressurized (Figure 11.3(b)) vessels.

The second step of the FEM modeling consists in the calculation of the transient stress field as a function of the local temperatures and of the other loads present on the equipment shell. Weight, internal vapor pressure and hydraulic gradient are considered in the analysis. Figure 11.4 shows an example of the maps representing the stress intensity field acting on the equipment shell obtained from the temperature simulations in Figure 11.3 for the correspondent vessel.

The results obtained allow for the application of the failure criterion described in Section 5.4. Thus the time to failure of the vessels can be evaluated obtaining, respectively, 12 and 8 min for atmospheric and pressurized vessels. This is in

Table 11.1 Main Features of the Vessels Considered for the Case Study Discussed in Section 11.2

Item	Atmospheric Vessel	Pressurized Vessel
Nominal volume (m^3)	10,000	110
Geometry	Vertical cylinder, cone roof	Horizontal cylinder
External diameter (m)	30	3.0
Total height*/length† cylinder (m)	14	18.3
Wall thickness (mm)	20.5–6.5‡	20
Filling level (%)	50	50
Design gauge pressure (barg)	0.25	18.2
Steel type	Low carbon steel ASTM A570Grade33	High yield carbon steel P460NH
Exterior temperature (°C)	20	20
Type of fire exposure	Distant source radiation	Diesel pool fire engulfment
Thermal load in simulations (kW/m^2)	20	150

*Only for atmospheric vessel.
†Only for pressurized vessel.
‡Thickness decreases in upper boards.

agreement with what observed in accidents involving vessels exposed to fire radiation or full impingement (Birk et al., 2006a; Droste and Schoen, 1988; Townsend et al., 1974). Therefore, the presented example analysis shows the potentialities of the FEM tool in evaluating the conditions leading to vessel failure. FEM provides support for the escalation possibility evaluation of critical targets and it supports the planning of emergency by evaluating the residual time for mitigation prior to the escalation (e.g. the time to failure).

Another advantage of FEM application is the detailed design of eventual mitigation barriers, such as the installation of fireproofing materials on the outer surface of the target vessels or structures (Roberts et al., 2000; Di Padova et al., 2011; Tugnoli et al., 2012) or pressure relief devices (see Section 5.5 for more details). Several examples are available in the literature concerning the evaluation optimum design of thermal protections aimed at delaying or eventually avoiding the vessel failure (Birk, 2004, 2005, 2006; Landucci et al., 2009a,b,c). The FEM tools can also be used for the performance evaluation of a given protective barrier and eventually to verify the effects of the protection degradation during fire exposure (Gomez-Mares et al., 2012a,b, c). A further key issue in the effectiveness assessment of fireproofing by FEM tools is the influence of defects in the thermal protection. Defects may be caused both by ageing (e.g. corrosion and/or erosion) and by accidental impact. A decrement in the thermal protection action due to the presence of defects is shown by small-scale experiments on LPG tanks (VanderSteen and Birk, 2003; Birk et al., 2006b) and investigated by FEM in detailed studies of Birk (1999, 2005).

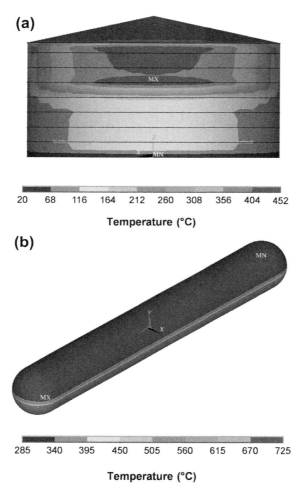

Figure 11.3 Temperature maps obtained by FEM for the analyzed vessels at the time to failure: (a) atmospheric vessel; (b) pressurized vessel. Temperature is expressed in °C. (For color version of this figure, the reader is referred to the online version of this book.)

11.4 CFD and FEM for the Assessment of Escalation Triggered by Blast Waves

The same type of approach shown in Section 11.2 may be extended for the analysis of equipment and pipes affected by shock wave (Robertson et al., 2000). In this case, a preliminary assessment of the blast load is instead needed for the setup of the mechanical FEM tool. Hence, a separate CFD analysis of the explosion is performed. Section 4.3 reports an example concerning a reaction section of a polypropylene unit

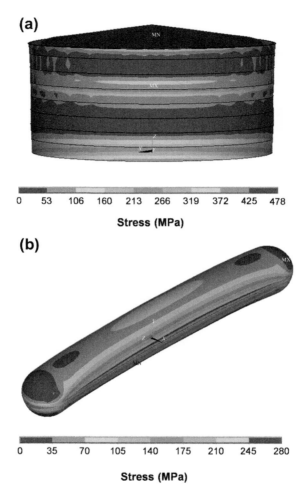

(a)

0   53   106   160   213   266   319   372   425   478

Stress (MPa)

(b)

0   35   70   105   140   175   210   245   280

Stress (MPa)

Figure 11.4 Stress intensity maps obtained by FEM implementing the temperatures evaluated in the thermal analysis (Figure 11.3): (a) atmospheric vessel; (b) pressurized vessel. Stress is expressed in MPa. (For color version of this figure, the reader is referred to the online version of this book.)

located in a complex industrial layout. The results of a commercial CFD code were implemented in a dynamic, nonlinear FEM. Displacements, relative displacements and the peak bending moments in the equipment and structures were calculated. The figures reported in Section 4.3 clearly evidence the level of detail that may be obtained by the approach.

An important issue in the combined method (CFD + FEM) regards the escalation effects for installation located at relatively long distance from the explosion source, for which CFD may not be adopted due to large computational effort. In this case, the analysis of shock wave propagation in the far field is typically coupled with simple methodologies as, e.g. the TNT-equivalency (for point-source explosion), the Multienergetic or the Baker–Strehlow Methodologies for VCE (American Institute of Chemical Engineers Center of Chemical Process Safety (AIChe-CCPS), 2010). These tools give the peak overpressure and impulse (i.e. the history, assuming a triangular shape of the wave) at any distance from the source, based on the energy of combustion or mass/energy of explosive and the maximum pressure at the source point, which may be obtained as result of CFD analysis. The pressure history reproduced at any distance may be used as an input for structural analysis based on Single Degree or Multidegree of freedom (Biggs's method) (Section 4.3), or FEM.

This type of analysis may be used for the design of blast wall in offshore unit, e.g. for residential unit protection or for sensitive structures, hence domino effects (Schleyer and Langdon, 2006).

11.5 Conclusions

The most detailed assessment of domino effect escalation scenarios needs at the same time the advanced 3D simulation of the physical phenomenon (the explosion, the fire, and the dispersion) and of its interaction with structures, either in the near field or in the far field. Quite clearly, the overall procedure is a complex and critical task, and can be rarely applied to actual full-scale problems of industrial relevance. Also, it should be considered that we are still far from the full comprehension of the chemistry and physics of most of the accidental phenomena typically occurring in the process industry as gas explosion, VCE (compressible and turbulent reaction system), flash fire, dispersion in complex geometry and others.

This chapter also presented the application of currently available advanced 3D models for the evaluation of process and storage equipment resistance to primary events impact.

In particular, FEM tools were described and applied to several case studies to predict the occurrence of escalation scenarios triggered by fire and explosion. The level of detail of the analysis and the results obtained evidenced the FEM potentialities. Nevertheless, as in the case of 3D tools for the assessment of accident consequences, the complexity of input information and data together with the computer time and man hours required to build the numerical domain pose important limitations to the application of such detailed models, which should be limited to the study of scenarios having particular relevance or criticality.

References

American Institute of Chemical Engineers Center for Chemical Process Safety, AIChE-CCPS, 2010. Guidelines for Vapor Cloud Explosion, Pressure Vessel Burst, BLEVE and Flash Fire Hazards. AIChE-CCPS, John Wiley & Sons Inc, New York, NY.
ANSYS Inc, 2007. ANSYS™ User Guide, vol. 11. ANSYS Inc, Canonsburg, PA.
Bi, M., Ren, J., Zhao, B., Che, W., 2011. Effect of fire engulfment on thermal response of LPG tanks. Journal of Hazardous Materials 192 (2), 874–879.

Birk, A.M., 1999. Tank-Car Insulation Defect Assessment Criteria: Thermal Analysis of Defects (TP 13518E Report). Transport Canada, Ottawa, ON.

Birk, A.M., 2004. Tank2004-A Computer Code for Modelling Pressure Vessels Containing Pressure Liquefied Gases in Fires: User Guide. Thermadyne Technologies Ltd, St. Louis, MO.

Birk, A.M., 2005. Thermal Model Upgrade for the Analysis of Defective Thermal Protection Systems. Transportation Development Centre, Transport Canada, Ottawa, ON.

Birk, A.M., 2006. Fire Testing and Computer Modelling of Rail Tank-Cars Engulfed in Fires: Literature Review (TP 14561E Report). Transport Canada, Ottawa, ON.

Birk, A.M., Poirier, D., Davison, C., 2006a. On the response of 500 gal propane tanks to a 25% engulfing fire. Journal of Loss Prevention in the Process Industries 19 (6), 527–541.

Birk, A.M., Poirier, D., Davison, C., 2006b. On the thermal rupture of 1.9 m^3 propane pressure vessels with defects in their thermal protection system. Journal of Loss Prevention in the Process Industries 19 (6), 582–597.

Chan, S.T., 1992. Numerical simulations of LNG vapor dispersion from a fenced storage area. Journal of Hazardous Materials 30 (2), 195–224.

Cook, R.D., 1995. Finite Element Modeling for Stress Analysis. Wiley, New York (NY).

Cross, M., Slone, A.K., 2005. FENET-multi-physics analysis (MPA) theme: a review of commercial MPA capability in 2005. In: NAFEMS World Congress PCD Proceedings 2005. Glasgow (UK), National Agency for Finite Element Methods and Standards NAFEMS.

Dharmavaram, S., Hanna, S.R., Hansen, O.R., 2005. Consequence analysis – using a CFD model for industrial sites. Process Safety Progress 24 (4), 316–327.

Di Padova, A., Tugnoli, A., Cozzani, V., Barbaresi, T., Tallone, F., 2011. Identification of fireproofing zones in oil and gas facilities by a risk-based procedure. Journal of Hazardous Materials 191 (1–3), 83–93.

Di Sarli, V., Di Benedetto, A., Russo, G., 2010. Sub-grid scale combustion models for large eddy simulation of unsteady premixed flame propagation around obstacles. Journal of Hazardous Materials 180 (1–3), 71–78.

Droste, B., Schoen, W., 1988. Full scale fire tests with unprotected and thermal insulated LPG storage tanks. Journal of Hazardous Materials 20, 41–53.

Ferziger, J.H., Peric, M., 2002. Computational Methods for Fluid Dynamics. Springer-Verlag, Berlin (DE).

Gavelli, F., Bullister, E., Kytomaa, H., 2008. Application of CFD (Fluent) to LNG spills into geometrically complex environments. Journal of Hazardous Materials 159 (1), 158–168.

Gomez-Mares, M., Larcher, S., Tugnoli, A., Cozzani, V., Barontini, F., Landucci, G., 2012a. Performance of passive fire protection for liquefied petroleum gas vessels: an experimental and numerical study. In: Bérenguer, C., Grall, A., Guedes Soares, C. (Eds.), Advances in Safety, Reliability and Risk Management. Taylor & Francis, London, pp. 1891–1899.

Gomez-Mares, M., Tugnoli, A., Landucci, G., Cozzani, V., 2012b. Performance assessment of passive fire protection materials. Industrial Engineering and Chemical Research 51, 7679–7689.

Gomez-Mares, M., Tugnoli, A., Landucci, G., Barontini, F., Cozzani, V., 2012c. Behavior of intumescent epoxy resins in fireproofing applications. Journal of Analytical and Applied Pyrolysis 97, 99–108.

Hanna, S.R., Hansen, O.R., Dharmavaram, S., 2004. FLACS CFD air quality model performance evaluation with Kit Fox, MUST, Prairie Grass, and EMU observations. Atmospheric Environment 38 (28), 4675–4687.

Health Safety Executive-HSE, 2012. Active and Passive Fire Protection. Health and Safety Executive, London, UK.

Hostikka, S., McGrattan, K.B., Hamins, A., 2003. Numerical Modeling of Pool Fires Using LES and Finite Volume Method for Radiation. Fire Safety Science Proceedings, Seventh (7th) International Symposium. In: Evans, D.D. (Ed.). International Association for Fire Safety Science (IAFSS), June 16-21, 2003, Worcester, MA, Intl. Assoc. for Fire Safety Science, Boston, MA, pp. 383–394.

I, Y.P., Chiu, Y.L., Wu, S.J., 2009b. The simulation of air recirculation and fire/explosion phenomena within a semiconductor factory. Journal of Hazardous Materials 163 (2–3), 1040–1051.

I, Y.-P., Shu, C.-M., Chong, C.-H., 2009a. Applications of 3D QRA technique to the fire/explosion simulation and hazard mitigation within a naphtha-cracking plant. Journal of Loss Prevention in the Process Industries 22 (4), 506–515.

Knudsen, J.G., Hottel, H.C., Sarofim, A.F., Wankat, P.C., Knaebel, K.S., 1999. Heat and mass transfer. Section 5. In: McGraw-Hill (Ed.), Perry's Chemical Engineers' Handbook, seventh ed. McGraw-Hill, New York.

Landesmann, A., Batista, E. de M., Drummond Alves, J.L., 2005. Implementation of advanced analysis method for steel-framed structures under fire conditions. Fire Safety Journal 40 (4), 339–366.

Landucci, G., Gubinelli, G., Antonioni, G., Cozzani, V., 2009c. The assessment of the damage probability of storage tanks in domino events triggered by fire. Accident Analysis and Prevention 41 (6), 1206–1215.

Landucci, G., Molag, M., Cozzani, V., 2009a. Modeling the performance of coated LPG tanks engulfed in fires. Journal of Hazardous Materials 172 (1), 447–456.

Landucci, G., Molag, M., Reinders, J., Cozzani, V., 2009b. Experimental and analytical investigation of thermal coating effectiveness for $3 \, m^3$ LPG tanks engulfed by fire. Journal of Hazardous Materials 161 (2–3), 1182–1192.

Landucci, G., Tugnoli, A., Busini, V., Derudi, M., Rota, R., Cozzani, V., 2011. The Viareggio LPG accident: lessons learnt. Journal of Loss Prevention in the Process Industries 24 (4), 466–476.

Launder, B.E., Spalding, D.B., 1972. Mathematical Models of Turbulence. Academic Press, London (UK).

Lea, C.J., Ledin, H.S., 2002. A Review of the State-of-the-Art in Gas Explosion Modelling (Report HSL/2002/02). Health and Safety Laboratory, Fire and Explosion Group, Buxton UK.

Liew, J.Y.R., 2008. Survivability of steel frame structures subject to blast and fire. Journal of Constructional Steel Research 64 (7–8), 854–866.

Luketa-Hanlin, A., Koopman, R.P., Ermak, D.L., 2007. On the application of computational fluid dynamics code for liquefied natural gas dispersion. Journal of Hazardous Materials 140 (3), 504–517.

Malalasekera, W.M.G., Versteeg, H.K., Gilchrist, K., 1996. A review of Research and Experimental Study on the Pulsation of Buoyant Diffusion Flames and Pool Fires. Fire and Materials 20, 261–271.

Maljaars, J., Soetens, F., Snijder, H.H., 2009. Local buckling of aluminium structures exposed to fire part 2: finite element models. Thin-Walled Structures 47 (11), 1418–1428.

Manu, C.C., Birk, A.M., Kim, I.Y., 2009. Stress rupture predictions of pressure vessels exposed to fully engulfing and local impingement accidental fire heat loads. Engineering Failure Analysis 16 (4), 1141–1152.

McGrattan, K., Baum, H.R., Rhem, R.G., 1998. Large eddy simulation of smoke movement. Fire Safety Journal 30 (2), 161–178.

McGrattan, K., Hostikka, S., Floyd, J., Baum, H., Rehm, R., 2010. Fire Dynamics Simulator Technical Reference Guide, (NIST Special Publication 1019–5). National Institute of Standard Technology, Gaithersburg, MD (USA).

Morris, M., Miles, A., Cooper, J., 1994. Quantification of escalation effects in offshore quantitative risk assessment. Journal of Loss Prevention in the Process Industries 7 (4), 337–344.

Najjar, S.R., Burgess, I.W., 1996. A nonlinear analysis for three-dimensional steel frames in fire conditions. Engineering Structures 18 (1), 77–89.

Noor, A.K., 1986. Survey of computer programs for heat transfer analysis. Finite Elements in Analysis and Design 2 (3), 259–312.

Paltrinieri, N., Landucci, G., Molag, M., Bonvicini, S., Spadoni, G., Cozzani, V., 2009. Risk reduction in road and rail LPG transportation by passive fire protection. Journal of Hazardous Materials 167 (1–3), 332–344.

Paris, L., Catala, C., Iddir, O., Salaün, N., 2010. Innovative design of an LDPE reactor bay against gas explosion using a risk-based approach coupled with high fidelity computer tools. Journal of Loss Prevention in the Process Industries 23 (5), 561–573.

Pontiggia, M., Derudi, M., Busini, V., Rota, R., 2009. Hazardous gas dispersion: a CFD model accounting for atmospheric stability class. Journal of Hazardous Materials 171 (1–3), 739–747.

Pontiggia, M., Derudi, M., Alba, M., Scaioni, M., Rota, R., 2010. Hazardous gas releases in urban areas: assessment of consequences through CFD modeling. Journal of Hazardous Materials 176 (1–3), 589–596.

Pontiggia, M., Landucci, G., Busini, V., Derudi, M., Alba, M., Scaioni, M., Bonvicini, S., Cozzani, V., Rota, R., 2011. CFD model simulation of LPG dispersion in urban areas. Atmospheric Environment 45 (24), 3913–3923.

Roberts, A., Medonos, S., Shirvill, L.C., 2000. Review of the Response of Pressurised Process Vessels and Equipment to Fire Attack (Offshore Technology Report – OTO 2000 051, HSL 2000). Health and Safety Laboratory, Manchester, UK.

Robertson, N.J., Fairlie, G.E., Draper, E.J., 2000. Gas explosion modelling and dynamic structural response. In: Proceedings of ERA Conference Major Hazards Offshore. ERA Technology, Leatherhead, UK, pp. 3.6.1–3.6.9.

Salzano, E., Marra, F.S., Russo, G., Lee, J.H.S., 2002. Numerical simulation of turbulent gas flames in tubes. Journal of Hazardous Materials 95 (3), 233–247.

Schleyer, G.K., Langdon, G.S., 2006. Pulse Pressure Testing of 1/4 Scale Blast Wall Panels with Connections – Phase II (Health and Safety Executive, Research Report 404). Health and Safety Executive, London, UK.

Sklavounos, S., Rigas, F., 2004. Validation of turbulence models in heavy gas dispersion over obstacles. Journal of Hazardous Materials 108 (1–2), 9–20.

Smith, P., Rawat, R., Spinti, J., Kumar, S., Borodai, S., Violi, A., 2003. Large Eddy Simulations Of Accidental Fires Using Massively Parallel Computers, Paper Number AIAA-2003-3697, 16th AIAA Computational Fluid Dynamics Conference, Orlando, Florida 23-26 Jun 2003

Spyrakos, C.C., 1996. Finite Element Modeling in Engineering Practice. West Virginia University Press, Morgantown.

Tan, D.M., Xu, J., Venart, J.E.S., 2003. Fire-induced failure of a propane tank: some lessons to be learnt. In: Proceedings of the Institution of Mechanical Engineers, Part E: Journal of Process Mechanical Engineering, 217 (2), pp. 79–92.

Townsend, W., Anderson, C.E., Zook, J., Cowgill, G., 1974. Comparison of Thermally Coated and Uninsulated Rail Tank-Cars Filled with LPG Subjected to a Fire Environment (Report FRA-OR&D 75–32). US Department of Transportation, Washington, DC.

Tu, J., Yeoh, G.-H., Liu, C., 2013. Computational Fluid Dynamics – A Practical Approach, second ed. Butterworth-Heinemann, Elsevier, Amsterdam (NL).

Tugnoli, A., Cozzani, V., Di Padova, A., Barbaresi, T., Tallone, F., 2012. Mitigation of fire damage and escalation by fireproofing: a risk-based strategy. Reliability Engineering and System Safety 105, 25–35.

Van de Wouwer, A., 2013. Modeling and simulation of distributed parameter systems. In: Control Systems, Robotics and Automation, vol. 6. EOLSS Publishers Co. Ltd, United Kingdom, pp. 85–108.

VanderSteen, J.D.J., Birk, A.M., 2003. Fire tests on defective tank-car thermal protection system. Journal of Loss Prevention in the Process Industries 16 (5), 417–425.

Wang, W.-Y., Li, G.-Q., 2009. Fire-resistance study of restrained steel columns with partial damage to fire protection. Fire Safety Journal 44, 1088–1094.

Wang, Z., Lu, Y., Hao, H., Chong, K., 2005. A full coupled numerical analysis approach for buried structures subjected to subsurface blast. Computers and Structures 83 (4–5), 339–356.

Yang, Z., 1997. Finite element simulation of response of buried shelters to blast loadings. Finite Elements in Analysis and Design 24 (3), 113–132.

Yeoh, G.H., Yuen, K.K., 2008. Computational Fluid Dynamics in Fire Engineering Theory, Modelling and Practice. Elsevier Ltd, Amsterdam (NL).

Part III

Prevention of Domino Effects from a Managerial Perspective

12 Managing Domino Effects from a Design-Based Viewpoint

Alessandro Tugnoli*, Valerio Cozzani*,
Faisal Khan†, Paul Amyotte‡

* LISES, Dipartimento di Ingegneria Civile, Chimica, Ambientale e dei Materiali, Alma Mater Studiorum – Università di Bologna, Bologna, Italy, † Safety and Risk Engineering Research Group, Faculty of Engineering and Applied Science, Memorial University of Newfoundland, St. John's, Newfoundland, Canada, ‡ Department of Process Engineering and Applied Science, Dalhousie University, Halifax, Nova Scotia, Canada

12.1   Introduction

The design of a plant influences in many ways the potential for domino effect. As discussed in Chapter 3, two basic elements are required for a domino escalation to take place: a primary scenario with enough energy to damage one or more than one "domino target unit" and the presence of at least one "domino target unit" within the reach of the primary scenario. In this context, a "domino target unit" is a unit that has the potential, once damaged, to generate a secondary scenario that results in an overall domino accident having consequences worse than those of the primary scenario (Chapter 3).

Plant design affects both the elements: the severities of the primary and secondary scenarios depend, among the other, on the substances, operative conditions and inventories defined in process and unit design; the presence and location of targets, instead, depend on the plant siting and layout design. Therefore, adequate measures implemented in the design phase may result in safer plants from the point of view of escalation.

Applicable safety measures implemented in process and unit design are mostly the same already implemented for conventional (i.e. non-domino-specific) safety strategies aimed at loss prevention, safety for the external population, safety for the employees, etc. However, the design of plant layout has a more distinctive influence on the potential for escalation and deserves a dedicated consideration.

Beside domino safety, plant layout design actually involves several different issues that have to be considered at the same time: constraints on process requirements, cost, safety, regulations, plant construction, services and utilities availability, etc. Suggested values for segregation distances for various equipment units are traditionally used in the layout design (Mecklenburgh, 1985; Mannan, 2005). Design optimization tools mainly focus on economic aspects, even if safety issues have been considered in some recent works (Díaz-Ovalle et al., 2010; Jung et al., 2010; Jung et al., 2011; Nolan and Bradley, 1987; Penteado and Ciric, 1996; Patsiatzis et al., 2004). A more detailed safety analysis, involving evaluation of possible accidental scenarios and consequence analysis (Section 12.2), is generally confined to the final stages of the design life cycle when the risk performance of the whole plant is verified. However, limited margins for layout improvement are left at this final stage. In order to anticipate safety issues related to the prevention of escalation in layout design, several indicators applicable in the early stages of layout design aimed at the identification of domino potential and optimal layout configuration were proposed (Tugnoli et al., 2007, 2008a,b, 2012). Some of these methods are presented in Section 12.4. In Section 12.3 the application of different levels of risk reduction strategies in safer design respect to domino hazards is discussed. In particular the crucial role of inherent safety design of layout plans is pointed out.

12.2   Safety Distances for Escalation: Application to Layout Definition

12.2.1   Thresholds for Escalation

Accidental scenarios involving intense energy releases are necessary to trigger the escalation sequences typical of domino accidents, where damage to atmospheric and pressurized target equipment takes place. Table 12.1 shows the primary scenarios that are more likely to trigger escalation effects, as emerged from the analysis in Chapters 2 and 3. Definition, modeling details and physics of the scenarios listed in Table 12.1 are widely described in Part 1. Table 12.1 also reports the physical effects responsible of escalation identified for each scenario, which were defined in Chapter 3 as the "escalation vectors" of the scenario. As shown in the table, three main escalation vectors were identified: heat radiation or flame impingement, overpressure and

Table 12.1 Specific Escalation Thresholds Considered for the Different Primary Scenarios

Scenario	Escalation Vector	Target Category	Escalation Threshold	Safety Distance
Flash fire	Heat radiation	Floating roof tanks	Flame envelope	Maximum flame distance
		All other units	Unlikely	–
Fireball	Heat radiation	Atmospheric	$I > 100 \text{ kW/m}^2$	Maximum flame distance
		Pressurized	Unlikely	–
Jet fire	Heat radiation	All	Flame envelope	–
		Atmospheric	$I > 15 \text{ kW/m}^2$	50 m from flame envelope
		Pressurized	$I > 45 \text{ kW/m}^2$	25 m from flame envelope
Pool fire	Heat radiation	All	Flame envelope	–
		Atmospheric	$I > 15 \text{ kW/m}^2$	50 m from pool border
		Pressurized	$I > 45 \text{ kW/m}^2$	20 m from pool border
VCE	Overpressure	Atmospheric	$P > 22 \text{ kPa}$	$R = 1.75$ (ME); 1.50 (BS)
		Pressurized; elongated (toxic)	$P > 20 \text{ kPa}$	$R = 2.10$ (ME); 1.80 (BS)
		Elongated (flammable)	$P > 31 \text{ kPa}$	$R = 1.35$ (ME); 0.85 (BS)
	Heat radiation	See flash fire	See flash fire	See flash fire
Confined explosion	Overpressure	Atmospheric	$P > 22 \text{ kPa}$	20 m from vent
		Pressurized; elongated (toxic)	$P > 20 \text{ kPa}$	20 m from vent
		Elongated (flammable)	$P > 31 \text{ kPa}$	20 m from vent
Mechanical explosion, BLEVE	Overpressure	Atmospheric	$P > 22 \text{ kPa}$	$R = 1.80$
		Pressurized; elongated (toxic)	$P > 20 \text{ kPa}$	$R = 2.00$
		Elongated (flammable)	$P > 31 \text{ kPa}$	$R = 1.20$
	Missile projection	All	Fragment impact	300 m (impact prob. $< 5 \times 10^{-2}$)

I, heat radiation intensity; P, maximum peak static overpressure; R, energy scaled distance; ME, multienergy method (Van den Berg, 1985); BS, Baker–Strehlow method (Strehlow et al., 1979; Tang and Baker, 1999).
Source: Adapted from Cozzani et al. (2006).

fragment projection. Toxic release was excluded from the present analysis because this physical effect does not result directly in a loss of containment (LOC) or in the damage of secondary equipment (Chapter 3).

Thresholds values can be effectively used to describe the minimum intensity of primary scenarios able to trigger escalation. As discussed in Chapter 3, the specific features of the accident scenario and of the target may play an important role in the propagation potential and deserve a throughout discussion. As a matter of facts, some scenarios may not be able to critically damage secondary equipment, either because not sufficiently intense (e.g. low-energy explosions) or because of their quick time evolution (e.g. flash fires). For other scenarios, vulnerability may depend on the characteristics of the target unit. For example, different thresholds for escalation can be defined for scenarios involving blast wave as escalation vector, depending on the kind of target equipment and the material contained. This is the combined result of the different toughness of the equipment to the escalation vector and, on the other hand, of the damage state (DS) required to initiate severe secondary scenarios. A conceptual framework based on the description of secondary target damage by a discrete number of structural DSs and of loss intensities (LI) was proposed to approach the problem (Cozzani et al., 2006). The analysis carried out evidenced that while severe DSs always lead to large LI and consequently to severe secondary scenarios, also minor structural damages can trigger severe secondary consequences in case of specific operative conditions and material hazards. This is, for example, the case of volatile toxic substances contained in pressurized equipment, which may form relevant toxic dispersions also in the case of relatively moderate leak diameters (e.g. failure of connection nozzles). Therefore, specific escalation thresholds, rather than structural damage thresholds, should be used in the analysis of propagation potential. Table 12.1 summarizes the main findings of Chapter 3 with this respect.

12.2.2 Safety Distances for Escalation

The intensity of each escalation vector depends on the total amount of energy (or of substance) which is possibly released from the primary system of containment (reactor, storage tank, etc.). The maximum distance at which escalation effects may be considered credible is called "safety distance", since it is the minimum segregation distance between units required to avoid an escalation event. If threshold values for escalation are available (Section 12.2.1; Chapter 3), the safety distance may be easily calculated by the application of standard literature models for consequence assessment (Center for Chemical Process Safety (CCPS), 1994; Uijt de Haag and Ale, 1999; Van Den Bosh and Weterings, 1997). Based on these considerations, Cozzani et al. (2007, 2009) proposed simplified normograms for the swift assessment of safety distances. Two examples are reported in Figure 12.1.

According to the discussion presented in Chapter 3, the safety distances closely reflect not only the nature of the escalation vector but also the primary accident scenario and the characteristics of the target unit (Table 12.1). For flash-fire scenarios, the extension of the flame envelope is the safety distance for potential sources of flammable vapors (e.g. floating roof tanks). Other kind of equipments were identified as unaffected by the

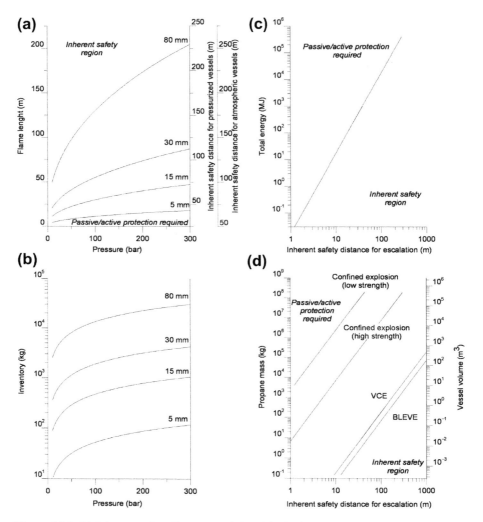

Figure 12.1 (a) Inherent safety distances and (b) critical primary vessel inventory for different release diameters in the case of jet fires from hydrocarbon gases at ambient temperature. (c) Explosion energy and (d) equivalent mass and volume of a propane vessel with respect to the inherent safety distances for escalation involving atmospheric vessels.
Source: Adapted from Cozzani et al. (2007, 2009).

scenario. For fireballs, the safety distance for atmospheric equipment may reasonably be assumed as to the radius of the fireball. In fact, recent results confirmed that the possibility of escalation may be reasonably excluded for pressurized vessels, even in the absence of passive protections (Cozzani and Salzano, 2004; Cozzani et al., 2006).

In the case of jet fires, the safety distance depends on the maximum flame length and direction. A study by Cozzani et al. (2006) suggests that time to failure (ttf) for atmospheric vessels is always higher than 15 min for targets located more than 50 m from the flame envelope. The value of 15 min was assumed as a reference time for activation of mitigation actions having a high probability of success (e.g. activation of water deluges at the arrival of the fire brigade). Since in general no thermal insulation is used on atmospheric tanks, this value may be assumed as the safety distance in the case of jet fires. With respect to pressurized vessels, the minimum value calculated by Cozzani et al. (2006) for the ttf is of about 13 min. These values are comparable to the time assumed as necessary for an effective mitigation. Thus, an escalation as a consequence of stationary radiation, without flame impingement, may be not excluded for an unprotected pressurized vessel. However pressurized vessels have usually passive fire protections (thermal insulation), as well as active protections (water deluges), that may raise the actual value of ttf up to safe values. Therefore, the escalation involving pressurized vessels is scarcely credible in the case of distant source radiation from jet fires. Furthermore, even in the case of unprotected vessels, the escalation is definitely not credible for distances higher than 25 m from the flame envelope in the jet direction.

The same analysis applied to pool fires evidenced that escalation may be considered possible for atmospheric vessels at distances lower than 50 m from pool border with a reference time of 15 min (Cozzani et al., 2006). In the case of pressurized vessels, a conservative safety distance of 20 m may be assumed, although this value may be further reduced taking into account the effect of thermal insulation (Section 12.3).

For scenarios involving overpressure as escalation vector, the safety distance depends primarily on the amount of energy released in the scenario, as well as on the type of scenario. The safety distance is related to the distance at which the peak overpressure of the blast wave equals the threshold values for damage by overpressure (Table 12.1). In the case of vapor cloud explosions (VCEs), this may be calculated by the standard approach used in Quantitative Risk Assessment (QRA) (CCPS, 1994; Uijt de Haag and Ale, 1999) estimating the explosion energy and evaluating the "strength" of explosion on the basis of plant layout and fuel reactivity, i.e. the "strength factor" (F) for the multienergy method (Van den Berg, 1985) or the flame Mach number (M_f) for the Baker–Strehlow methodology (Tang and Baker, 1999). The values of reduced distances reported in Table 12.1 can be considered for safety distance evaluation.

The explosion of industrial equipment due to the internal combustion of gases, vapors or dust is in general destructive even for high-strength enclosures. Hence, venting devices are often introduced for mitigation purposes. When the vent section opens, the rapid depressurization through the vent section may produce a blast wave. External explosions due to the combustion of unburnt gases released after the vent opening were also observed (Forcier and Zalosh, 2000; Whitham, 1956). The analysis of safety distances by Cozzani et al. (2006) evidenced as failures due to overpressure effects is not possible for targets located more than 20 m from the vent outlet.

Missiles are an escalation vector generated in the case of mechanical explosions and boiling liquid expanding vapor explosions (BLEVEs). The escalation is caused by the missile impact on a target vessel, causing a LOC. This requires two conditions to be verified: the distance of the target vessel must be lower than the maximum credible projection distance and the impact must be followed by an LOC at the target vessel. The latter requirement is usually assumed to be verified in the conservative approach used in the assessment of missile damage (Gubinelli et al., 2004; Mannan, 2005). The distance of projection is mainly dependent on the initial missile velocity, direction and on the drag factor (CCPS, 1994). For most mechanical explosions, the maximum projection distance of fragments is usually higher than 1000 m and projection distances up to 900 m were observed in past accidents (Gubinelli et al., 2004; Holden and Reeves, 1985; Westin, 1971). Since these distances are well beyond typical interunit distances of a plant, defining an escalation threshold based solely on possible projection distances is therefore impractical. However, the probability of impact quickly decreases with the projection distance. A study by Gubinelli et al. (Gubinelli et al., 2004; Cozzani et al., 2006) shows that the impact probability of a fragment form a mechanical explosion is about 3×10^{-1} at 100 m and of 5×10^{-2} at 300 m. In the case of BLEVE, conservative values of impact probability are of 2.5×10^{-1} at 100 m and of 2.5×10^{-2} at 300 m. Once defined a tolerability criteria for impact probability, these findings can be used to evaluate likely affected areas.

12.3 The Role of Safety Barriers

12.3.1 Categories of Safety Barriers

The risk of domino accident can be reduced during design activities by adopting different strategies. According to the schematization provided by CCPS (Bollinger et al., 1996), the domino propagation can be eliminated by inherently safer design, limited by engineered barriers (passive or active systems) and/or managed by appropriate procedural safeguards. Current practice mainly relies on active and passive safety strategies. The passive safety approach to domino prevention consists in the proper design of physical barriers and protection systems (e.g. thermal insulation of process equipment) whose effect, when needed, is available without any external intervention. This strategy is widely used for the reduction of accident consequence, although the cost of passive protection systems may be relevant (Hendershot, 1997). Active strategies to prevent escalation events are usually considered less reliable in the hierarchy of safety but, at least for some primary scenarios as pool or jet fires, these approaches may be effective and are often compulsory in the national legislation as well as in the international design standards.

12.3.2 Inherent Safety Barriers

The inherent safety approach aims at the elimination of the domino hazard (i.e. the potential to trigger an escalation sequence leading to a domino event). The advantages of applying an inherent safety philosophy throughout the whole plant life cycle have been clearly highlighted by Kletz (Kletz, 1978, 1991; Kletz and Amyotte, 2010) and other studies (Bollinger et al., 1996; Hendershot, 1997; Hurme and Rahman, 2005; Khan and Amyotte, 2003). An inherent approach to domino prevention may be easily applied in early plant design, taking into account the possibility of domino events during layout definition. In this case, escalation events may be avoided simply by introducing appropriate safety distances between the more hazardous process units (those having large inventories of flammable or toxic substances) and other process installations. The spatial arrangement of process units influences the ability of an accidental event to propagate from one unit to another (domino effect), resulting in escalation of the magnitude of the accident consequences (Cozzani et al., 2007; Khan and Abbasi, 1999). The position of populated targets (e.g. buildings) with respect to possible sources of hazard is, as well, of major concern due to the possibility of exposure and fatalities. Moreover, layout design affects the accessibility of the different areas in a plant, which is a critical element for both accident frequency (e.g. affecting easy, regular operations and maintenance) and accident management (e.g. firefighting operations and evacuation). Finally, inherently safer systems can reduce the high costs usually associated with the full plant safety life cycle—from hazard management to regulatory liabilities and safety system maintenance (Gupta et al., 2003; Kletz, 1984, 1998; Khan and Amyotte, 2005).

Thus plant layout design is the best stage to carry out the implementation of specific preventive and protective inherent safety measures aimed to prevent escalation. However, practical constraints limit the full application of inherent safety strategies. As a matter of fact, the integration of active and passive measures with an inherent safety approach seems to be the more promising route toward an efficient reduction of the risk due to escalation events. Therefore domino hazard reduction through inherent safety will make add-on measures less critical and more effective.

The principles of inherent safety can be summarized by a list of guide words (Kletz, 1998; Hendershot, 1999). Although the terminology of inherent safety varies somewhat throughout the process safety community, there exists a general commonality of thought on the meaning of the different principles when expressed with alternate labels. Cozzani et al. (Cozzani et al., 2007, 2009; Tugnoli et al., 2008a) analyzed the application of a set of classical inherent safety guide words (intensification, moderation, substitution, simplification and limitation of the effects) to the layout definition activities. They concluded that safer layouts can be achieved by the integrated application of inherent safety guide words to both the process design and the actual layout design. Clearly enough, the same inherent safety guide word has a different implementation in these two design stages, since different constraints are in place.

The process design phase is characterized by the possibility to change parameters; for example, operative and storage conditions, equipment design, and inventories. Here, the guide word "intensification" can be effectively referred to the reduction of the inventory in single equipment items or of the number of equipment items. Since the inventory involved is often a significant parameter in determining the escalation of a primary scenario (e.g. for BLEVEs or VCEs), the minimization of quantities stored or processed is an effective measure for the reduction of hazard. Also actions related to "moderation" guide word would lead to important reduction in escalation possibilities. The use of less hazardous conditions, as the shift to safer storage technologies (as, in general, the use of cryogenic instead of pressurized storages), is effective in reducing, on one hand, the hazard of the primary event and, on the other, the vulnerability of equipment to escalation as well as the severity of the possible secondary scenarios. Actions falling under the "intensification" and "moderation" guide words, when introduced specifically for issues related to the prevention of escalation, may be considered particularly for equipment items having relevant inventories, such as storage tanks (e.g. reducing storage capacity or changing storage conditions). Finally, the "substitution" of substances with others having less hazardous proprieties, and the "simplification" of processes, would be other inherent safety guide words implemented in the process design phase. However, these usually require relevant modifications of the design that may conflict with other drivers of the design phase (costs, use of well-known technologies, standardization, etc.).

In the later phase of plant layout design, the inherent safety guide words should be applied to the layout plot as a system. For example, the "simplification" guide word calls for the reduction of unnecessary complexity in the layout plot. Complexity of layouts can easily arise as the disposition of units diverts from the logical process flow order, or as further items (e.g. walls, equipment of other production lines, and buildings) are added to the plan. Therefore, it is possible that layout design choices, even if oriented to satisfy the other inherent safety guide words, eventually result in a negative feedback with respect to "simplification" of the layout.

The "limitation of the effects" guide word is fully applied in the layout design phase. "Limitation of effects" is sometimes considered as a "minor" guide word, as it accepts that a negative effect will somehow take place. However, in the perspective of escalation, this guide word should assure that no secondary event will be caused by domino effects, thus pursuing the prevention of domino accidents by an effectively inherent approach. This includes providing appropriate segregation distances among units, as well as choosing a safer spatial arrangement of the plant units (e.g. avoid sensitive targets near hazardous units, use low-hazard units as "hazard buffer").

12.3.3 Passive Safety Barriers

Inherent safety strategies are applicable mainly in basic design activities. Passive protection systems are usually implemented in more detailed design stages. Passive strategies consist in general in the use of physical barriers (e.g. thermal insulation of process equipment, fire-resistant walls, blast walls, etc.). They can be applied both as prevention and/or mitigation barriers. In the case of domino prevention, passive barriers have a distinctive value as mitigation safeguards, since, by their very nature, are immediately effective at the instant of the primary accident. This null response time makes passive barriers effective in preventing escalation also for scenarios that are rapidly evolving or that are difficult to detect. For example, in case of heat radiation originated from a fireball scenario only passive mitigations systems (e.g. thermal insulation) should be considered effective, since the rapid evolution of the scenario (in general of the order of 5–20 s) excludes the possibility of active protection systems. The same applies to protection from shock waves originated by explosion scenarios. In some cases (e.g. fragment projection) passive barriers can be the only applicable mitigation measure, since inherent strategies may not be applied in practice (e.g. excessive segregation distances). The CSB Report on the Bayer CropScience accident of 2008 (U.S. Chemical Safety and Hazard Investigation Board (US-CSB), 2011) presents an analysis of the protection provided by a blast blanket in the case of fragment impact. Though the report refers to a specific case, it evidences as blast protection can be an effective solution for mitigating damage from both small and large missiles. However, the design of the supporting structure must specifically account for the expected dynamic loads, since deformations originated by the impact may as well damage piping and process equipment.

Passive protections from heat radiation or flame impingement (e.g. fireproofing, firewalls, etc.) may be very effective in slowing the heat-up of the structures in a plant caused by a primary fire scenario, delaying secondary failures or even preventing them, if appropriately integrated with active or procedural actions (e.g. insulation, firefighting, and depressurization of possible target units). However, in the case of flame impingement, the integrity of the protection is critical, since hot spots may be formed in presence of defects of the insulation, leading to local stress, overheating and failure (Mannan, 2005; Roberts, 2004a,b; Shirvill, 2004).

Catch basins are a particularly effective form of passive mitigation barriers when dealing with flammable liquid spills: they contribute both to limit the extension of possible impingement scenarios (pool fires) as well as the evaporation rate (cloud formation and subsequent possible VCE). Trenches and impounding basins positively contribute in a similar way.

Passive protection can also be effective for mitigation in domino chains involving blast waves or fragments as propagation vector. However, the installation of barriers such as blast walls, mounding, etc. can introduce significant costs and maintenance issues that have to be carefully considered: as a matter of fact, the use of these barriers is usually limited to the protection of the most critical structures in the plant (control room, critical equipment, storage of explosive material, etc.).

12.3.4 *Active and Procedural Safety Barriers*

As well as passive barriers, active safety barriers may be divided according to their role as prevention or mitigation barriers. Prevention barriers are mainly aimed at the reduction of the frequency of occurrence of the primary scenario (e.g. alarm-activated process shout-down). These barriers, while surely effective in reducing the domino risk, do not have a specific effect limited to domino scenarios, but are part of the general loss prevention measures of the plant. As such, they are usually implemented based on considerations or requirements independent of their role in domino risk reduction. The active mitigation barriers, instead, can be specifically targeted at the reduction of the domino propagation (e.g. depressurization and water deluge). However, active mitigations typically have a significant time lag of intervention. Time effectiveness criteria require comparing the activation time with the characteristic time of the accident scenario.

For example, in the case of fireball, the characteristic time ranges typically between 1 and 20 s and active mitigation on target vessel (e.g. water deluge) is generally ineffective. The same consideration applies to overpressure and fragment projection scenarios. For example, in a typical VCE, the total duration of the explosion may range typically from few tens to hundreds of milliseconds (or even more in the case of very low Mach deflagrations). These times are typically less than characteristic response times of any protection equipment (Hoiset et al., 1999). Clearly enough some mitigation measures, such as unit isolation by emergency shutdown, can still be effective in reducing the severity of consequences in those scenarios which involve delayed ignition (e.g. VCE).

In the case of stationary fires (jet fire and pool fires) active barrier may be very effective in reducing the probability of domino escalation. For example, water deluge can be used to minimize the temperature increase of equipment exposed to fire. This is however generally limited to the heat radiation zone, since the intense heat load in the impingement area may make water deluge ineffective, via formation of local hot spots (Roberts, 2004a,b; Shirvill, 2004). A close integration of passive and active measures is advised in these cases: passive protections (primarily fireproofing) may be crucial to provide sufficient time for intervention of active measures. The study by Cozzani et al. (2006) evidences that in the case of pressurized vessels, the ttf may be considerably extended by passive protections, allowing active or procedural measures to be implemented.

The same considerations about active barriers generally apply to procedural emergency measures for accident mitigation. For procedural measures the characteristic response time may be longer by an order of magnitude compared to active measures. Therefore no procedural measures are usually applicable to fast evolving scenarios (fireball, mechanical explosions, VCE, etc.). On the other side, the effectiveness in emergency management of scenarios involving stationary fires (e.g. use of fire monitors) can be crucial in preventing escalations. As discussed above, a close integration with passive and active protections is crucial in these cases. Similarly, the firefighting operations aimed at the extinction of the flame eliminate the escalation vector, though they may be limited only to some stationary fires (e.g. pool fire). It is worth reminding that not properly managed firefighting operations may introduce new scenarios for accident escalation, as in the case of environmental damage from spill of contaminated firefighting water. Adequate design standards and management procedures should be implemented to prevent this kind of postaccident fallouts (Mannan, 2005).

Finally evacuation should be mentioned as an effective measure to prevent harm to people in domino accidents, especially for the internal population where characteristic times of egress may be shorter than some of the propagation times. However, it must be remarked that this measure do not mitigate the domino escalation itself, but only the adverse effects to people, leaving unaltered the consequences on the assets and the environment.

12.4   Advanced Methods for Layout Definition

12.4.1   Approach to Domino Prevention in Layout Definition

Design experience shows that the use of conceptual tools alone frequently fails in solving the tradeoffs related to the conflicting needs in the design improvement. A solution consists in coupling the conceptual tools with appropriate performance metrics. Metrics for design analysis and computer-aided tools have been developed to assist the various steps of plant design and layout definition (see, for example, Mannan,

2005; Mecklenburgh, 1973, and the references cited therein). However, current research worldwide is focused primarily on optimization of the economic aspects of the facility plot (see, for example, Barbosa-Pòvoa et al.,2002; Deb and Bhattacharyya, 2005; Georgiadis et al., 1999; Papageorgiou and Rotstein, 1998) rather than on prevention of domino accidents. From the safety point of view, early layout design is mainly based on industrial practice and simple guidelines or empirical rules (e.g. tables of conventional segregation distances, Mecklenburgh, 1985; Mannan, 2005). Some attempts to include safety aspects in economic optimization of layout design have been made (e.g. accounting for cost of safety devices and of losses) (Nolan and Bradley, 1987; Penteado and Ciric, 1996; Patsiatzis et al., 2004). A layout and siting approach coupling consequence simulation of selected accident scenarios and mixed-integer nonlinear program optimization was proposed by Mannan et al., (Jung et al., 2010, 2011; Diaz-Ovalle et al., 2010). Tugnoli et al. (Cozzani et al., 2009; Landucci et al., 2008; Tugnoli et al., 2008a,b,c, 2012) have recently proposed several metrics applicable to support inherently safer layout design.

12.4.2 Key Performance Indicators for the Assessment of Hazard Potential

The Inherent Safety Key Performance Indicators (IS-KPI) method was initially developed by Tugnoli et al. (2007, 2012) to support the inherently safer design of industrial processes. The method was later extended to the assessment of domino potential, and consequently, to the support of layout design activities (Cozzani et al., 2009; Landucci et al., 2008; Tugnoli et al., 2012). The output of the methodology is a set of key performance indicators (KPIs) capturing different aspects of the inherent safety performance and framing an integrated safety profile of the technology, that can support decision making in design and/or communicate the performances to the stakeholders. The recursive application of the IS-KPI method can provide continuous improvement in the design.

The IS-KPI methodology consists of four main steps, as illustrated in Figure 12.2: (I) identification of process units; (II) definition of accident scenarios; (III) calculation of escalation distances; and (IV) calculation of KPIs. The input data required for the application of the method are preliminary data on process equipment and operational parameters: process flow diagram, substances, operating conditions, preliminary material flows, general technical specifications of the equipment, estimation of inventories, and preliminary layout.

In step I of the methodology, the plant is divided in process units. A process unit is an equipment unit or item with a definite function and geometrical taxonomy. The units identified at this stage represent the basic elements of the assessment. The equipment units are classified based on their geometrical structure.

In step II of the methodology, the critical events (CEs) leading to primary scenarios are identified. Several well-known techniques are available for this activity (American Petroleum Institute (API), 2004; Center for Chemical Process Safety (CCPS), 2008; Delvosalle et al., 2006; International Organization for Standardization (ISO), 2000; Paltrinieri et al., 2011; Mannan, 2005; Stör-fall

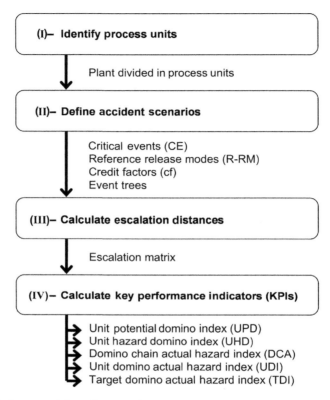

Figure 12.2 Conceptual flow diagram of the IS-KPI method applied to layout hazard assessment.

Kommission (SFK), 2002). The use of an integrated array of different techniques may be necessary in this step when complex hazards are present.

Reference release modes (R-RMs) are then identified for the possible CEs. A R-RM is characterized by release geometry (e.g. instantaneous release, leak through a breach of a given equivalent release diameter, etc.), material released (substance or mixture), condition of release (phase/phases, pressure, and temperature), and duration (duration of release and quantity released). The release categories suggested by API 581 standard (American Petroleum Institute (API), 2000), the MIMAH procedure (Delvosalle et al., 2006) or the "Purple Book" (Uijt de Haag and Ale, 1999) are examples of release modes suitable for this step.

Design, structural characteristics, and operation mode can make a unit more susceptible to a specific R-RM rather than to another. Design and operation features may result, as well, in one unit to be much more robust than another unit toward the same R-RM (Heikkilä, 1999). A "credit factor" may be determined for each R-RM, reflecting the inherent performance of the unit design in maintaining the integrity of containment. The quantification of the credit factor is based on the estimation of the credibility of each R-RM. Reference failure frequency data may be easily used to evaluate the "credit factor" linked to each class of equipment (see e.g. API, 2000; Delvosalle et al., 2006; Ministerie van de Vlaamse Gemeenschap, 2004; OREDA Project, 2002; Uijt de Haag and Ale, 1999). Specific failure frequency data, derived for example from available statistical data or from conventional fault-tree analysis (Delvosalle et al., 2006), may be introduced to account for the adoption of technologies with higher safety standards. Clearly enough, the effect of engineered barriers (active and procedural) on frequency reduction is neglected in the assessment.

Finally, a set of event tree diagrams is associated to each CE, featuring the possible dangerous phenomena or consequence scenarios (e.g. fireball, pool fire, etc.). A set of reference event trees, derived from conventional approaches proposed in the technical literature (Delvosalle et al., 2006; Mannan, 2005; Uijt de Haag and Ale, 1999), was defined to identify the expected accident scenarios relevant to each CE.

In step III, the areas potentially affected by the dangerous phenomena following any possible R-RM are calculated. The characteristic dimension of these areas (e.g. the maximum distance where the consequences of an accident can damage targets severely enough to result in domino propagation) is critical for the quantification of KPIs, since it quantifies the intensity of each escalation vector. This parameter is named "escalation distance".

Different types of physical effects (thermal radiation, overpressure or fragment projection) can be caused by the accident scenario. The thresholds discussed in Section 12.2 can be adopted as reference for each physical effect of concern. Several widely accepted models and commercial software tools are available for consequence analysis, and may be used for the current purpose. However, the use of the same model in the assessment of similar scenarios is advised in order to obtain consistent results.

Modeling the dangerous phenomena associated to each R-RM yields a matrix of escalation distances called the "escalation matrix". Each element of the escalation matrix for the k-th unit, $n_{i,f,k}$, is the expected damage distance calculated for the f-th dangerous phenomenon of the i-th R-RM:

$$n_{i,f,k} = \max(d_{i,f,k}, c) \tag{12.1}$$

where $d_{i,f,k}$ is the calculated damage distance and c is a constant, considered equal to 5 m in this study. The use of the constant c in Eqn (12.1) allows the definition of a "near-field" zone, in which the consequences of the event are neglected. This takes into account the unreliability of conventional consequence assessment models in accurately describing the consequences of the events in the "near field", thus avoiding biases in the analysis due to these uncertainties.

A vector, named "unit escalation vector", $\mathbf{e}_{i,k}$, is defined for each unit from the escalation matrix. It consists of the maximum damage distances calculated for each i-th R-RM considered for the k-th unit:

$$\mathbf{e}_{i,k} = \max_{j}(n_{i,f,k}). \tag{12.2}$$

Finally, in step IV escalation distances may be used for a preliminary evaluation of the inherent domino hazard in a layout. An extensive set of KPIs is available for this purpose:

- A "unit potential domino index" (UPD) that may be used for a preliminary ranking of the worst-case hazard of the primary units independently of the definition of a layout.
- A "unit hazard domino index" (UHD) that may be used for a preliminary ranking of the expected hazard of the primary units independently of the definition of a layout.
- A "domino chain actual hazard index" (DCA) that may be used to assess the hazard of a specific escalation between a primary and a secondary unit.
- A "unit domino actual hazard index" (UDI) that expresses the hazard due to escalation scenarios triggered by a given primary unit.
- A "target domino actual hazard index" (TDI) that expresses the hazard for a given target unit due to escalation scenarios triggered by other primary units.

The UPD is defined as the area potentially affected by worst-case escalation effects, estimated on the basis of the intensity of the escalation vector.

$$\text{UPD}_k = \max_i(\mathbf{e}_{i,k})^2 \tag{12.3}$$

where UPD_k is the unit domino potential index of the k-th unit and $\mathbf{e}_{i,k}$ is the corresponding escalation vector. The unit potential domino index (UPD) is therefore representative of the maximum impact area that may be derived from the worst-case accident scenario considered for the unit. The index represents a leading indicator of the domino hazard potential of the unit generating the escalation vector. It allows a preliminary screening of domino hazard and the identification of critical potential sources of escalation events, independently on the actual layout configuration.

The UHD is defined similarly to UPD. However in this case the safety performance (robustness to CEs) of the unit equipment is taken into account, quantifying the effects that may derive from the worst credible accidents.

$$\text{UHD}_k = \sum_i \text{cf}_{i,k} \cdot \mathbf{e}_{i,k}^2 \tag{12.4}$$

where UHD_k is the unit domino hazard index of the k-th unit, $\text{cf}_{i,k}$ and $\mathbf{e}_{i,k}$ are, respectively, the credit factor and the corresponding escalation vector element for the i-th R-RM. The value of the index is higher for units having higher potential hazards or a higher record of LOC events. The index represents a leading indicator of the domino hazard of the unit that accounts for the credibility of the loss and, thus, the inherent safety of the containing equipment. Also in this case, the index value is a reference independent of the actual layout configuration.

In order to evaluate the specific escalation hazard between two units in a given layout, the DCA is defined as the ratio of the escalation vector to the actual distance between the primary scenario and the target unit:

$$\text{DCA}_{i,k,j} = \frac{\mathbf{e}_{i,k}}{D_{k,j}} \tag{12.5}$$

where $\text{DCA}_{i,k,j}$ is the hazard index for the i-th R-RM of the k-th unit, considered as a domino trigger toward the j-th unit. The actual equipment distance ($D_{k,j}$) is available only if a layout plan, either in a preliminary form, is defined. If separation distances and plant layout are not available (e.g. in the early steps of plant design), conventional safety distances may be used to provide at least an estimation of the expected chain propagation hazard. As a matter of fact, these distances reflect the current practice in layout design and are more likely adopted in a standard layout design. Such distances are reported in several technical publications and standards (Mannan, 2005; Mecklenburgh, 1973; and references cited therein).

The critical primary units with respect to domino hazards may be identified by the UDI:

$$\text{UDI}_k = \sum_{j=1}^{u_k} \max_{i=1}^{m_k}(\text{DCA}_{i,k,j} \cdot \alpha_{i,k}) \tag{12.6}$$

where u_k is the total number of units considered for possible escalation caused by the k-th unit, m_k is the total number of primary escalation scenarios of the k-th unit that may trigger escalation and $\alpha_{i,k}$ is an inventory parameter, ranging from 0 to 10. The inventory parameter takes into account that the escalation hazard may depend on the inventory of the primary unit (Cozzani et al., 2009). This is of particular importance for escalations triggered by stationary radiation, where a reference time was assumed to define the escalation thresholds.

The TDI is related to the hazard of a target unit:

$$\text{TDI}_j = \sum_{k=1}^{q_j} \max_{i=1}^{m_k}(\text{DCA}_{i,k,j} \cdot \alpha_{i,k}) \tag{12.7}$$

where TDI_j is the target domino actual hazard index for j-th target unit, q_j is the total number of units considered for possible escalation scenarios having j-th unit as a target and the other parameters are defined as in Eqns (12.5) and (12.6). The index is aimed at the definition of the actual hazard for a target unit within the assessed layout. Clearly enough it may be calculated as well for external units, in order to assess the escalation hazard toward other facilities.

The numeric values of the UPD and UHD depend on the measurement units used for the damage distance and credit factors, while DCA, UDI and TDI are nondimensional numbers. Further details on the assessment of these indicators and examples of application may be found elsewhere (Tugnoli et al., 2007, 2012; Cozzani et al., 2009; Landucci et al., 2008).

12.4.3 Integrated Inherent Safety Index for Layout Assessment

Khan and Amyotte (2004, 2005) proposed the integrated inherent safety index (I2SI) as an indexing method for the swift assessment of alternative process options in the light of the inherent safety guide words. The method was expanded and applied to the analysis of the inherent safety and safety cost performance of layout design expanded by Tugnoli et al. (2008a,b). The resulting "integrated inherent safety index for layout assessment (I2SI-LA)" can be considered an effective integrated tool for domino prevention that introduces inherent safety concepts into the early stages of layout design by means of an easy-to-use scoring approach. When compared to the IS-KPI approach described in the previous section, the I2SI-LA allows a more swift and comprehensive assessment, though it resorts more extensively to expert judgment in the definition of the contributors to index scores. An example of application of I2SI-LA in the comparison of alternative layout plans is reported by Tugnoli et al. (2008b).

The conceptual framework of the I2SI is given in Figure 12.3. The input information necessary to perform the assessment of a layout option are a preliminary definition of the plant plot and of the parameters necessary to describe the hazard of

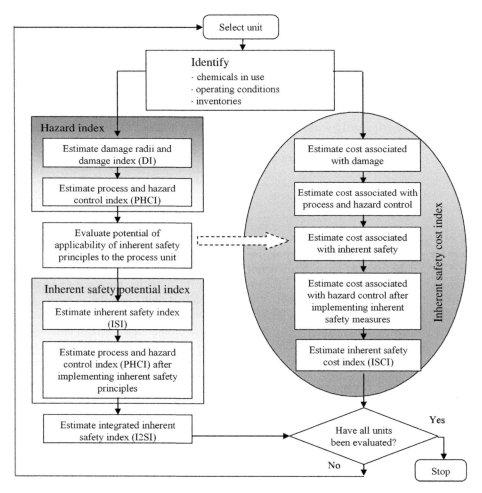

Figure 12.3 Conceptual flow diagram of the I2SI assessment method. (For color version of this figure the reader is referred to the online version of this book.)
Source: Adapted from Khan and Amyotte (2005).

the pieces of equipment (chemicals and their proprieties, operative conditions, reactions, material balances, inventories, and control system).

The I2SI-LA assessment compares alternative options for a plant layout. One of the options—called the "base case"—is assumed as an arbitrary reference in the assessment. The other options are assessed in relative terms to the base case, evaluating the improvements in the light of inherent safety principles. Each plant option is divided in units. For each unit, the I2SI-LA is composed of two main subindices: a hazard index (HI) and an inherent safety potential index (ISPI):

$$I2SI = \frac{ISPI}{HI} \tag{12.8}$$

The HI is a measure of the damage potential of a single unit after taking into account the process and hazard control measures. The ISPI, on the other hand, accounts for the applicability of the inherent safety principles (or guide words) to the unit. The HI is calculated for the units of the "base case" and the values remain the same for the corresponding units in all the other possible options. A value of I2SI greater than unity denotes a positive response of the inherent safety guide word application (i.e. an inherently safer option). The higher the value of I2SI, the more pronounced the inherent safety impact.

To evaluate alternative layout options, the I2SI values for the N units constituting a plant are combined according to Eqn (12.9):

$$I2SI_{system} = \left(\prod_{i=1}^{N} I2SI_i \right)^{1/2} \tag{12.9}$$

The HI is calculated from two subindices: a damage index (DI) and a process and hazard control index (PHCI).

$$HI = \frac{DI}{PHCI} \tag{12.10}$$

The DI is an expression of the damage radius expected from the unit. It considers for four hazard parameters (fire and explosion, acute toxicity, chronic toxicity, and environmental damage). These are estimated as a function of the expected damage radii for each scenario, as obtained by using simple indexing approaches such as the Safety Weighted Hazard Index (Khan et al., 2001). Further details on the calculation of DI can be found in the original version of the method (Khan and Amyotte, 2004).

The PHCI accounts for the add-on process and hazard control measures that are required or are present in the system. This index is quantified on a scale mutually agreed upon by process safety experts. The index ranges from 1 to 10 for any control arrangement (e.g. temperature control, level control, blast wall, sprinkler system, etc.) and is quantified based on the necessity of this control arrangement in maintaining safe operation for the unit. Further details on PHCI can be found in the original version of the method (Khan and Amyotte, 2005).

The ISPI is composed, similarly to the HI, of two subindices: an inherent safety index (ISI) and a hazard control index (HCI).

$$ISPI = \frac{ISI}{HCI} \tag{12.11}$$

The denominator in Eqn (12.11) is defined in I2SI-LA as "HCI after the implementation of safety measures". In the assessment of HCI, the requirement to install further add-on hazard control measures after the previous analysis and implementation of safety measures in the layout option is assessed (Khan and Amyotte, 2004, 2005). The ISI in Eqn (12.11) is calculated using scores based on the applicability of the inherent safety guide words. The evaluation of ISI follows the same procedure as an hazard and operability (HAZOP) study in which inherent safety guide words are applied to the assessed system. Based on the extent of the applicability and on the ability to reduce the hazard, an index value is computed for each guide word.

The inherent safety guide words were revised in I2SI-LA for their applicability to layout design and evaluation (Tugnoli et al., 2008a). The analysis evidenced that constraints related to previous design steps (e.g. chemical route choice, process design, equipment selection, etc.) limit the applicability of measures aimed at enhancing inherent safety in the layout options. In particular the guide words "minimization" and "substitution" are generally not applicable because materials, equipment characteristics and inventories have already been selected in previous design phases.

"Attenuation", "simplification" and "limitation of effects" were instead identified as the relevant guide words in the framework of layout design. For each guide word a specific value of ISI is estimated and combined together according to Eqn (12.12):

$$ISI = [\max(25, ISI_a^2 + ISI_{si}^* \|ISI_{si}\| + ISI_l^2)]^{1/2} \tag{12.12}$$

where the subscripts refer to the considered guide words (a for attenuation, si for simplification and l for limitation of effects). Equation (12.12) allows negative values for the simplification parameter, although limiting to 5 the lowest value of the final ISI. This yields ISPI equal to 1 for base case units (i.e. ISI = 5) that do not require any hazard protection device (i.e. HCI = 5). The values of ISI for "attenuation" and "limitation of effects" are estimated by conversion monograph (Figure 12.4) as a function of the extent of applicability of the guide word to the assessed option. For "simplification" a reference table is proposed for the direct assessment of the index value by linguistic guidelines (Figure 12.4).

For the guide word "attenuation" the extent of applicability is assessed mainly as the ability of the layout option to reduce the hazard potential from domino effects. A unit having a lower potential to affect other units, triggering less severe domino consequences, scores a higher values of E_a. The protection provided by passive devices is not accounted for applicability of the attenuation guide word, as the focus is on the domino escalation potential which can be reduced only by inherent measures. Since the evaluation of the extent of applicability is admittedly subjective, specific guidelines were introduced to facilitate the quantification of this parameter (Tugnoli

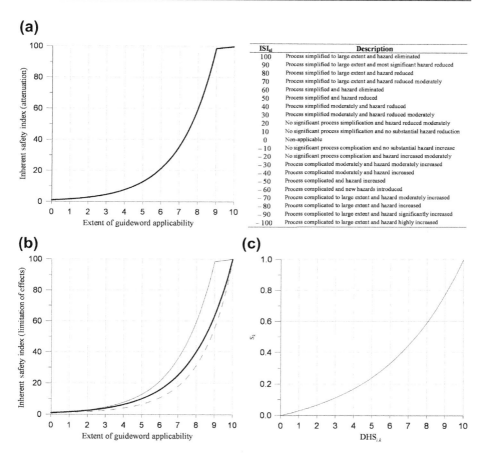

Figure 12.4 Diagrams and table for definition of "inherent safety index (ISI)" and "credit factor for domino escalation" in the I2SI-LA method. (a) ISI for the "attenuation" guide word. (b) ISI for the guide word simplification. (c) ISI for the "limitation of effects" guide word: "limitation of effects of domino escalation" (thin solid line), "limitation of the potential to target buildings" (thick solid line), "limitation of the affected area" (dashed line). (d) Credit factor for domino escalation as a function of $DHS_{i,k}$.
Source: Adapted from Tugnoli et al. (2008a,b).

et al., 2008a,b). In particular an auxiliary index, the DHI is proposed to quantify the domino effect hazards caused by a unit in a specific layout (Tugnoli et al., 2008b).

For the guide word "simplification", an arbitrary reference table is proposed for the direct assessment of the index value by linguistic guidelines (Figure 12.4). The ISI_{si} may score negative values since layout options aiming at safer performance with respect to attenuation and limitation of effects may incur an increase in layout complexity. Since complexity can be defined as negation of simplicity, pursuing "simplification" in layout design means to limit the increase in complexity up to values that are overbalanced by the positive effects from the application of other guide words.

"Limitation of effects" is a guide word that deals with the reduction of the extent of negative consequences arising from accidental events. Accepting that a negative effect may somehow occur, this guide word implies a consideration of the measures aimed to limit consequences. In early layout design both inherent and passive strategies can be implemented to pursue this goal. Thus, the limitation of effects guide word has been considered in the safety analysis of both inherent and passive measures. The extent of applicability of this guide word involves three different elements: (1) the "limitation of the effects of domino escalation" (ISI_{le}), (2) the "limitation of the damage potential to target buildings" (ISI_{lb}), and (3) the "limitation of the affected area" (ISI_{la}).

"Limitation of the effects of domino escalation" considers the reduction of the effects and consequences of domino escalation events, resulting from the integrated action of inherent and passive strategies.

"Limitation of the damage potential to target buildings" accounts for the appropriate location of buildings (workshops, administrative buildings, trailers, etc.) and control or emergency structures (control room, medical center, etc.) in the layout plan so as to limit harm to people and impairment of accident response. The assessment of extent of applicability is related to the presence of buildings into hazard-susceptible areas (i.e. areas affected by fire, explosion and acute toxic effects). The combined effect of different primary units on the same building must be taken into account, since they may change from one layout option to another.

"Limitation of the affected area" assesses the reduction of the spatial area susceptible to dangerous consequences from an accidental event. Such reduction is generally achieved by passive measures (e.g. blast walls and fire resistance walls). The affected area is therefore an indicator of the hazard present, no matter if particular structures are located there (e.g. units or buildings), but simply because final targets (e.g. people and environment) can potentially be present in it.

Monographs for converting the extent of applicability of each parameter to an ISI value are defined in Figure 12.4, following a procedure similar to the attenuation guide word. Auxiliary indexes (e.g. domino hazard score (DHS)) are available to guide the evaluation of extent of applicability (Tugnoli et al., 2008a).

The three contributions are finally combined in Eqn (12.13)

$$ISI_l = \min\left\{ 100, [(ISI_{le})^3 + (ISI_{lb})^3 + (ISI_{la})^3]^{1/3} \right\} \tag{12.13}$$

The cost indexing procedure of I2SI-LA gives an insight on the economic aspects of inherent safety. The costing system (right-hand side of Figure 12.3) is composed of three subindices for each unit in the plant: the conventional safety cost index (CSCI), the inherent safety cost index (ISCI), and the loss saving index (LSI).

The CSCI compares the cost of achieving an acceptable safety level for a unit by conventional safety measures and the expected economic consequence of an accident:

$$CSCI = \frac{C_{ConvSafety}}{C_{Loss}} = \frac{C_{ConvSafety}}{C_{PL} + C_{AL} + C_{HHL} + C_{ECC} + C_{DEC}} \tag{12.14}$$

The numerator in Eqn (12.14), $C_{ConvSafety}$, is the sum of the costs of process control measures and add-on (end-of-pipe) safety measures (i.e. $C_{ConvSafety} = C_{Control} + C_{Add-on}$).

It can be estimated by the number of measures required and their representative reference costs (see, e.g. Khan and Amyotte, 2005). The denominator in Eqn (12.14), C_{Loss}, represents the dollar value of expected losses caused by accidental events in a unit. It is composed of five components: the production loss (PL), which is the economic consequence of production shutdown (i.e. business interruption); the direct asset loss (AL), which is the value of the physical unit itself, depleted by the accidental event; the human health loss (HHL), which is the social cost of fatalities/injuries directly caused by the accident at the unit; the environmental cleanup cost (ECC), which is the cost of restoration of the volume of soil, water and air that were contaminated by the accidental event; and the domino escalation cost (DEC), which is the expected cost of the losses deriving from the possible domino propagation.

The DEC can be expressed as the sum of the loss related to the secondary units involved, weighted by a parameter that features the probability of being involved, as expressed by Eqn (12.15):

$$C_{DEC} = \sum_{j} s_j (C_{AL,j} + C_{HHL,j} + C_{ECC,j}) \tag{12.15}$$

where $C_{AL,j}$, $C_{HHL,j}$ and $C_{ECC,j}$ are, respectively, the additional direct AL, HHL and ECC costs for the failure of each j-th secondary unit, as a result of escalation from the primary unit under assessment. The PL cost is not accounted for the second time in C_{DEC} because the target units are considered to be in the same production line as the primary unit. The factor s_j is a number between 1 and 0 that accounts for the credibility that the failure of the considered unit affects the j-th secondary unit (Tugnoli et al., 2008a,b) for its calculation).

The ISCI features an economic balance between cost of safety for an inherently safer option (marginal cost of inherent safety + cost of additional measures) and the expected cost of losses in case of accident:

$$\text{ISCI} = \frac{C_{InhSafety}}{C_{Loss}} = \frac{C_{InhSafety}}{C_{PL} + C_{AL} + C_{HHL} + C_{ECC} + C_{DEC}} \tag{12.16}$$

The denominator in Eqn (12.16), C_{Loss}, is the same as in Eqn (12.14) for the CSCI. However, the numerator, $C_{InhSafety}$, is the sum of the costs of inherent safety implementation, process control measures, and add-on (end-of-pipe) safety measures still required in the inherently safer layout option (i.e. $C_{InhSafety} = C_{Inherent} + C_{Control} + C_{Add-on}$). The costs of process control and add-on safety measures are calculated following the same procedure as for CSCI. The costs for inherent safety implementation are estimated considering the extent of application of the inherent safety guide words and the costs associated with their application. A marginal cost (i.e. capital cost difference of the assessed option relative to the base case) is calculated for the application of each guide word. For example, the cost of extra space required for increased unit segregation is estimated and referred to as the cost of implementing the guide word attenuation, as earlier discussed. This cost is divided by the extent of applicability of the inherent safety guide words, which denotes the extent to which the guide word will eliminate/reduce the hazards. Hence, the total cost of inherent safety implementation may be calculated as follows:

$$C_{\text{Inherent}} = C_A/E_A + C_{Si}/E_{Si} + C_L/E_L \tag{12.17}$$

where the C variables are the costs and E, the extents of applicability of, respectively, attenuation (A), simplification (Si), and limitation of effects (L).

The possibility of escalation by domino effects, assessed by C_{DEC}, is frequently a prevailing term within the cost of loss. This value can have significant variation for different layout options because of the choices specifically aimed at inherent safety improvement. The LSI index is proposed to map out the economic effect of escalation reduction deriving from inherently safer layout design:

$$\text{LSI}_{\text{option } n} = \frac{C_{\text{InhSafety,option } n} + (C_{\text{Loss,option } n} - C_{\text{loss,base option}})}{C_{\text{loss,base option}}} \tag{12.18}$$

This index compares inherent safety costs with a parameter that represents the savings from avoided loss by domino escalation, since it considers loss variations between the base case (an arbitrary chosen layout option) and the n-th option under assessment.

12.5   Conclusions

In this chapter, the role of design in reducing domino hazard has been explored. Layout design was identified as a key stage in this reduction. The review of the escalation vectors allowed for the identification of threshold values to be used in design activities. In particular, specific thresholds are identified with respect to the primary scenario, the type of target unit and the materials involved. Appropriate safety distances can be estimated and implemented in layout design once the effect thresholds are defined. The hierarchy of risk reduction strategies (inherent, passive, active and procedural) was analyzed respect to their role in domino prevention. The integration of different strategies in accident mitigation was identified as the key in achieving safer plants as regard domino propagation. Inherent safety application in layout definition emerged as a primary strategy to reduce domino hazards. Since conceptual tools alone frequently fail in solving the tradeoffs related to the conflicting needs in the design improvement, inherent safety metrics were introduced to support design activities. Both the metrics presented here (IS-KPIs and I2SI-LA) provide insight on several key parameters relevant for the definition of layouts more effective for the prevention of escalation and domino accidents.

References

American Petroleum Institute (API), 2000. Risk-Bases Inspection Base Resource Document. API Publication 581, first ed. API Publication, Washington.
American Petroleum Institute (API), 2004. Security Vulnerability Assessment Methodology for the Petroleum and Petrochemical Industries, second ed. API Publication, Washington.

Barbosa-Pòvoa, A.P., Mateus, R., Novais, A.Q., 2002. Optimal design and layout of industrial facilities: a simultaneous approach. Industrial and Engineering Chemistry Research 41, 3601–3609.

Bollinger, R.E., Clark, D.G., Dowell III, A.M., Ewbank, R.M., Hendershot, D.C., Lutz, W.K., Meszaros, S.I., Park, D.E., Wixom, E.D., 1996. Inherently Safer Chemical Processes: A Life Cycle Approach. American Institute of Chemical Engineers. CCPS, New York, NY.

Center for Chemical Process Safety (CCPS), 1994. Guidelines for Evaluating the Characteristics of VCEs, Flash Fires and BLEVEs. AIChE, New York, NY.

Center for Chemical Process Safety (CCPS), 2008. Guidelines for Hazard Evaluation Procedures, third ed. Wiley/AIChE, New York, NY.

Cozzani, V., Salzano, E., 2004. Threshold values for domino effects caused by blast wave interaction with process equipment. Journal of Loss Prevention in the Process Industries 17, 437.

Cozzani, V., Gubinelli, G., Salzano, E., 2006. Escalation thresholds in the assessment of domino accidental events. Journal of Hazardous Materials A129, 1–21.

Cozzani, V., Tugnoli, A., Salzano, E., 2007. Prevention of domino effect: from active and passive strategies to inherently safer design. Journal of Hazardous Materials A139, 209–219.

Cozzani, V., Tugnoli, A., Salzano, E., 2009. The development of an inherent safety approach to the prevention of domino accidents. Accident Analysis and Prevention 41, 1216–1227.

Deb, S.K., Bhattacharyya, B., 2005. Solution of facility layout problems with pickup/drop-off locations using random search techniques. International Journal of Production Research 43, 4787–4812.

Delvosalle, C., Fievez, C., Pipart, A., Debray, B., 2006. ARAMIS project: a comprehensive methodology for the identification of reference accident scenarios in process industries. Journal of Hazardous Materials 130, 200–219.

Díaz-Ovalle, C., Vázquez-Román, R., Mannan, S., 2010. An approach to solve the facility layout problem based on the worst-case scenario. Journal of Loss Prevention in the Process Industries 23 (3), 385–392.

Forcier, T., Zalosh, R., 2000. External pressures generated by vented gas and dust explosions. Journal of Loss Prevention in the Process Industries 13, 411.

Georgiadis, M.C., Schilling, G., Rotstein, G.E., Macchietto, S., 1999. A general mathematical programming approach for process plant layout. Computers and Chemical Engineering 23, 823–840.

Gubinelli, G., Zanelli, S., Cozzani, V., 2004. A simplified model for the assessment of the impact probability of fragments. Journal of Hazardous Materials 116, 175.

Gupta, J.P., Hendershot, D.C., Mannan, M.S., 2003. The real cost of process safety – a clear case for inherent safety. Process Safety and Environmental Protection 81, 406–413.

Heikkilä, A., 1999. Inherent Safety in Process Plant Design (Dissertation n. 384). VTT Publications, Espoo, Finland.

Hendershot, D.C., 1997. Inherently safer chemical process design. Journal of Loss Prevention in the Process Industries 10, 151–157.

Hendershot, D.C., 1999. Designing safety into a chemical process. In: Proceedings of the 5th Asia Pacific Responsible Care Conference, Shanghai, China.

Hoiset, S., Hjertager, B.H., Solberg, T., Malo, K.A., 1999. Properties of simulated gas explosions of interest to the structural design process. Process Safety Progress 17, 278.

Holden, P.L., Reeves, A.B., 1985. Fragment Hazards from Failures of Pressurized Liquefied Gas Vessels. IChemE Symposium Series 93 p. 205.

Hurme, M., Rahman, M., 2005. Implementing inherent safety throughout process lifecycle. Journal of Loss Prevention in the Process Industries 18 (4–6), 238–244.

International Organization for Standardization (ISO), 2000. Petroleum and Natural Gas Industries – Offshore Production Installations – Guidelines on Tools and Techniques for Hazard Identification and Risk Assessment. ISO standard 17776. ISO, Geneva, Switzerland.

Jung, S., Ng, D., Lee, J.-H., Vazquez-Roman, R., Mannan, M.S., 2010. An approach for risk reduction (methodology) based on optimizing the facility layout and siting in toxic gas release scenarios. Journal of Loss Prevention in the Process Industries 23 (1), 139–148.

Jung, S., Ng, D., Diaz-Ovalle, C., Vazquez-Roman, R., Mannan, M.S., 2011. New approach to optimizing the facility siting and layout for fire and explosion scenarios. Industrial and Engineering Chemistry Research 50 (7), 3928–3937.

Khan, F.I., Abbasi, S.A., 1999. The world's worst industrial accident of the 1990s. Process Safety Progress 18, 135–145.

Khan, F.I., Amyotte, P.R., 2003. How to make inherent safety practice a reality. Canadian Journal of Chemical Engineering 81, 2–16.

Khan, F.I., Amyotte, P.R., 2004. Integrated inherent safety index (I2SI): a tool for inherent safety evaluation. Process Safety Progress 23, 136–148.

Khan, F.I., Amyotte, P.R., 2005. I2SI: a comprehensive quantitative tool for inherent safety and cost evaluation. Journal of Loss Prevention in the Process Industries 18, 310–326.

Khan, F.I., Husain, T., Abbasi, S.A., 2001. Safety weighted hazard index (SWeHI): a new, user-friendly tool for swift yet comprehensive hazard identification and safety evaluation in chemical process industries. Process Safety and Environmental Protection 79, 65–80.

Kletz, T.A., 1978. What you don't have, can't leak. Chemistry and Industry 6, 287–292.

Kletz, T.A., 1984. Cheaper, Safer Plants, or Wealth and Safety at Work. Institution of Chemical Engineers, Rugby, UK.

Kletz, T.A., 1991. Plant Design for Safety, a User-Friendly Approach. Hemisphere – Taylor & Francis, New York, NY.

Kletz, T.A., 1998. Process Plants: A Handbook for Inherent Safer Design. Taylor & Francis, Bristol, PA.

Kletz, T.A., Amyotte, P., 2010. Process Plants: A Handbook for Inherent Safer Design, second ed. Taylor & Francis, Boca Raton, FL.

Landucci, G., Tugnoli, A., Cozzani, V., 2008. Inherent safety key performance indicators for hydrogen storage systems. Journal of Hazardous Materials 159 (2–3), 554–566.

Mannan, S., 2005. Lees' Loss Prevention in the Process Industries, third ed. Elsevier, Oxford, UK.

Mecklenburgh, J.C., 1973. Plant Layout. John Wiley & Sons, New York, NY.

Mecklenburgh, J.C., 1985. Process Plant Layout. George Goodwin, London, UK.

Ministerie van de Vlaamse Gemeenschap, 2004. Handboek kanscijfers: voor het opstellen van een veiligheidsrapport. ver. 2.0. Heirman JP, AMINAL, Brussel, Belgium.

Nolan, P.F., Bradley, C.W.J., 1987. Simple technique for the optimization of lay-out and location for chemical plant safety. Plant/Operations Progress 6, 57–61.

OREDA Project, 2002. Offshore Reliability Data Handbook (OREDA), fourth ed. SINTEF Technology and Society, Høvik, Norway.

Paltrinieri, N., Tugnoli, A., Bonvicini, S., Cozzani, V., 2011. Atypical scenarios identification by the DyPASI procedure: application to LNG. Chemical Engineering Transactions 24, 1171–1176.

Papageorgiou, L., Rotstein, G.E., 1998. Continuous domain mathematical models for optimal process plant layout. Industrial and Engineering Chemistry Research 37, 3631–3639.

Patsiatzis, D.I., Knight, G., Papageorgiou, L.G., 2004. An MILP approach to safe process plant layout. Chemical Engineering Research and Design 82, 579–586.

Penteado, F.D., Ciric, A.R., 1996. An MILP approach for safe process plant layout. Industrial and Engineering Chemistry Research 4, 1354–1361.

Roberts, T.A., 2004a. Linkage of a known level of LPG tank surface water coverage to the degree of jet fire protection provided. Journal of Loss Prevention in the Process Industries 17, 169.

Roberts, T.A., 2004b. Directed deluge systems designs and determination of the effectiveness of the currently recommended minimum deluge rate for the protection of LPG tanks. Journal of Loss Prevention in the Process Industries 17, 103.

Shirvill, L.C., 2004. Efficacy of water spray protection against propane and butane jet fires impinging on LPG storage tanks. Journal of Loss Prevention in the Process Industries 17, 111.

Stör-fall Kommission (SFK), 2002. Report of the German Hazardous Incident Commission. SFK–GS–38. Stör-fall Kommission: Bonn, Germany. Available at: www.sfk-taa.de (accessed 13.01.13.).

Strehlow, R.A., Luckritz, R.T., Adamczyk, A.A., Shimp, S.A., 1979. The blast wave generated by spherical flames. Combustion and Flame 35, 297.

Tang, M.J., Baker, Q.A., 1999. A new set of blast curves from vapour cloud explosion. Process Safety Progress 18, 235.

Tugnoli, A., Cozzani, V., Landucci, G., 2007. A consequence based approach to the quantitative assessment of inherent safety. AIChE Journal 53 (12), 3171–3182.

Tugnoli, A., Khan, F., Amyotte, P., Cozzani, V., 2008a. Safety assessment in plant layout design using indexing approach: implementing inherent safety perspective. Part 1—guideword applicability and method description. Journal of Hazardous Materials 160, 100–109.

Tugnoli, A., Khan, F., Amyotte, P., Cozzani, V., 2008b. Safety assessment in plant layout design using indexing approach: implementing inherent safety perspective. Part 2—domino hazard index and case study. Journal of Hazardous Materials 160, 110–121.

Tugnoli, A., Khan, F., Amyotte, P., 2008c. Inherent safety implementation throughout the process design lifecycle. In: Book of Abstracts of PSAM9, Hong Kong, China, p. 24.

Tugnoli, A., Landucci, G., Salzano, E., Cozzani, V., 2012. Supporting the selection of process and plant design options by inherent safety KPIs. Journal of Loss Prevention in the Process Industries 25 (5), 830–842.

Uijt de Haag, P.A.M., Ale, B.J.M., 1999. Guidelines for Quantitative Risk Assessment (Purple Book). TNO – Committee for the Prevention of Disasters, The Hague, The Netherlands.

U.S. Chemical Safety and Hazard Investigation Board (US-CSB), 2011. Investigation Report – Pesticide Chemical Runaway Reaction Pressure Vessel Explosion – Bayer CropScience, LP, Report No. 2008-08-I-WV. Available at: http://www.csb.gov (accessed 13.08.12.).

Van den Berg, A.C., 1985. The multi-energy method—a framework for vapor cloud explosion blast prediction. Journal of Hazardous Materials 12, 1.

Van Den Bosh, C.J.H., Weterings, R.A.P.M., 1997. Methods for the Calculation of Physical Effects (Yellow Book). TNO – Committee for the Prevention of Disasters, The Hague, The Netherlands.

Westin, R.A., 1971. Summary of Ruptured Tank Cars Involved in Past Accidents. Report No. RA-01-2-7. Railroad Tank Car Safety Research and Test Project, Chicago, IL.

Whitham, G.B., 1956. On the propagation of weak shock waves. Journal of Fluid Mechanics 1, 290.

13 Managing Domino Effects in a Chemical Industrial Area

Genserik Reniers*,†, *Robby Faes*‡

*Centre for Economics and Corporate Sustainability (CEDON), HUB, KULeuven, Brussels, Belgium, †Centre for Economics and Corporate Sustainability (CEDON), HUB, KULeuven, Brussels, Belgium, ‡3M Environmental Health & Safety, Zwijndrecht, Belgium

13.1   Introduction

Roughly, three types of risks can be distinguished: risks where a lot of historical data are available (type I), risks where little or extremely little historical data are available (type II), and risks where no historical data are available (type III). Consequences of type I risks mainly relate to individual employees (e.g. most work-related accidents), the outcome of type II risks may affect a company or large parts thereof (e.g. large explosions, domino effects, etc.), and type III risks may, e.g. have an unprecedented and unforeseen impact upon the organization and society.

For preventing type I risks turning into accidents, risk management techniques and practices are widely available. Statistical and mathematical models based on past accidents can be used to predict possible future type I accidents, indicating the prevention measures that need to be taken to prevent such accidents.

Type II risks and related accidents are much more difficult to predict. They are extremely difficult to predict via commonly used mathematical models since the frequency with which these events happen is too low and the available information is not enough to be investigated via, e.g. regular statistics. The errors in probability estimates are very large and one should thus be extremely careful while using such probabilities. Hence, managing such risks is based on the scarce data that are available and on extrapolations, assumptions and expert opinions (see, e.g. Casal, 2008). Such risks are also investigated via available risk management techniques and practices, but these techniques should be used with much more caution, since the uncertainties are much higher for these types of risks than for type I risks. A lot of risks (and latent causes) are present which never turn into large-scale accidents due to adequate risk management, but very few risks are present that turn into accidents with huge consequences. Hence, highly specific mathematical models and software should be employed for determining such risks. It should be noted that the complex calculations lead to the appearance of these risks to be accurate; nonetheless, they are not accurate at all (due to all the assumptions that have to be made for the calculations) and they should be regarded and treated as relative risks (instead of absolute risks).

The related accidents with the third type of risks are simply impossible to predict. No information is available about the risks and the risks only extremely rarely turn into accidents. They are the result of pure coincidence and they cannot be predicted by past events in any way, they can only be predicted or conceived by imagination. Such accidents can also be called "black swan accidents" (Taleb, 2007) or highly "atypical accidents" (Paltrinieri et al., 2012). Such events can truly only be described by "the unthinkable" (which does not mean that they cannot be thought of, but merely that people are not capable of (or mentally ready to) realizing that such event really may take place).

In summary, type I unwanted events can be regarded as "occupational accidents" (for example, accidents resulting in the inability to work for several days, accidents requiring first aid, etc.), whereas both type II and type III accidents can be categorized as "major accidents" (e.g. multiple fatality accidents, accidents with huge economic losses, etc.). Thus, domino accidents, being type II or III events, cannot be managed and treated in the same way as occupational accidents. Domino risks should thus be managed with approaches and models which are not based on casuistic of occupational accidents, but based on the expertise and the mindset related to disaster prevention.

Based on the oil and gas producers model for human factors (OGP, 2005), Reniers and Dullaert (2007) discern three dimensions in which measures can be taken to avoid and to prevent unwanted events, and to mitigate their consequences. The three dimensions are People, Procedures and Technology. Applied to domino risk management within an organization, the first dimension, People, indicates how people (individually and in group) deal with cascade risks and think about domino risks in the organization (including domains such as, for example, training for major disaster, competence, behavior and attitude, etc.). The second dimension, Procedures, concerns all management measures taken in the organization to tackle domino risks in all possible situations and under all conceivable circumstances (including topics such as work instructions, procedures, guidelines, etc.). The third component, Technology, comprises all technological measures and solutions taken and implemented with respect to domino risk management (including, e.g. domino effects software, safety instrumented functions, etc.).

The first dimension to influence domino risk management is thus being defined as "People". According to some estimates, human error contributes 80–90% of all accidents. This number considers all possible sources of error, including front-line operating personnel, engineers, and supervision. In case of security, even all incidents are human made. That is why creating (safety and security) risk awareness among all employees is essential for adequately managing domino risks, as well as providing proper training, providing safety and security incentives, creating a safety-driven and security-driven organizational community, enhancing competences of employees at all levels, etc.

The second dimension, "Procedures", is being managed by a safety management system (SMS) (in case of safety), and a security management program (in case of security). These management systems revise the existing procedures used to maintain a good safety culture and/or security culture. The term "procedures" can be interpreted very broadly. It concerns procedures to operate safely and securely, to safely store hazardous substances, to manage the competences of employees, to manage emergency situations, to have determined, detect, and delay procedures in place, etc. Logically, the organizational structure and culture play a large role in this.

One needs little argumentation that the technological dimension is indispensable to ensure a good domino risk management policy. Following the as low as reasonably achievable principle, it should be noted that no risk can be reduced to zero without expenses that cannot be justified economically. That is why the technological dimension has to be designed in a way that the resulting domino risk lays between socially accepted boundaries. Governments will impose these bounds, but often organizations will surpass the required measures.

Every dimension (People, Procedures and Technology) can be looked upon from a single chemical plant's viewpoint as well as from a multiorganizational perspective. This way, these three dimensions can be used to explain how internal domino effects (single-plant's viewpoint, see Section 13.2) and external domino effects (multiple plant viewpoint, see Section 13.3) can be managed within an industrial area.

13.2   Managing Internal Domino Effects

13.2.1   Dimension 1: People (for Managing Internal Domino Effects)

13.2.1.1   Systems Thinking

For understanding how to manage domino effects, it is very important to gain insights into the "logic behind the system". Therefore, some general insights in systems thinking are needed to be followed by risk managers. Four systems thinking insights are given hereafter.

13.2.1.1.1   Systems Thinking Insight 1: Reaction Time or Retardant Effect
Every safety measure, taken based on risk management, has a certain "reaction time", that is, it takes a certain amount of time before the effect(s) of a measure become(s) apparent. It is important to know this reaction time or at least to have an idea of this time, in order to avoid taking new measure(s) too quickly. Hence, a long-term vision

while taking safety and health measures and interpreting the results needs to be supported by the insights of the working of the system, and of the long-term effects of measures on the system.

13.2.1.1.2 Systems Thinking Insight 2: Law of Communicating Vessels

The law of communicating vessels simply states that in physics matters are often linked to one another, and that in some way they have an impact on each other. This is no different for socioeconomic systems such as organizations. Every measure or change within a system leads to changes and shifts within other parts of the system, and these domino changes need to be identified and, if necessary, additional measures need to be taken.

The whole system at once, instead of different parts of the system, needs to be considered, and the relationships between the parts of the system, needs to be taken into account when making risk decisions on domino effects prevention. Hence, awareness of possible unexpected changes and continuous vigilance always need to be present.

13.2.1.1.3 Systems Thinking Insight 3: Nonlinear Causalities

Linear causality is very hard to find in real life and in real industrial practice: almost never an accident is the result of one cause. In the case of a domino accident, it is always caused by the concurrence of circumstances and a variety of factors. A factor in itself frequently does not suffice to cause an accident, and if another order of events had happened, the consequences of a domino accident might have been completely different. Moreover, cause and consequence regularly are not closely linked in space and time. Nonetheless, the urge to "think in linear causalities" by humans is very strong, also so for risk managers, and it certainly has been in the past.

Instead of viewing reality as a static picture and, based on this picture, taking preventive measures, risk managers need to discern change patterns, looking at positive and negative feedback loops between events, and based on this improved perception of reality, take health and safety measures.

13.2.1.1.4 Systems Thinking Insight 4: Long-Term Vision

Research indicates that business failures more often originate from bad adaptation of a business to slowly emerging threats, rather than from sudden threats. Insidious, latent (long-term) problems usually do not receive the attention they deserve; on the contrary, short-term failures leading to problems often do.

Due to the fact that analyses are limited in space and time, gradually emerging failures and problems are much more difficult to detect. Increasing space and time while analyzing a certain part of reality improves the perception of this reality piece. An improved perception of reality leads to better decisions.

13.2.1.1.5 Systems Thinking Conclusions

All previously mentioned "laws of systems thinking" indicate the need for insights in a system, thereby considering all relevant system parts, their relationships, and their interdependences. Often, "easy solutions" are sought, looking for symptoms instead of for underlying causes and structures. This way, the "visible problem" is solved for a short time, cannot be seen anymore, and all seems to be well. Of course, in reality, the problem is still present, and it will reappear later in time, often with more persistence and possibly somewhere else in a system. Also, the problem has often become that of another person, who has more difficulty to solve the problem due to the superficial actions taken earlier in time.

Such a non-systems-thinking behavior is strongly present in people, and knowledge and reasoning are therefore necessary to implement true systems thinking behavior in organizations, especially to adequately manage domino effects.

13.2.1.2 High Reliability Theory

High reliability theory (HRT), a school of thought on how to prevent major accidents such as domino effects, believes that with intelligent organizational design and management techniques a company can compensate for weaknesses within the organization and guarantee accident-free operations. According to the HRT, four aspects lead to zero-accident safety: leadership that prioritizes safety objectives as an organizational goal, high levels of human and nonhuman redundancy, an organizational setting of high reliability with decentralized authority and a line level culture of reliability and training, and an approach to trial-and-error learning where organizations learn from (internal and external) experiences.

Organizations capable of gaining and sustaining high reliability levels are called high reliability organizations (HROs). Such organizations identify and correct risks very efficiently and effectively.

A typical characteristic of HROs is collective mindfulness. Hopkins (2005) also indicates that HROs organize themselves in such a way that they are better able to notice the unexpected in the making and halt its development. Hence, collective mindfulness in HROs implies a certain approach of organizing themselves. Five key principles are used by HROs to achieve such mindful and reliable organization.

The first three principles mainly relate to anticipation, or the ability with which organizations can cope with unexpected events. Anticipation concerns disruptions, simplifications, and execution, and requires means of detecting small clues and indications, with the potential to result in large, disruptive events. Of course, such organizations should also be able to decrease, to diminish or to stop the consequences of (a chain of) unwanted events. Anticipation implies the ability to imagine new, noncontrollable situations, which are based on little differences with well-known and controllable situations. HROs take this into account by principles 1, 2, and 3.

Whereas the first three principles relate to proaction, the fourth and fifth principles focus on reaction. It is evident that if unexpected events happen despite all precautions taken, the consequences of these events need to be mitigated. HROs take this into account by principles 4 and 5.

13.2.1.2.1 HRO Principle 1: Targeted at Disturbances

This principle points out that HROs are very actively and in a proactive manner looking for failures, disturbances, deviations, inconsistencies, etc. because they realize that these phenomena can escalate into larger problems and system failures. They achieve this goal by urging all employees to report (without a blame culture) mistakes, errors, failures, near-misses, etc. HROs are also very much aware that a long period of time without any incidents or accidents may lead to complacency with the employees of an organization, and may thus further lead to less risk awareness and less collective mindfulness, eventually leading to accidents. Hence, HROs rigorously see to it that such complacency is avoided at all times.

13.2.1.2.2 HRO Principle 2: Reluctant for Simplification

When people—or organizations—receive information or data, there is a natural tendency to simplify or to reduce it. Parts of the information considered as nonimportant or irrelevant are omitted. Evidently, information which may be perceived as irrelevant might in fact be very relevant in order to avoid incidents or accidents. HROs will therefore question the knowledge they possess from different perspectives and at all times. This way, the organizations try to discover "blind spots" or phenomena which are hard to perceive. To this end, extra personnel (as a type of human redundancy) are used to gather information.

13.2.1.2.3 HRO Principle 3: Sensitive toward Implementation

HROs strive for continuous attention toward real-time information. All employees (from front-line workers to top management) should be very well informed about all organizational processes, and not only about the process or task they are responsible for. They should also be informed about the way that organizational processes may fail and how to control or repair such failure.

To this end, an organizational culture of trust between and among all employees is an absolute must. A working environment in which employees are afraid to provide certain information, e.g. to report incidents, will result in an organization being poor of information, and in which efficient working is impossible. A so-called "engineering culture", in which quantitative data/information are much more appreciated than qualitative knowledge/information, should also be avoided. HROs do not distinct between qualitative and quantitative information.

HROs are also sensitive toward routines and routine-wise handling. Routines can be dangerous when leading to mindlessness and distraction. By installing job rotation and/or task rotation in an intelligent way, HROs try to prevent such routine-wise handling.

Furthermore, HROs view near-misses and incidents as opportunities to learn. The failures going hand in hand with the near-misses always reveal potential (otherwise hidden) hazards, hence such failures serve as an opportunity to avoid future similarly caused incidents.

13.2.1.2.4 HRO Principle 4: Devoted to Resiliency

HROs define resiliency as the capacity of a system to retain its function and structure, regardless of internal and external changes. The system's flexibility allows it to keep on functioning, even when certain system parts do not function as required anymore. An approach to ensure this is that employees organize themselves into ad hoc networks when unexpected events happen. Ad hoc networks can be regarded as temporary informal networks capable of supplying the required expertise to solve the problems. When the problems have disappeared or are solved, the network ceases to exist.

13.2.1.2.5   HRO Principle 5: Respectful for Expertise

Most organizations are characterized by a hierarchical structure with a hierarchical power structure, at least to some degree. This is also the case for HROs. However, in HROs, the power structure is no longer valid in unexpected situations in which certain expertise is required. The decision process and the power are transferred from those with the highest hierarchy (in normal situations) toward those with the most expertise regarding certain topics (in exceptional situations).

13.2.2   Dimension 2: Procedures (for Managing Internal Domino Effects)

13.2.2.1   Safety Management System

A chemical plant's Safety Management System (SMS) aims to ensure safety for the various risks posed by operating the facility. Effective management procedures adopt a systematic and proactive approach to the evaluation and management of the plant, its products and its human resources.

To enhance safety regarding internal domino risks, the SMS considers safety features throughout scenario selection and process selection, inherent safety and process design, industrial activity realization, commissioning, beneficial production and decommissioning. Arrangements are made to guarantee that the means provided for safe operation of the industrial activity are properly designed, constructed, tested, operated, inspected and maintained and that persons working on the site (contractors included) are properly instructed.

Five indispensable features for establishing an organizational SMS can be listed as follows:

1. Parties involved
2. The policy—objectives
3. List of actions to be taken
4. Implementation of the system
5. Monitoring and continuous improvement of the system.

The essence of accident prevention practices consists of safety data, hazard reviews, operating procedures and training. These elements need to be integrated into a safety management document which is implemented in the organization on an ongoing basis. To enhance implementation efficiency, the latter topics can be divided into 11 subjects (Reniers, 2006):

1. Safe work practices

A system should be installed to guarantee that safe work practices are carried out in an organization through procedural and administrative control of work activities, critical operating steps and critical parameters, through prestartup safety reviews for new and modified plant equipment and facilities, and through management of change procedures for plant equipment and processes.

2. Safety training

The necessity to periodically organize training sessions emerges from the continuously changing environment of plants, installations, and installation equipment.

Employees and contractors at all levels should be equipped with the knowledge, skills and attitudes relating to the operation or maintenance of organizational tasks and processes, so as to work in a safe and reliable manner. Safety training sessions should also lead to a more efficient handling of any incident or accident.

3. Group meetings

An organization should establish a safety group meeting for the purpose of improving, promoting and reviewing of all matters related to safety and health of employees. This way, communication and cooperation between management, employees and contractors is promoted, ensuring that safety issues are addressed and appropriate actions are taken to achieve and maintain a safe working environment.

4. Pursuing in-house safety rules and complying with regulations

A set of basic safety rules and regulations should be formulated in the organization to regulate safety and health behaviors. The rules and regulations should be documented and effectively communicated to all employees and contractors through promotion, training, or other means, and should be made readily available to all employees and contractors. They should be effectively implemented and enforced within the organization. The company rules should be in conformance with the legislative requirements and rules that are nonstatutory should conform to international standards and best practices.

5. Safety promotion

Promotional programs should be developed and conducted to demonstrate the organization's management commitment and leadership in promoting good safety and health behaviors and practices.

6. Contractor and employee evaluation, selection and control

The organization should establish and document a system for assessment and evaluation of contractors to guarantee that only competent and qualified contractors are selected and permitted to carry out contracted works. This way, personnel under external management, but working within the organization, are treated, evaluated and rewarded in the same manner (concerning safety issues) as internally managed personnel.

7. Safety inspection, monitoring and auditing

The organization needs to develop and implement a written program for formal and planned safety inspections to be carried out. The program should include safety inspections, plant and equipment inspections, any other inspections (including surprise inspections), and safety auditing. This way, a system is established to verify compliance with the relevant regulatory requirements, in-house safety rules and regulations and safe work practices.

8. Maintenance regimes

A maintenance program needs to be established to ensure the mechanical integrity of critical plant equipment. In fact, all machinery and equipment used in the organization needs to be maintained at all times so as to prevent any failure of these equipments and to avoid unsafe situations.

9. Hazard analysis and incident investigation and analysis

All hazards in the organization need to be methodically identified, evaluated and controlled. The process of hazard analysis should be thoroughly documented.

Written procedures should also be established to ensure that all incidents and accidents (including those by contractors) are reported and recorded properly. Furthermore, procedures for incident and accident investigation and analysis so as to identify root causes and to implement effective corrective measures or systems to prevent recurrence should be installed.

10. Control of movement and use of dangerous goods

A system should be established to identify and manage all dangerous goods through the provision of material safety data sheets and procedures for the proper use, storage, handling, and movement of hazardous chemicals. To further ensure that all up-to-date information on the storage, use, handling and movement of dangerous goods in the organization reaches the prevention and risk management department, a continuously adjusted database with information should be established.

11. Documentation control and records

An organization should establish a central documentation control and record system to integrate all documentation requirements and to ensure that they are complied with.

13.2.2.2 Business Continuity Planning

Effective business continuity planning (BCP) is essential if, despite all measures and precautions taken, a domino effect would manifest. BCP as a corporate activity has the fundamental strategic objectives of ensuring corporate survivability and economic viability when business profits and/or continuity are threatened by external or internal potentially destructive events.

Three characteristics can be expressed as separating a domino accident from an occupational accident or any type I event based on (Meyer and Reniers, 2013):

- Surprise—Domino events come at a time or a level of intensity beyond everyone's expectations.
- Threat—All domino events create threatening circumstances that reach beyond the typical problems organizations face.
- Short response time—The threatening nature of domino events means that they must be addressed quickly.

Hence, adequate BCP is an essential aspect of managing internal domino effects. Anticipation, the master word of BCP, is (Meyer and Reniers, 2013)

- imagining and studying the different incurred scenarios in the company, by integrating also the known catastrophic scenarios;
- to equip oneself with "thinking" and "acting" in a structured way when the moment comes, known as the crisis cell;
- prepare a minimum of the functioning logistics and procedures;
- prepare the crisis communication so as not to add an aggravating factor in case of mediatization.

- identify in advance the right people: those who have the knowledge and expertise, those who are united and can work in team, those who will not "flinch" when there are decisions to make and finally those whose legitimacy would not be questioned.

Anticipating and preparing for a domino event is answering the following questions:

- How to react and be organized to face an unpredictable disaster?
- Which crisis management device should be deployed?
- Which collaborators should be implicated? What should they be attributed and what would their responsibilities be?
- Which logistic should be planned?
- Which crisis communication should be put into place? And for which addressees?
- Which procedures and which operation should be implemented to reduce crisis impact and proceed with the restarting of the activities?

13.2.2.3   Hazard and Operability Analysis for Domino Effects

In the chemical industry, a strong diversification of the use of risk analysis methodologies can be detected. The two most popular techniques for hazard identification within a chemical process (Reniers, 2006) are Hazard and Operability (Hazop) analysis and What-if analysis (see, e.g. CCPS, 2008). Safety advisors prefer these techniques because they are systematic and analytical, user-friendly, qualitative and have an excellent track record.

The Hazop procedure and the What-if procedure may be adapted to investigate domino hazards (Reniers, 2010). Hazop users apply so-called "guide words". The original guide word-based Hazop approach was developed by ICI Chemicals, in mid-1970s. In the ICI approach, each guide word is combined with relevant process parameters and applied at each point (study node, process section, or operating step) in the process that is being examined.

Over the years many organizations have modified the Hazop analysis technique to suit their special needs (Ford and Brown, 1990). These various industry-, company-, or facility-specific approaches may be quite appropriate in the applications for which they are intended. For example, to guide teams more quickly to specific process safety areas, the original guide words and the original process parameters have been modified and specialized lists of guide words have been set up (Greenberg and Cremer, 1991). In order to investigate off-premises hazards, domino effects-specific guide words and parameters can thus be developed and used. Acikalin (2003) suggests that domino accident escalation may take place due to three different effects: overpressure, radiation and missile projection. Considering domino effects, literature (Fievez, 1996) also makes a distinction between seven different accident scenarios. Based on these specific off-site data, guide words and domino parameters can be listed as in Tables 13.1and 13.2.

Domino Hazop parameters focus attention upon a particular consequence aspect of the design intent toward a domino accident scenario. Domino Hazop guide words, when combined with domino Hazop parameters, suggest possible deviations on the process point under consideration possibly leading to a domino effects scenario. The

Table 13.1 Domino Hazop Analysis Guide Words and Meanings

Domino Guide Words	Meaning
Overpressure effects?	Possibility of overpressure effects inducing secondary (escalation) effects
Radiation effects?	Possibility of radiation effects inducing secondary (escalation) effects
Missile projection?	Possibility of missile projection inducing secondary (escalation) effects

Table 13.2 Domino Hazop Analysis Parameters

Boiling liquid expanding vapor explosion (BLEVE)
Vapor cloud explosion (VCE), confined and unconfined
Poolfire
Jetfire
Tankfire
Boilover
Explosion

following example can illustrate investigating domino deviations using domino guide words and domino parameters:

Domino Hazop Guide Word		Domino Hazop Parameter		Deviation
Missile projection	+	BLEVE	=	Missile projection upon this process point might cause deviations (on this process point), possibly resulting in a (secondary) BLEVE.
Radiation effects	+	VCE	=	Radiation effects upon this process point might cause distortions (on this process point), possibly resulting in a (secondary) VCE.
Overpressure effects	+	Poolfire	=	Overpressure effects upon this process point might cause distortions (on this process point), possibly resulting in a (secondary) poolfire.

13.2.3   Dimension 3: Technology (for Managing Internal Domino Effects)

Technology specifically addressing domino events, transpires, amongst others, into domino software. There are very few software packages available to study domino accidents in complex industrial areas and to forecast potential domino catastrophes with secondary order, tertiary order or higher order accidents. Available domino software focuses mostly on risk assessment and on consequence assessment. Software available for domino effects is discussed in Chapter 14.

Other technological solutions that can be used for managing and/or preventing domino effects are listed in Table 13.3. The table shows possible measures that can be applied in the chemical and process industries, in order to prevent domino effects or to reduce their effects. As such, these measures do not necessarily prevent the initial event or reduce the consequences of an initial event, but they do affect the possible domino effect (with cardinality 1).

13.3 Managing External Domino Effects

Although the fact that external domino effects are comparable with internal domino effects with respect to the technical matters of such accidents, they cannot be managed in the same way or with the same approach as explained in the previous section. Internal domino effects and external domino effects indeed represent different situational contexts. Managing domino effects within a multiorganizational context requires much more agreement, collaboration, fine-tuning, and caution than managing internal domino effects. How the three dimensions from the risk management model—People, Procedures, and Technology—can be adapted to a multi-plant context is explained and elaborated in this section.

13.3.1 Dimension 1: People (for Managing External Domino Effects)

13.3.1.1 Getting the People from Different Neighboring Plants Involved in a Cluster Initiative

As a chemical company nowadays it is obvious to conduct safety studies to detect the risks of major accidents. It is indeed a fact that these risks threaten not only the people within the company but also the people in the surrounding neighboring companies, and also the residents. The neighboring companies generally have more resources to protect themselves than the average population. However, this does not automatically mean that they could deploy their resources effectively and immediately when necessary.

When a crisis situation occurs, many of the decisions often are taken until after the first hour and by the emergency services. Very well-intentioned, but, by far, too late. Gas clouds, pressure and radiation effects do not wait to assert their influence. In 1985 the "Responsible Care" thought was developed. In this idea, the companies themselves take responsibility to decrease the effects of "major accidents" to a minimum. This should be a sufficient motivation for companies to take the responsibility instead of acting upon the measures imposed by any authorities.

Bringing all of these under control of this matter is a moral duty of anyone who intends to work with hazardous materials.

Due to this motivation, 3M Company EHS decided to start a safety cluster in the Antwerp Port area (in Belgium) in January 2010. It involved moral obligation toward

Table 13.3 Prevention and Protection Measures for Dealing with Possible
Domino Effects

Damage Preventing	Design or Control	Measure
Damage preventing	Design equipment layout	• Choice of the right materials. • Pressure resistance of the equipment. • Mechanical overpressure protection, rupture disks, pressure relief valves, and blowdown tanks. • Corrosion resistance. • Supporting equipment for recipients and pipes. • Vent sizing. • Inherent safety design (can tolerate some equipment failures). • Emergency stop buttons. • Flame arrestors.
Damage preventing	Design preventing fire	• Thermal protections. • Differential protections. • Use of nonflammable materials. • Nonflammable gaskets and valves (containment). • Avoiding unstable products. • Preventing oil getting into substances as rock wool. • Gas detectors. • Smoke detectors. • The permanent leakage detection system is incorporated in the inspection program. • Radiation shields. • Pressure shields and explosion films.
Damage preventing	Design process control	• Safety integrity levels and redundancy. • Preventive maintenance logging and investigations. • Automatic monitoring and continuous monitoring. • Safety studies installation and product: Hazop, what if, Swift, … • Incidents/dangerous situations/near misses, logging and investigations. • Steering of installations by remote control. • Installation of independent interlocks. • Instrumental protections. • Alarm on abnormal level changes in a (storage)tank. Stop product transfer if amounts are larger than considered normal.

Table 13.3 Prevention and Protection Measures for Dealing with Possible
Domino Effects—*cont'd*

Damage Preventing	Design or Control	Measure
Damage limiting or preventing	Design explosion related	• Explosion hatches. • Emergency cooling systems.
Damage preventing	Design ATEX related	• Avoiding flammable liquids. • Covering with foam blankets (aut/man). • Forced ventilation. • Avoiding ignition sources and grounding. • Nitrogen blanketing. • Intrinsic safe.
Damage limiting	Design containing	• The bund walls keep the product/fire away from other tanks. • Sloped fire trench keep the fire under control and away from other tanks/equipment. • Prevent spreading of flammable products via sewers. • Release volume containing. Keep gas releases inside for safe scrubbing. • Dubbel wall tanks and pipes to prevent leakage. • Safety couplings (unique for one product).
Damage limiting	Design Compartmentation	• Emergency stops. Immediate stop and go to a safe position. • The break-away coupling prevents spills in case of a separation of the temporary connection. • Detection of a loss of containment in combination with an automatic compartmentation. Divide in smaller volume sections that when a loss of containment take place only the small volume is lost. • Check valves. Stop product going back with an unwanted effect as a consequence. • Fire-activated valves. Spring-loaded valve closes if a fire melts a substance that keeps the valve open.
Damage limiting	Design limit fire damage	• Automatic sprinklers with or without foam addition. • Fire-resistant materials. • Fire-retardant materials. • Fire-proof due structure sprinklers. • Automatic or semiautomatic deluges with sprinkler heads as detection or other detection system. • Fire protection for electrical installations.

(Continued)

Table 13.3 Prevention and Protection Measures for Dealing with Possible
Domino Effects—*cont'd*

Damage Preventing	Design or Control	Measure
Damage limiting and/or preventing	Design evacuation safety shelter	• Area protection against pressure waves. • Area protection against toc gases (air tight), manually or (semi)automatic system. • Area protection against radiation IR, ... • Decreasing staffing possibilities in the still active zones. • Alarm on the air quality intake.

the neighboring companies, the public and several other actors. However, mapping scenarios and domino effects separately company by company is almost an impossible task. Moreover, one gets no overview of all possible scenarios so it is impossible to detect all external domino effects. So people—safety and risk managers—from neighboring companies need to get involved in the safety cluster initiative. It is important that the one who wants to start the safety cluster is the one who can get enough people at the table to get the ball rolling.

A critical mass of EHS people is necessary who have sufficient responsibility and "weight" to realize that this approach is needed to be able to change things permanently. Based on the experience of the Antwerp safety cluster, some pitfalls for trying to make people from different chemical companies collaborate on the prevention of external domino effects can be identified. These are summarized hereafter.

13.3.1.2 Pitfalls

1. Admit to third parties that the own company may be a hazard is very difficult.

The will and conviction can be there, but finally it should be recorded which companies could affect each other, and in which ways. This could trigger the necessary resistance against the initiative.

2. Confidentiality of data generated.

It is very important that the safety cluster works democratically. To the extent possible, the people of the cluster take unanimous decisions.

Anyone can customize "his company input" at any time. Making the company-specific information audit-proof, anyone can keep company info up-to-date at any time. To remember to do so, the corrections of the company-specific information needed by the safety cluster are a fixed item on the cluster agenda, triggering continuous improvement and updates.

Before the initial release of the emergency planning matrix content (Section 13.3.2) to the various authorities, a separate release to all companies concerned is necessary.

For a release after adjustment of the emergency planning matrix by one company, there is a release of the company or companies that made the adjustment.

3. Imposition of unilateral measures (how well-intentioned they may be) to the companies by the authorities.

It is important that the safety cluster can operate in a serene way. Measures "coined" by the companies on the one hand will favorably affect the enforceability, while on the other hand measures imposed (by authorities) will deprive the companies the chance to acting upon their "responsible care" or sustainability. Given the difficulty of the matter, it is important that the companies can perform this analysis thoroughly.

External pressure should in no case be the driving force leading to a thought-out and achievable result. Of course, one should be given an acceptable deadline to find a solution.

4. The insufficient understanding of each other's definitions and criteria.

To avoid misunderstandings, it is very important that there is agreement about the applied definitions. It will strongly influence the final result.

Example: Determination of the scenarios.

- To be listed in the emergency planning matrix, the substance has to be a "Seveso" product.
- Fatal rupture of a container will not be considered. However, the fatal fracture of an appendage with an internal diameter of 10 cm with a continuous release will be considered.
- Ordinary warehouse fires will not be included in the scope. Warehouse fires with an "increased risk" due to the presence of certain products or product combinations are included in the scope.

13.3.1.3   The Multi-Plant Council

To overcome confidentiality issues, we suggest using an independent supraplant body, called the cluster council or the multi-plant council (MPC) as suggested by Reniers (2010), to implement the model in a chemical cluster. The MPC is divided into two parts. The first part consists of plant representatives who mainly have a counseling function, formulating recommendations as a result of brainstorming sessions. The second part consists of independent and external consultants who are responsible for gathering, assessing and analyzing all relevant and confidential risk information from the chemical plants in the cluster. By dividing the MPC into two parts, a balance between confidentiality and data information is targeted. Figure 13.1 illustrates the different parts of the MPC.

Part 1 of the MPC itself consists of plant safety representatives and plant security representatives. It has a typical counseling function, formulating safety and security recommendations as a result of joint think-tank brainstorming and communication sessions. The other part of the MPC, the MPC data administration, is composed of independent consultants (i.e. impartial knowledgeable personnel) responsible for administering all necessary (confidential) safety- and security-related information gathered from the different plants of the cluster. For more information on the MPC concept, the reader is referred to Reniers (2010).

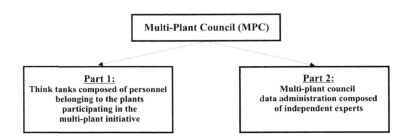

Figure 13.1 Constitution of the multi-plant council.

13.3.1.4   Scope of a Safety Cluster Initiative

The cooperation of the safety cluster should at least encompass the areas of

- sharing information regarding incidents;
- mutual assistance;
- approach process safety management-related topics;
- action points from audits and how to comply;
- working with third parties;
- drafting a multi-plant emergency planning matrix;
- etc.

13.3.2   Dimension 2: Procedures (for Managing External Domino Effects)

13.3.2.1   Multi-Plant Safety Management System

For a better understanding of the additional feature recommendations required to elaborate a multi-plant SMS when compared with the single-plant SMS topics as discussed in Section 13.2.2, these topics are more thoroughly described hereunder.

1. Multi-plant safe work practices

Gradual agreement over work practices, used in the different plants, should be established. By achieving this, best work practices are standardized step-by-step wherever possible, useful and feasible and their effectiveness is continually optimized through evaluation of the past experiences in the plants forming part of the cluster.

2. Multi-plant safety training

The necessity to periodically organize cluster training sessions emerges from the continuously changing environment of plants at subcluster level and of the cluster vicinity at supracluster level. Employees working in a hazardous environment are trained together with their colleagues in nearby plants. Such group safety training sessions should lead to a more efficient handling of an ongoing domino accident due to a collective know-how regarding how to act correctly. Coordinating the organization of these training sessions belongs to the responsibilities of the cluster council.

Depending on the circumstances and on the subjects, employees from different plants are trained either separately or jointly. The cluster training programs are reviewed periodically by the cluster council. In the view of this, each company reports changes made to operational procedures.

3. Multi-Plant group meetings

The MPC is responsible for organizing group meetings. To achieve a smooth procedure at meetings, every plant may first organize an internal group meeting about cluster safety issues to which the plant employees concerned are invited. Afterward, a plant representative can be delegated to the cluster group meeting to report the plant's own particular point of view.

4. Multi-plant pursuing in-house safety rules and complying with regulations

In-house safety rules take into account the possibility of a domino accident caused by a neighboring company. Furthermore, the European Seveso-III Directive includes a list of liabilities for top-tier chemical companies:

- Every company potentially able to cause a domino accident has an obligation to exchange information with external companies concerning such a risk.
- The exchanged data have to be used to create an emergency plan, to formulate a safety report and to establish a safety policy concerning the prevention and controlling of major accidents.
- An internal emergency plan and a contingency plan have to be drafted regarding domino risks.

5. Multi-plant safety promotion

Safety training and plant group meetings are perfect tools to increase employees' awareness of the danger of hazardous installations and domino risks. However, to create a continuous awareness of the importance of cluster safety, other tools such as signs and posters will indicate the cross-company danger of some installations and substances.

6. Multi-plant contractor and employee evaluation, selection and control

The participating companies establish an equivalent contractor assessment system based on the same basic safety performance indicators. This way, personnel under external management, but working within different plants belonging to the chemical cluster, are treated, evaluated and rewarded in the same manner (concerning safety issues) as internally managed personnel.

7. Multi-plant safety inspection, monitoring and auditing

The MPC is responsible for cluster safety inspection and cluster safety auditing. First, the council monitors plant compliance with cluster safety regulations (e.g. concerning domino accident prevention). Second, the council inspects the effectiveness of the individual plants' safety inspection programs (described in the plant's SMS).

8. Multi-plant maintenance regimes

A plant maintenance program plays a crucial role in the prevention of major incidents. Therefore, such plant programs are reviewed regularly by the MPC and experience-based recommendations are made by the council.

9. Multi-plant hazard analysis

To perform a cross-company hazards analysis, it is vital that the MPC has access to all relevant information from the different participating companies. Information such as hazardous chemicals material safety data sheets, the location of hazardous chemicals, etc. have to be given to the MPC (data administration). Based on these confidential data, cluster domino hazard analyses can be executed by an independent and certified organization. The MPC also informs other stakeholders, such as the public and the public authorities, about the consequences of the identified scenarios on health (including public health) and the environment. The results of the analysis are discussed in group meetings with safety managers from the different companies involved. If necessary, these managers will then transfer the information to their own employees to be sure that it is clearly communicated to all parties involved.

10. Multi-plant control of movement and use of hazardous chemicals

To ensure that all information on the use, handling and transfer of hazardous chemicals in the cluster reaches the affected companies, a continuously adjusted database with nonconfidential information should be accessible to all the neighboring plants. The database is for example located on an information network shared by the different plants so that every company can carry out hazardous chemicals changes while safety managers of the other plants belonging to the cluster are informed in real time.

11. Multi-plant documentation control and records

In order to ensure that all companies have the same major accident prevention know-how, a central database managed by the MPC for gathering nonconfidential as well as confidential safety information on individual plant level about potential (theoretical and experienced) incidents and accidents possibly affecting other plants of the cluster can prove very useful.

13.3.2.2   A Framework for Preventing External Domino Effects

Cross-company management essentially differs from single company management by the amount of information available to managers. Another problem is that different organizational perspectives have to be combined. As Hovden (1998) explains, each frame or perspective provides a different way of interpreting events and actions, and each implies a different focus with consequences for choice of strategies and approach to effective management.

Hence, bringing these viewpoints together is not an easy assignment. There should be no communication problems or misunderstandings between the parties involved. Therefore, a safety risk management strategy has to be worked out to be sure that responsible personnel from different plants communicate with the same know-how, on the same level, and about the same safety issues. To provide a situation where communication conflicts are minimized, the risk analysis procedures and their results need to be understood by the different experts concerned with the study. This can be achieved by "speaking the same risk analysis language", and hence to use one risk analysis technique for certain major risks in the cluster.

Combining Hazop analysis, what-if analysis and the risk matrix, the three most popular and most widely used risk assessment techniques in the process industries, into one framework allows for constituting a tool for optimizing the discussion of process hazard analysis performances by employees of neighboring companies in an industrial area. Reniers (2010) elaborates on such a framework, called "Hazwim", and the reader is therefore referred to this reference for more information.

13.3.2.3   Multi-Plant Emergency Planning Matrix

The objective of drafting a multi-plant emergency planning matrix is to create an overview of how the companies within the safety cluster should respond due to a calamity in a preagreed upon way. The measures should be expected to be executed easily, since experience indicates that complex rules usually fail. Moreover, implementations and operations as a result of the matrix are guaranteed by the matrix simplicity.

It is the intention of the emergency planning matrix to limit the damage in the very early stages of an accident or a crisis situation, before the community emergency services become operational.

In order to raise awareness and recognition of respective problems and those of the neighboring companies, each company presents an overview of its "risks on major accidents" to the other companies belonging to the cluster.

An explanation is given regarding the interpretation of emergency planning rules and legislation by the authorities and also the concrete operation of community emergency services is presented to all members of the cluster.

Afterwards, a draft of the emergency planning matrix "layout" is made, and the criteria for the accident impact levels are determined.

A user-friendly and widely used program should be employed, because of a low-threshold access (e.g. Excel), to bring all data in one overview. The different major accident scenarios of every company of the cluster are then placed on the Y-axis. The companies possibly affected by the major accident scenarios, including other nonchemical companies, are placed on the X-axis of the matrix. The influence and the influence level can then be found on the coordinate axis. The influence level and the necessary actions to be taken by a company based on this level are further determined through agreement between the cluster members on accurate definitions.

Table 13.4 provides an overview of easily applicable criteria.

To achieve the objective of collaboration between the members of the cluster and in order to finish the emergency planning matrix, one company should set a good example by providing its potential effects (based on major accident scenarios) to the other companies and distributing the example to all members of the safety cluster. To be able to do this, a risk manager or prevention advisor should obtain as much internal support as possible. Once one of the companies initiated the drafting of a matrix, other companies are likely to follow suit.

Table 13.4 Levels and Related Actions to be Taken

Severity	A	Very severe	Outside the cluster	>> level 1
	B	Severe	Inside the cluster	>> level 2
	C	Important	Inside the company fence	>> level 3

	Level 3*	Informative "notification type of alarm"	No action required by affected companies.
	Level 3	Prealarm "notification type of alarm"	Increased alertness by affected companies, company where accident originates, and assesses own situation.
	Level 2	Alarm "advice stay inside"	Affected companies shut windows and doors, and put the ventilation off.
	Level 1	Alarm "advice go to a shelter"	Shelter = ventilation under control, medical and sanitation under control.
	Level 0	Alarm "plant evacuation"	Decision reserved for the leadership of the community emergency services.

It is important that every company calculates major accident effects in a similar way. Therefore, a cross-diagonal check has to be carried out by controlling if—and if so, understanding why—the effect distances of a certain substance can vary from company to company. Differences in calculations may, for example be caused by the stored quantities, the layout of the installation, the process conditions, the implemented prevention measures and the harm reduction measures taken.

Furthermore, the preventive and protective measures eligible for improvement need to be identified, and shortcomings are adjusted wherever possible. Related to this, cross-company mutual aid possibilities are investigated to eliminate the deficiencies, for example companies offering a shelter to other companies in emergency situations, potential interventions are provisioned, potential joint interventions are worked out, a quick and easy alert/communication system is developed, etc.

Regarding external domino effects, it is very important that the definition of external domino effects is understood and agreed upon by all members of the safety cluster. Otherwise discrepancies may occur that will negatively affect the final model. Measures to prevent or stop external domino effects need to be formulated and implemented by the cluster members.

A simple method to notify neighboring companies about a possible major accident will need to be introduced as well. As an example, in the Antwerp cluster initiative, it was agreed by the companies participating in the cluster that a fire (even though there is no hazardous substance involved) will be reported to the other members of the safety cluster from the moment there is a clearly visible or nonvisible effect to neighboring companies.

Such "alert notification" is regarded as advice to the potentially affected companies. One can focus on this advice without obligation until the moment the leadership of the community emergency services becomes operational. From that moment on, their advice will become mandatory.

To elaborate a multi-plant emergency planning matrix, the following stepwise approach is suggested:

1. Discuss and set the objectives.
2. Set the operating modalities and meeting moments.
3. Each company presents a review of its "major accident hazards".
4. Analyze the legal framework. Presentation.
5. Discuss an approach plan.
6. Create a proposal for the emergency planning matrix and discuss its functions.
7. Define the severity of possible accidents and its relationship to the levels of Table 13.4.
8. All companies put their information in the emergency planning matrix, starting with one initiative-taker.
9. Process all data and visualize in a general overview.
10. Cross-diagonal check of the data to guarantee a coherent content.
11. Discuss possibilities for mutual aid to eliminate the deficiencies of preventive and protective measures.
12. Build and agree upon the definition of an external domino effect.
13. Develop a method to detect external domino effects.
14. Determine measures to prevent or stop external domino effects.

13.3.3 Dimension 3: Technology (for Managing External Domino Effects)

Available major accident software usually does not facilitate an objective ranking of installations to determine where to carry out risk analyses concerning external domino effects. Nonetheless, dividing a chemical industrial area into smaller subareas between which no domino effects are possible is a possible technique to decrease external domino safety and security risk impacts.

Optimizing the decision process for taking prevention measures (for possible technological measures to deal with external domino effects, see Table 13.3, which is the same table that is applicable to internal domino effects) in a multi-plant chemical industrial surrounding requires quantifying the danger (with respect to external domino effects) of the entire industrial area, creating subareas, succeeded by domino danger quantification of possible paths giving rise to domino effects in these subareas. Available data should allow for user-friendly and objective safety and security management decisions. The computer-automated decision support tool to this end should be developed by the MPC of the multi-plant initiative. Using a holistic approach and thus looking at the multi-plant area as a whole, the tool should allow taking preventive measures which minimize the impacts from potential intentional attacks and accidental escalating accidents within the area.

Furthermore, it should be investigated whether dividing the entire multi-plant area into smaller subareas can be achieved by eliminating a certain percentage of the installations belonging to the multi-plant area, i.e. making these installations extremely safe or secure to the extent that they can be regarded as nonexistent within the network representing the industrial area. The latter should be based on objective criteria and should be mathematically demonstrable. Since the different subareas emerging from this operation based on quantitative data cannot "affect" each other with domino effects, they form isolated domino effect islands, completing the first objective of the domino decision support software.

Confidentiality issues about input and output info provided by the multi-plant software should be taken into account. The software should also be very straight-forward, simple to use and to interpret, and should not discuss that way how to deal with risks. The approach and the external domino safety policy of the cluster should only be the result of agreement between the cluster members.

It is true that between 2001 and 2010, a variety of computer-automated tools have been developed to determine the likely occurrence of domino effects and their consequences (Chapter 14). However, these tools do not offer transparent answers in terms of prioritizing the taking of domino prevention measures in a complex multi-plant surrounding of chemical installations. Moreover, these tools only aim at taking precautions in terms of domino safety risks, and fail to address domino security risks. Nevertheless, managerial decisions to prevent such catastrophic major accidents have to be made as efficiently and as effectively as possible. Decisions must be based not only on safety and security requirements but also on economic constraints. Therefore, Reniers (2010) proposes a methodology for the relative ranking of sequences of chemical installations in terms of their liability to produce escalating effects. The philosophy behind this approach is to address questions arising from the domino risk analysis, thus determining the relative significance from a safety and a security point of view, before performing additional and much more costly hazard evaluation or risk analysis studies. The mathematical requirements and the working procedure for actually developing such domino computer program are given by Reniers (2010), and the reader is therefore referred to this reference.

13.4   Conclusions

Managing single-company and cross-company domino effects is not identical. Dealing with internal (single company) domino effects only requires the involvement of company personnel (management, maintenance, engineers, etc.) of a single plant. This can be done by elaborating adequate measures and attitudes to tackle cascade risks on three domains: human-related, procedural, and technological. External (cross-plant) domino effects need the involvement of personnel (management, maintenance, engineers, etc.) from different plants, and thus the organization and implementation of such domino effects prevention is much more difficult. However, it is not impossible: if, again on the domains "People, Procedures, and Technology", the necessary measures, concepts and organizational structures (as explained in this chapter) are realized and implemented, external domino effects can be effectively dealt with.

References

Acikalin, H.A., 2003. Dynamische Simulation thermisch initiierter domino-Effecte. Doctoral Dissertation (in German). Berlin University, Berlin, Germany.

Casal, J., 2008. Evaluation of the Effects and Consequences of Major Accidents in Industrial Plants. Elsevier, Amsterdam, The Netherlands.

CCPS, 2008. Guidelines for Hazard Evaluation Procedures, third ed. American Institute of Chemical Engineers, New York, USA.

Fievez, C., 1996. Effets domino dans l'industrie chimique: recherché d'une méthodologie de prevention sur base d'une analyse accidentologique. Facultés Polytechniques de Mons, Mons, Belgium.

Ford, K.A., Brown, W.H., 1990. Innovative Applications of the Hazop Technique. Presentation at the AIChE Spring National Meeting, Orlando, FL, USA.

Greenberg, H.R., Cremer, J.J., 1991. Risk Assessment and Risk Management for the Chemical Process Industry. J. Wiley and Sons Inc., New York, USA.

Hopkins, A., 2005. Safety, Culture and Risk. The Organizational Causes of Disasters. CCH Australia Ltd., Sydney, Australia.

Hovden, J., 1998. Models of organizations versus safety management approaches: a discussion based on studies of the "internal control of HSE" reform in Norway. In: Hale, A., Baram, M. (Eds.), Safety Management. The Challenge of Change. Pergamon, Oxford, UK.

Meyer, T., Reniers, G., 2013. Engineering Risk Management. De Gruyter, Berlin, Germany.

OGP, 2005. Human Factors. International Association of Oil and Gas Producers, London, UK.

Paltrinieri, N., Dechy, N., Salzano, E., Wardman, M., Cozzani, V., 2012. Lessons learnt from Toulouse and Buncefield disasters: from risk analysis failures to the identification of atypical scenarios through a better knowledge management. Risk Analysis 32, 1404–1419.

Reniers, G., 2006. Shaping an Integrated Cluster Safety Culture in the Chemical Process Industry. Doctoral Dissertation, University of Antwerp, Antwerp, Belgium.

Reniers, G., Dullaert, W., 2007. Gaining and Sustaining Site-Integrated Safety and Security in Chemical Clusters. Nautilus Academic Books, Zelzate, Belgium.

Reniers, G., 2010. Multi-Plant Safety and Security Management in the Chemical and Process Industries. Wiley-VCH, Weinheim, Germany.

Taleb, N.N., 2007. The Black Swan. The Impact of the Highly Improbable. Random House, New York, USA.

14 Decision Support Systems for Preventing Domino Effects

Bahman Abdolhamidzadeh

Center for Process Design, Safety and Loss Prevention (CPSL),
Sharif University of Technology, Tehran, Iran

Chapter Outline

14.1   Introduction

It has been almost 50 years from the genesis of the first decision support systems (DSSs). Although the main focus of the early-generation DSSs was on supporting business decisions, nowadays DSS software is used more than ever in almost every field of science and engineering. Actually there is theoretical possibility to apply DSS in any knowledge domain. Process safety and risk assessment cannot be considered as exceptions in this regard.

A DSS is generally defined as an integrated computerized information system that supports the decision-making process for individuals and organizations. A properly

Domino Effects in the Process Industries. http://dx.doi.org/10.1016/B978-0-444-54323-3.00014-2

designed DSS helps decision makers to compile useful information from raw/input data. DSS is mainly used when a user is faced with an unstructured problem, in which several alternatives are possible and making a decision is not an easy task. There are numerous analogous situations in science and engineering where DSS toolkits are developed in order to overcome the above-mentioned difficulties.

In process safety, similarly to many other engineering domains, DSSs have been used increasingly. Among all the diverse applications of DSSs in process safety, utilization for managing and preventing domino effects seems promising. Domino accidents are inherently complex phenomena with usually numerous possible alternative scenarios and interacting parameters. Moreover, the amount of numerical calculations needed to assess the risk due to domino effect is huge. Thus, the necessity of applying expert systems and automated tools to assess the risk posed by domino scenarios is unquestionable.

In this chapter, we aim at thoroughly reviewing all available computerized tools and software packages presented up to now for managing domino effects. Application scope, approach, technical features, advantages, shortcomings and limitations of each DSS are analyzed and discussed. However, earlier in this section the definition, architecture, benefits and essential characteristics of DSS toolkits are described to give the reader an insight into a perfectly designed DSS and a comparison with the characteristics of available ones. Finally, a framework for an integrated DSS to prevent domino effects based on all existing advancements and potential possibilities for improvement is introduced.

14.2   Decision Support System

In order to discuss the role of the DSS toolkits in domino accident prevention, it is necessary to have a perspective on the definition, the components and especially the mechanism by which these tools help decision makers.

There is no general agreement on the definition of DSS, as it is a broad concept. DSS definitions vary depending on the point of view of the authors (Druzdzel and Flynn, 1999). For Keen and Scott Morton (1978), a DSS couples the intellectual resources of individuals with the capabilities of the computer to improve the quality of decisions. They defined DSS as a computer-based support for decision makers who are dealing with semistructured problems. There are several other definitions such as the liberal one presented by Power (2002) that defines DSS as a useful and inclusive term for many types of information systems that support decision making.

In the present context, a DSS is considered as an integrated set of computerized tools that support and reinforce user judgments while making decisions. Application of such automated systems is quite helpful when a user is faced with several choices and decisions that are hard to specify in advance.

Gorry and Scott Morton (1971) coined the phrase DSS and during the 1970s the concept of DSS became a notable field of research. DSS evolved a lot through all these years. New generations of DSSs shifted from conventional record keeping and file drawer systems toward optimization and representational models. The later generations are the so-called Intelligent Decision Support Systems which make extensive use of artificial intelligence techniques. In the new millennium, web-based DSSs have emerged and are increasingly used (Power, 2007).

Nowadays DSS toolkits are widely used in almost every field of engineering and science. The scope of DSS application is quite vast and ranges between different management disciplines such as emergency management (Mendonça et al., 2006), financial management (Zopounidis et al., 1997) or even hospital management (Forgionne and Kohli, 1996) to sophisticated engineering application such as design, control and optimization of petrochemical plants (Stephanopoulos and Han, 1996).

The application of DSS in process safety has been initiated in the past decade, although less intensively with respect to other engineering disciplines. At the present time, safety specialists have the opportunity to apply the most recent advancements and achievements in DSS structure and architecture. They only need to adopt these available tools for their own purposes. Reniers et al. (2006) have discussed the use and functionality of the different DSS toolkits available to major accident prevention in the chemical process industry. In the present chapter, 30 available software tools are investigated, based on the feedback from developers. Several evidences of the application of DSS toolkits to risk management can be found elsewhere (Khan and Abbasi, 1997a, 1999; Papazoglou et al., 2000; Contini et al., 2000; Gheorghe and Vamanu, 2004).

14.2.1 Classification

Similar to what took place for its definition, different authors have proposed different classifications for DSSs. These differences are mainly due to classifications based on different points of view. One of these classifications is categorizing based on user level. A DSS can be considered *passive, active* or *cooperative*. A *passive* DSS is one that only facilitates the process of decision making but does not offer suggestions or solutions. However, an active DSS can produce such suggestions or solutions. The third class is *cooperative* DSSs that provide a set of suggestions that should be later refined or modified by the user. The majority of existing DSS toolkits that are used in process risk assessment are of *passive* or *cooperative* category.

14.2.2 Components

There is a relative agreement on the three basic components of every DSS. These three components are: (1) the user interface, (2) the model and (3) the database (Sprague and Carlson, 1982; Power, 2002). Some researchers tend to include the user as one of the components of a DSS. They support their idea by the justification that especially in unstructured or semistructured problems the function of the user is beyond entering data, since it is needed to apply judgment and intuition during the decision-making process. So the user should be considered as an element in a DSS. As a matter of fact, in decision-making processes regarding safety and risk assessment, the human factor (human interaction, human decisions, etc.) still plays an undeniable role.

Figure 14.1 shows the different components of a typical DSS and the interconnections between these elements. Two different defined boundaries for DSSs are shown in Figure 14.1.

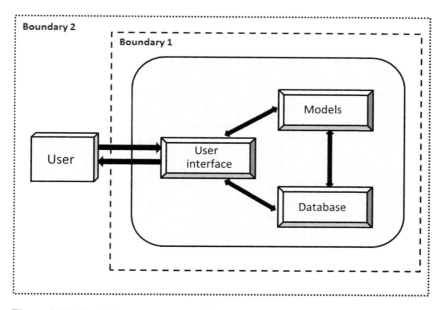

Figure 14.1 The distinct components of a typical DSS with two different boundaries.

14.2.3 Benefits

There are several advantages associated with the application of DSS software in every decision-making process. Some of the main benefits are the following (Keen, 1981; Sprague and Watson, 1993):

- Increase in user efficiency
- Accelerated problem solving
- Speeding up in the process of decision making
- Promotion of learning and training
- Facilitation of communications
- Increased number of alternatives to be examined
- Saving of time and money.

Some disadvantages are also observed. The most frequent one is the user tendency to accept output solutions and suggestions without any interpretation or postprocessing, because it is generated by a so-called intelligent system.

14.2.4 Characteristics

Sprague and Watson (1993) have mentioned user-friendliness, accessibility and a wide range of applications as the essential characteristics of a well-designed and efficient DSS toolkit. Some other specifications could be added to these features, such as

- Completeness
- Accuracy
- Cost
- Communication capabilities.

A balance between all the above-mentioned features is necessary. For a DSS software that should be used in process safety and risk assessment decision making, the possible application to a wide range of applications is very important as the type and specifications of processes, chemicals and equipment involved are highly diverse. Completeness means providing all the necessary output data to support decision making. For a domino DSS toolkit, completeness can be mainly defined as covering the whole process of risk assessment, not only some specific part of that such as consequence assessment. Since there are usually simulations and probabilistic calculations involved in these DSS toolkits, model accuracy is vital to prevent the user from making wrong judgments and decisions. The software cost can also be a determining factor, particularly for small organizations.

As it was mentioned earlier, the application of DSS toolkits in making risk-based decisions has grown a lot in recent years. Some software packages become industry standards. But when it comes to dealing with domino effects, even some of the most widely used software tools for safety assessment in the process industries are not able to handle these scenarios at all. PHAST, EFFECTS, SAFETI, and FRED/Shepherd are widely used in current risk assessment projects but none of these software tools is able to take into account, calculate and represent domino scenarios directly.

Actually the information about DSS toolkits which support domino scenarios are scattered and mainly published as scientific papers, while most of the industrial safety managers rely on internal company standards, internet search and software developers' websites for choosing their toolkits. Thus there is a need to introduce readily available software capable of dealing with domino scenarios and escalation hazard. In the next section, in addition to introducing the DSS software available for domino risk management, an overview and analysis of the capabilities, features and shortcomings of tools was provided, considering different points of view.

14.3   DSS Application to the Prevention of Domino Effect

14.3.1   State of the Art in DSS for Domino Effect Assessment

According to several statistical analyses of past accidents (Kourniotis et al., 2000; Darbra et al., 2010; Abdolhamidzadeh et al., 2011) the share of domino accidents among the process accidents is notable in sheer numbers as well as with respect to their severity. That is why importance of domino effects is highlighted even in regulations such as the Seveso Directives (Directive 96/82/EC; Directive, 2012/18/EU). Due to the congestion, complexity, higher production capacities and vicinity to population centers, the potential of domino effects' occurrence increased in process plants. Nowadays, no risk assessment can be considered complete without including the analysis of domino effect. Several studies such as those by Cozzani et al. (2006b) or Abdolhamidzadeh et al. (2010) have revealed that neglecting domino risk in quantitative risk assessment (QRA) leads to risk underestimation.

The assessment of domino effects is complex and this is the reason behind the frequent absence of domino risk studies in industrial QRAs. Normally, there are a lot of scenarios with too many interacting parameters involved, forcing the risk analyst or the decision maker to face a so-called semistructured problem. In such problems there is some agreement on the data, process and evaluation criteria to be used in decision making but some level of user judgment is required in the decision-making process. Thus, the application of DSS in this regard seems quite promising.

At the moment, computer-automated DSSs for risk assessment are being used more than ever. Broadly speaking, every risk assessment software can be considered as a DSS for preventing domino effects because the results after processing can be used in some steps of domino risk assessment. However, the focus in this section is on the DSS toolkits developed merely for the purpose of domino effect prevention. The majority of the available DSSs for domino effect prevention only allow assessing domino risks or elements of risks such as consequences. There are only few software packages for domino prevention which have chosen different approaches. All the DSS tools for domino risk management which are discussed in scientific literature are reviewed in this section. Some of them are merely the result of academic research and should be viewed as theoretical exercises rather than tools for industrial practice. Others have a higher degree of maturity and were already used or are ready to use in the assessment of industrial processes.

14.3.2 DOMIFFECT (DOMIno eFFECT) Software

Khan and Abbasi (1998) presented a computer-automated tool which was possibly the first software package specifically developed for dealing with domino effect risk assessment. This software tool enables its user to understand (Khan and Abbasi, 1998)

1. Whether domino effects are likely to occur in a given setting
2. What are the likely accident scenarios
3. What would be the likely impacts of the different scenarios.

The developers believe that the generated outputs help the user in decision making toward strategies needed to prevent domino accidents.

14.3.2.1 Approach

The approach of the developers was based on a conjunction of deterministic and probabilistic models to forecast the occurrence probability and damage potential of different domino scenarios. Deterministic models have been used to estimate the consequences of a wide range of scenarios. These models enable DOMIFFECT to assess several leak, dispersion, fire and explosion accident scenarios. Probabilistic models have been incorporated mainly to assess the event frequencies in addition to damage and domino probabilities.

The backbone algorithm for development of this software is a conceptual framework called domino effect analysis (DEA) presented earlier in a scientific paper by Khan and Abbasi (1998). Later on, in order to perform an automated DEA, these authors have developed DOMIFFECT. Figure 14.2 outlines the sequence of steps which is followed in DOMIFFECT to perform a DEA.

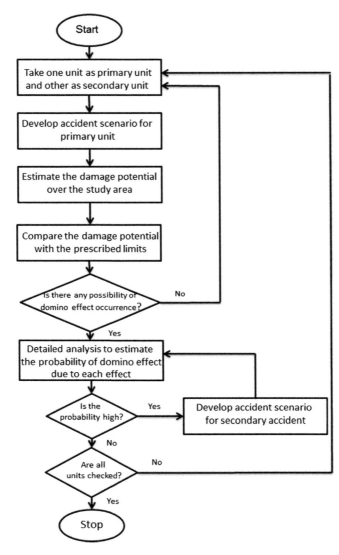

Figure 14.2 DOMIFFECT sequential steps (Khan and Abbasi, 1998).

As shown in Figure 14.2, after the selection of a primary unit, the damage potential should be estimated for all other units. This step is carried out by performing a detailed consequence assessment for the primary event at the location of target (secondary) units. To facilitate the DEA process, Khan and Abbasi (2001a) have proposed the use of a two-tier sequence. At the first level, damage potential is compared with the prescribed limits (threshold values) to identify those units that undergo a domino accident. If the estimated damage potential values are higher than the selected threshold values, a detailed analysis is performed in the second tier of the assessment. In the second tier, based on the damage potential and on the characteristics of the target unit, the probability of a domino accident is estimated. If the probability is high then accident scenarios are developed for the secondary accident. As outlined in Figure 14.2, the process based on the selection of a unit as "primary" and checking its damage potential on the other units is continued until all units are examined.

In addition to the damage potential and domino probabilities for different scenarios, the software generates individual risk (IR) contours. By combining the IR for primary accidents and the IR for accidents caused by primary events, the software comes up with an "overall IR contour". By the application of DOMIFFECT, the so-called hot spots or the units that pose higher risk can be identified. A decision maker can use these outputs to consider strategies to prevent domino accidents.

14.3.2.2   Technical Features

DOMIFFECT is coded in C++ and the program size is 60 MB. It is a personal computer (PC)-based, menu-driven and interactive software. This tool consists of six main modules: DATA, ACCIDENT SCENARIO, ANALYSIS, DOMINO, GRAPHICS and USER INTERFACE (Khan and Abbasi, 1998). Three of these modules, i.e. DATA, ACCIDENT SCENARIO and ANALYSIS, are similar to the modules in a previously presented software called MAXCRED-II (Khan and Abbasi, 1997b). The DOMINO module makes this software different from a typical consequence assessment software. In this module, the damage potential of the primary event is estimated at the location of all secondary units. By this module, the probability of secondary accidents is checked and if it is high enough, the appropriate accident scenarios are developed and analyzed. Probit equations are used to estimate the probability of domino scenarios. Similar to any DSS, this tool has a USER INTERFACE module. This module provides necessary input information for other modules, stores the results and handles different files. The GRAPHICS module enables DOMIFFECT to generate output risk contours and also execute input site layouts.

14.3.2.3   Advantages vs. Disadvantages

Developers of DOMIFFECT have tried to demonstrate the wide applicability and flexibility of this software by utilizing this tool in domino risk assessment of different industrial case studies. These include refineries, petrochemicals plants (Khan and Abbasi, 2001a) and chemical or fertilizer production units (Khan and Abbasi, 2001b). This wide range of potential applications can be considered as one of the basic advantages of this software.

Completeness is one of the necessary characteristics of any DSS. Regarding risk assessment software, completeness can be mainly defined as covering the whole process of risk assessment, not only some specific steps. As DOMIFFECT is a software which covers almost all of the main steps in risk assessment, completeness can be pointed out as a strong point for this software as a DSS.

A further issue of DOMIFFECT is that its developers have claimed that the software has been studded with state-of-the-art models capable of handling a wide variety of accident situations with precision and accuracy (Khan and Abbasi, 1998). Although this statement seemed true at the moment of the development of the software, this seems no more true. Some of the models used for estimation of damage probability in this software are only valid in special cases. To name a few, the relations used for the estimation of vapor fraction generated inside a vessel due to temperature rise caused by heat load is only valid for limited operating conditions and special equipment. Also, in damage probability estimation due to overpressure, a generic probit model was used. Further studies have revealed that equipment-specific probit models are more accurate (Cozzani and Salzano, 2004).

On the other hand, DOMIFFECT is not capable of dealing with potential synergic effect of secondary accidents. Actually this complex mechanism is impossible to be analyzed using classical consequence and damage assessment models. Although some graphical applications are incorporated in DOMIFFECT for loading maps and site layouts and also for generating output risk contours, the quality of the generated graphical output is low due to computational limitations present at the time of the software development. Thus, these features cannot be categorized as one of the advantages of this software. In recent years, more sophisticated tools are available to increase the exploitation of results.

Finally, from the point of view of flexibility, it seems that the modular structure of DOMIFFECT makes it easy to link this tool with other risk assessment software.

14.3.3 Domino XL 2.0

As a response to the emphasis given to domino effects in the Seveso II Directive (Directive 96/82/EC), several research projects started in Europe in the late 1990s. Although the importance of domino effects was highlighted in the Seveso II Directive, a methodology that explains how to assess—or even a clear definition of—domino effects was not provided. Thus several projects, mostly funded by control authorities, were carried out to overcome these ambiguities.

As one of the teams engaged in this topic, Delvosalle et al. (2002) have proposed a methodology to assess the probability of domino effects inside an industrial site (internal domino effect) or between different establishments (external domino effect). The methodology gave birth to a software called Domino XL 2.0, which aimed to assess possible domino effects in the Seveso industries and which can also be used as a safety tool in these industries (Delvosalle et al., 2002).

14.3.3.1 Approach

The methodology which has been used in development of Domino XL 2.0 includes four main steps. These basic steps are outlined in Figure 14.3.

The preliminary step is the gathering of the necessary information, which shall be done prior to using the software. In this step, general information about the establishments under study, the site layout, as well as information about all the equipment

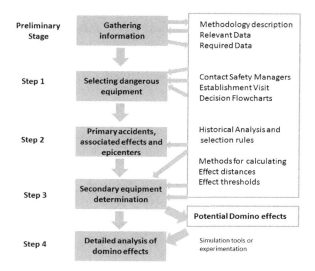

Figure 14.3 Summary of the steps followed in Domino XL 2.0 methodology (Delvosalle et al., 2002).

such as inventory, operating condition and design specifications are required. The other steps are the following (Delvosalle et al., 2002):

- The first step consists of listing, categorizing and locating on a site layout all dangerous equipment likely to be involved in a domino effect.
- In the second step, all retained equipment items likely to be involved in a primary accident must be studied. Resulting effects and the epicenter location must be associated with each primary accident.
- By means of simple criteria (effect distance and threshold), the third step allows to determine equipment zones likely to be damaged by a primary accident and to trigger off a secondary accident (domino effect).
- In the last step, the relevance of the results obtained in the second and third steps needs to be analyzed.

The first step of this methodology is aimed at identifying all the dangerous equipment likely to initiate a domino accident or become a target in cascading accidents. The developers have proposed grouping all the equipment items with similar behavior in case of accident into "equipment zones". The pieces of equipment in each equipment zone are studied together. This may solve the problem of dealing with numerous equipment items in each establishment. The methodology gives useful grouping criteria. In the following steps, the location and safety systems which are present in each group need to be introduced.

In the second step, primary accidents, potential effects and the epicenter should be associated with each equipment group. Based on a historical survey of past accidents, some decision flowcharts have been generated by Delvosalle et al. (2002) to be used in this methodology in order to help the user to select proper primary scenarios and potential effects. Another part of this step consists in associating an epicenter to each accident scenario.

The criterion to determine equipment groups likely to be damaged by primary accidents in this methodology is to compare the damage potential with threshold values. In the third step, in order to estimate the damage potential caused by each primary scenario, consequence assessment is carried out. Domino XL 2.0 developers have mainly used the classical approaches proposed by the TNO Yellow Book (Van Den Bosh and Weterings, 1997) to assess the effect distances. Finally, in the fourth step, the relevance of pairs of equipment or equipment zones selected in the third step as likely to be involved in a domino effect is verified. Due to the numerous calculations required—especially in the third step—when this methodology is applied to a real industrial case, the Domino XL 2.0 software was developed to facilitate the process of assessing domino effects.

14.3.3.2 Technical Features

Domino XL 2.0 has been built so that the user must follow all the steps considered in the proposed methodology one by one. An online help is provided that links the instructions of the methodology with the user's guide.

In order to enter the required information while performing the first and second step of the methodology, an input interface called DATA MANAGER is considered. With the DATA MANAGER, the user is able to input the sections present in each establishment, the equipment items included in these sections and their specific characteristics. The user needs to group equipment items in equipment zones and specifies the pipes linking these equipment zones. Then the user must associate primary accidents for each selected equipment zone or pipe, and finally, congested zones shall be defined. Several specific pages were developed in the DATA MANAGER.

Several decision flowcharts are provided to help the user with data input. Domino XL 2.0 also includes a substance database. Finally, the user can obtain a clear synthesis of the input data by a reporting tool embedded in the software.

The third step of the methodology is fully automated and the user is not involved in carrying out this step. Normally numerous calculations are done in this step to generate the results. The software produces three kinds of results. The first type (called GENERAL RESULTS) consists of tables showing the secondary items that may be the targets of escalation of primary accidents. Number and color coding helps the user to identify different outcome scenarios such as direct, indirect, internal and external domino effects. A graphical display of the results is also available in this section. There are two other types of results available at the end of the analysis: WEIGHED RESULTS and SPECIFIC WEIGHED RESULTS. These outputs are generated to help the analyst or decision maker dealing with domino assessment. By defining two factors called "dangerousness factor" and "vulnerability factor" different scenarios are weighted and compared. The first factor focuses on primary items and classifies these items according to the severity of domino consequences. The second factor (vulnerability factor) focuses on secondary items and ranks these items according to their vulnerability with respect to escalation. SPECIFIC WEIGHED RESULTS gives the comparison of the above-mentioned factors for a specific equipment zone or pipe. To facilitate the comparison and analysis of results, the output of these sections is reported in a bar chart format.

14.3.3.3   Advantages vs. Disadvantages

Domino XL 2.0 is a user-friendly software and instructions are given step by step to help the user to input correct data and make decisions. During the data input, the user is helped with several decision flowcharts.

Domino XL 2.0 aims to identify critical equipment and to evaluate the relevance of safety systems with respect to domino effects (Delvosalle et al., 2002). Thus, software output and analysis results are mainly in the format of relative factors. This makes it easy for decision makers to identify "hot spots" causing domino effects or vulnerable zones against such effects. As the main goal of Domino XL 2.0 is categorizing and prioritizing equipment zones and items, in some steps of the analysis, simplified approaches are introduced. For example, in order to assess whether an accident escalates to a secondary equipment or not, the escalation criterion is the comparison between the consequences of the primary event at the location of the target equipment item and predefined threshold values. This means that in the approach it is assumed that damage can occur or not (0 or 1) depending on the value of the physical effects with respect to escalation thresholds, and no damage probability is calculated. This approach reduces the accuracy of the analysis. Even if damage or escalation threshold values are applied, specific thresholds taking into account the characteristics of different categories of target vessels such as the ones presented by Cozzani et al. (2006a) shall be applied rather than general values.

As it was mentioned earlier, the tool does not include any frequency estimation for the domino scenarios and thus no risk calculation. Thus, with respect to completeness (coverage of the main steps in risk assessment), this tool cannot be considered as a complete risk assessment software.

Based on the models used by Domino XL 2.0 (classical consequence modeling approach), it can be realized that this software is not capable of assessing the consequences of multiple contemporary events that may take place during higher level domino accidents.

There are no graphical applications incorporated in Domino XL 2.0 for loading maps and site layouts. However, the software is capable of generating output results in a graphical format to facilitate communication.

Taking into account all the above-mentioned pros and cons, since Domino XL 2.0 identifies dangerous and vulnerable items in a process plant and also is able to compare the effect of safety systems on the domino effect occurrence probability, it is a helpful risk ranking tool for preliminary domino risk decision making.

14.3.4   Domino Version of Aripar-GIS Software

Cozzani et al. (2006b) have developed and added a specific domino software package to Aripar-GIS software in order to boost its capabilities and handle domino QRA. Aripar-GIS software was developed in the framework of the Aripar project (Egidi et al., 1995) and is a risk assessment tool. With the application of Aripar-GIS, individual and societal risk assessment can be performed for fixed installations and also for transport systems (Spadoni et al., 2000). Since this software is equipped with all the necessary databases, risk calculation modules and a geographical user interface, Cozzani et al. (2006b) have benefited from this opportunity and added extra features and models to obtain a version of Aripar-GIS software that can assess the individual and societal risk due to domino scenarios.

14.3.4.1 Approach

The developers of the domino version of Aripar-GIS software have implemented in the tool a systematic procedure for the quantitative assessment of domino effect that they developed a year before (Cozzani et al., 2005). By the application of this methodology, domino risk is assessed quantitatively. The procedure requires the estimation of the domino frequencies and the evaluation of the consequences of primary and secondary events. This goal is achieved carrying out a multistep procedure, which is outlined in Table 14.1.

The early steps are similar to many other domino risk assessment software tools. After the identification of primary scenarios and of relevant escalation vectors, the credible escalation events shall be selected. Using threshold values in this step is

Table 14.1 Main Steps Followed in Domino Version of Aripar-GIS for Domino Effect QRA (Adapted from Antonioni et al., 2009)

Stage	Step	Tools Needed
1. Risk identification	1.1 Escalation assessment of the primary event considered	Consequence analysis models
	1.2 Identification of possible target units	Threshold values
	1.3 Identification of each secondary scenario that may follow the damage of each target unit	
2. Frequency calculation	2.1 Estimation of the damage probability for each target unit	Damage probability models
	2.2 Identification of each possible combination of secondary scenarios	Specific software tools
	2.3 Conditional probability calculation for each possible combination of secondary scenarios	Specific software tools
3. Consequence assessment	3.1 Consequence assessment of the primary scenario	Consequence analysis models
	3.2 Consequence assessment of each secondary scenario considered	Consequence analysis models
4. Risk recomposition	4.1 Calculation of vulnerability maps for each possible combination of scenarios	Human vulnerability models
	4.2 Calculation of individual risk for each possible combination of scenarios and for each primary event considered	Risk recomposition software
	4.3 Calculation of overall risk indexes	

a common practice and has been used in several methodologies, although there is not a general agreement on the damage threshold values in literature. Cozzani et al. (2006a) have derived a set of equipment-specific threshold values to be used in the application of the methodology. They have proposed to use the threshold values to identify the credible escalation scenarios. The selected scenarios would be analyzed in detail in further steps. As any primary event may cause more than one secondary event, frequency and consequence shall be estimated for all the possible combinations of events.

For frequency estimation of domino scenarios, at first equipment-specific damage models in the format of Probit equations are used to estimate the damage probability of possible secondary targets. These damage probabilities and the expected frequency of the primary event shall be combined to yield the domino scenario frequency. An analytical approach using probability rules is defined for the estimation of the overall damage and escalation probabilities.

In the original version of the method, escalation has been limited to the secondary scenarios or in other words "first-level" domino scenarios. This is a strategy to keep the calculation load reasonable but of course the accuracy is reduced. Recent developments, discussed in Chapter 10, may allow the extension of the methodology to the assessment of higher level domino effects.

Frequency assessment of domino scenarios involving the contemporary damage of more than one secondary unit is a substantial difference with respect to previous methodologies, which only consider the possible damage of a single secondary unit (Cozzani et al., 2006a,b).

For the consequence analysis of the primary and secondary events, the approach applies conventional models (CCPS, 2000; Van Den Bosh and Weterings, 1997) similar to the models which had been used in the previously discussed software tools. However, the innovative approach in this methodology is the way that "damage maps" resulting from the conventional models are combined to yield the overall consequences of each domino scenario. Since damage maps are not homogenous and cannot be easily combined, "vulnerability maps", which are matrices yielding the death probability due to the event as a function of the position with respect to the source of the event (Leonelli et al., 1999), are generated and then combined. In order to generate vulnerability maps, a set of selected human vulnerability models are also introduced in this methodology. The methodology was discussed in Chapter 10 of this book.

Finally, the overall contribution of domino effect to IR, societal risk and potential life loss index are calculated by a specific procedure, taking into account all the credible combinations of secondary events that may be triggered by each primary scenario (Cozzani et al., 2005). As the number of contributing scenarios even in simple layouts is very high, Cozzani et al. (2006b) have developed a software tool to facilitate the application of their methodology. The available platform of Aripar-GIS software and its capabilities gave them the opportunity to build this domino risk assessment tool.

14.3.4.2   Technical Features

As the domino version of Aripar-GIS software is an extension of an existing software, almost all the technical features are the same as that of the main software (Aripar-GIS), which is described in detail elsewhere (Spadoni et al., 2000). In brief, this software consists of three main parts: DATABASE, RISK CALCULATION MODULES and GEOGRAPHICAL USER INTERFACE.

A data input phase is necessary before starting the analysis. Due to the GIS technology capabilities, different types of maps can be imported and used as input data. In this software, after importing the maps of the impact area, a grid has to be defined for risk quantification and representation. The user can specify grid sizes based on the required accuracy desired for risk calculation. The risk will be calculated in each grid point in the following steps. The application of GIS technology has enormously simplified the data input phase. The graphical description of "vulnerability centers" (hospitals, churches or commercial centers) requires only a simple click of the mouse at the point where the center is located (Bonvicini et al., 2012). Definition of "risk sources" can also be easily done by a simple "click-and-drag" operation of the mouse.

At the end of the input phase, modules for calculating the risk of a generic source on all points of the grid and on vulnerability centers can be run. The consequences on all points of the impact area are calculated by means of interpolating functions (Spadoni et al., 2000).

Finally, results are generated and displayed. In addition to a numerical output, a graphical representation of results can easily be done. Aripar-GIS generates different types of graphical results in the format of IR values, isorisk curves, and societal risk plots such as $F–N$ curves and $I–N$ histograms.

In the domino version of Aripar-GIS software, some additional tables and checkboxes are provided to allow the user considering domino scenarios. For domino risk calculation, additional input data is required mainly regarding the target equipment, relevant equipment damage models and secondary event occurrence probabilities. Proper tables are provided for the user to specify these data.

14.3.4.3   Advantages vs. Disadvantages

The domino version of the Aripar-GIS software is suitable for the assessment of risks posed by fixed installations as well as by transport systems. This makes a "wide range of applicability" as one of the advantages of using this DSS. As all the steps of risk assessment are covered in the domino version of the Aripar-GIS software, "completeness" is another characteristic of this DSS.

One of the most notable superiorities of this tool is the integration of a domino risk assessment methodology with a GIS platform. This allows an automatic identification of possible escalation targets and a straightforward application of the proposed methodology. The GIS technology not only leads to simplification of the data input phase but also makes the representation of the result much more efficient.

Technically speaking, using state-of-the-art and equipment-specific threshold values and damage models makes this software distinct compared to other domino software available. The ability to consider contemporary damage of more than one secondary unit is one of the other significant improvements done in this software toward a more realistic assessment.

However, as obvious, there is still room for improvement. As it was mentioned earlier for the sake of simplicity, second or higher level domino scenarios are not considered in the frequency estimation. However, a comprehensive past accident analysis (Abdolhamidzadeh et al., 2011) has revealed that almost half of the past domino accidents went beyond the first level. Even if an extension of the methodology to consider further escalation levels is possible, due to the mathematical approach selected for frequency estimation in this methodology, this extension would increase the computational load and subsequently the computational time significantly. This makes questionable the possibility of such extension without the use of specific domino frequency estimation while using normal PCs.

Similar to other available software in this field, the domino version of the Aripar-GIS software uses conventional models for consequence assessment. Thus, the analysis of the potential synergic effect of secondary accidents which has been evidenced in some past domino accidents is strongly simplified. Taking into account this complex behavior is hard even using advanced computational fluid dynamics (CFD) tools, although for sure excluding it affects the precision of the analysis.

In general, the development of a domino version of Aripar-GIS software gives decision makers and risk analysts the opportunity to easily assess the contribution of domino scenarios to industrial risk indexes such as individual and societal risk. Application of this tool also reduces the effort and additional resources required to carry out the analysis of domino effect within a QRA, due to the availability of the GIS platform.

14.3.5 DomPrevPlanning

A common disadvantage of the available domino DSS that bans widespread application of such software in industrial risk assessments is the requirement both of numerous diverse input data and of extensive user knowledge and expertise. In order to tackle this barrier, Reniers and Dullaert (2007) have developed a computer-automated tool called DomPrevPlanning which, by using simple input data, generates a simple ranking of domino sequences as an output. Using a holistic approach, this software determines the prioritization of domino sequences in an industrial area in order to facilitate decisions concerning the prevention and mitigation of domino effects (Reniers and Dullaert, 2007).

14.3.5.1 Approach

Reniers and Dullaert (2007) have elaborated a 10-step methodology to prioritize domino effects in an industrial area. The main steps are outlined in Figure 14.4 in two different levels. DomPrevPlanning is steered and carried out by a supraplant council. Some steps are done on plant level, while others shall be done on the multiplant level or at the so-called cluster level.

Similar to the few available domino risk DSSs, Step 1 of the methodology begins with the supraplant council gathering information for all the individual establishments. As shown in Figure 14.4, this step is carried out at plant level. In Step 2 of the proposed methodology, the actual distance between each couple of installations in the area under

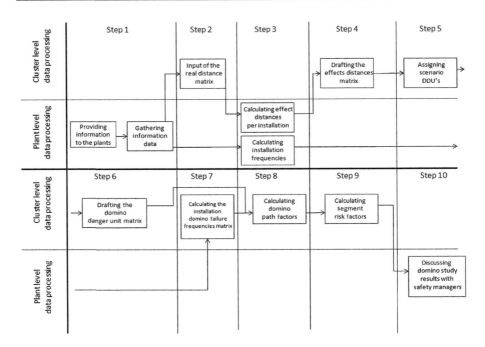

Figure 14.4 Main steps followed in DomPrevPlanning (Reniers and Dullaert, 2007).

study shall be calculated. This step drafts a matrix called "real distances". It is obvious that the second step is performed at the cluster level.

In Step 3, effect distances and installation failure frequencies are calculated per installation at plant level. Reference documents for calculations in this step are Instrument Domino Effects (RIVM, 2003), the "manual for failure frequency figures" (AMINAL, 2004) and "guidelines for quantitative risk assessment", i.e. the so-called Purple Book (Uijt de Haag and Ale, 1999). This implies that although some frequently used mathematical models are embedded in the procedure, detailed consequence modeling and frequency estimation is not performed in this methodology.

A matrix called "effect distances" is drafted in Step 4, containing all the possible scenario effect distances between each pair of installations in the area. In Step 5, comparing the real distance between each pair of installations with the possible scenario effect distance, a categorization of scenario severity is drawn up and a scenario domino danger unit (DDU) is assigned. As DDU values are specified by the user of this procedure, taking into account the preventive and protective measures per installation is possible. This means that DDU values are not dependent only on the outcome of the risk analysis. Afterwards, all scenario DDUs for every unidirectional pair of installations are added and a DDU matrix is calculated in step six. For the calculation of the "installation domino failure frequencies" matrix, the causing installation-specific failure frequency shall be multiplied with the probability of direct ignition of the vulnerable installation. This task is done in Step 7 on cluster level. In

Step 8, drafted matrices in Steps 6 and 7 are used to calculate the so-called domino danger path factor (DDPF). DDPF is determined by multiplying the overall DDU of the path and the overall probability of the path. Finally, segment risk factors are calculated in order to prioritize three-sequence installations of a whole complex cluster area in Step 9. Step 10 is dedicated to the communication of results.

When it comes to the application of the proposed methodology, the number of possible scenarios for real industrial cases are too high to be handled manually. For this reason, Reniers and Dullaert (2007) have developed the DomPrevPlanning software package to perform all the rigorous calculations associated with application of the procedure.

14.3.5.2 Technical Features

DomPrevPlanning has been developed in Visual Basic for Applications (VBA) to allow for a user-friendly input procedure and output analysis in Microsoft Excel (Reniers and Dullaert, 2007). In practice, installation data and real distance data are entered by the user into the data manager for each site considered. Except for Steps 1 and 2, the other steps are carried out automatically.

14.3.5.3 Advantages vs. Disadvantages

DomPrevPlanning is a useful DSS for installation ranking with respect to domino effects in a cluster of industrial plants. Since it is not always possible to perform highly complicated QRAs especially in the early stages of design, there is a need for a software such as DomPrevPlanning which prioritizes installations based on domino risk. This tool uses relatively simple data to obtain simple domino effect output data. DomPrevPlanning provides fast and ready-to-use output results for risk analysts. This software enables the risk analyst to investigate the relative importance of couples of installations from a domino safety viewpoint, before performing additional and much more costly domino risk assessment studies.

DomPrevPlanning is an easy-to-handle and user-friendly decision support tool that provides its users with straightforward information ready for implementation (Reniers and Dullaert, 2007). In this software, the amount of data needed is restricted to a minimum. In order to facilitate the input data step, suggestions are given by the software to assign required values based on simple criteria. Thus, the users of this tool do not require a detailed knowledge about the system and domino risk assessment.

Not performing detailed consequence modeling, frequency estimation or accurate calculation of the risk indexes cannot be considered as one of the DomPrevPlanning shortcomings as this software is designed mainly for ranking installations with respect to the hazard of domino effect. This obviously limits the application of the software to the early stages of design and decision making, when the mapping of all possible installation-specific domino effect scenarios and their consequences is not yet available.

Another simplification that is used in the methodology is that only the serial domino effects or the so-called domino chains are considered (see Chapter 3). Since different past accident analyses have revealed that in many cases more than one secondary accident may occur simultaneously, the selected approach reduces the accuracy.

14.3.6   STARS Domino

STARS Domino is an improved version of STARS Software developed for the automatic assessment of domino effects. Ballocco et al. (2001) have tried to establish a full integration between deterministic and probabilistic models in this tool. The development of this tool was part of a research project called I.T.E.R.E which was aimed to study new techniques to be applied in risk assessment on a local scale. While STARS Software was oriented to reliability and safety analysis using a probabilistic approach, within the I.T.E.R.E project a tentative was carried out to complete the toolkit by adding deterministic models for accident simulation (Ballocco et al., 2001).

14.3.6.1   Approach

The approach used in the development of STARS Domino is simple and straightforward. After specifying the location of the installation creating a layout or importing drawings, the user can define accident scenarios for each item. Accident scenarios consist of connected objects in an event tree fashion, starting with the initiating event. The initiating event is followed by other objects such as a physical phenomenon (e.g. release, fire or explosion), systemic events (e.g. an operator intervention) or a safety protection systems (Ballocco et al., 2001). A probability can be associated with each object. Thus, based on an event tree logic, the probability of all the sequences can be estimated. The phenomenological events of the accident sequences are simulated by models which are selected by the user from those proposed by the STARS Domino software. After the calculation of consequences, the system executes a spatial analysis to identify any target object with the possibility of damage caused by a primary accident. This task is done by comparing the damage effect at the location of the target installation with a damage threshold. If a secondary accident is identified, the system automatically adds a new initiating event within the accidental sequence which is already developed.

14.3.6.2   Technical Features

STARS Domino is composed of four main modules: KNOWLEDGE BASE, SYSTEM MODELS, FAULT TREE and EVENT TREE. Specific functions, called *Feeds*, have been considered to connect the different modules and exchange data between those (Ballocco et al., 2001).

For layout specification in addition to importing CAD drawing, the user can easily create a layout as the software is linked to Microsoft Visio. For result visualization STARS Domino has the option of using GIS technology. Damage curves may be automatically displayed on territorial maps.

Results can be visualized as text, tables, charts and if applicable isocurves on area maps. All the outputs can be easily exported to any of the other commonly used Microsoft Office software tools (Ballocco et al., 2001).

14.3.6.3   Advantages vs. Disadvantages

One of the most significant advantages of STARS Domino is the ability of easily interfacing with different software. For example, in the data entry phase, data can be imported from Excel software. In order to load an area map, the software can be easily linked to Microsoft Visio or may handle CAD drawings. Another innovative feature of this toolkit is that it allows the possibility to be connected to external calculation models for accident simulation. Users can import their own consequence assessment models by interfacing with a MATLAB tool or using any external executable model. Different options are provided for the visualization of results. Results can be provided in the format of text, tables, charts and curves.

The major disadvantage of this package is its rather oversimplified approach. This makes the STARS Domino a suitable tool for quick calculation and identification of potential targets for domino effect. However, the software does not provide the risk analyst with detailed risk estimation or risk ranking of different installations. To mention only some of these simplifications, in this approach the escalation criterion is based on the comparison of the primary accident consequences with fixed damage thresholds which do not seem accurate at all. Ballocco et al. (2001) have not mentioned the possibility of higher level domino effects. Thus, STARS Domino is not capable of dealing with synergic effects due to contemporary events. The software is also not able to consider the estimation of the probability of contemporary domino events.

14.3.7   FREEDOM

Different methods are proposed to estimate domino accident frequencies. Although there are some differences among these methods, the fundamental elements are common and are based on analytical formulations. Developing analytical relations based on probability rules is mathematically cumbersome for complex systems with large numbers of equipment. This complexity increases when considering the possibility of domino effect escalation to second or higher level domino effects (see Chapter 3). Abdolhamidzadeh et al. (2010) have proposed a new method for assessing the domino frequencies with a completely different approach which overcomes the limitations of analytical methods. This method is based on the application of simulation techniques instead than on analytical formulations.

14.3.7.1   Approach

This method is based on Monte Carlo Simulation (MCS). MCS is a method for iteratively evaluating a deterministic model using sets of random numbers as inputs. This method is often used when the model is highly complex, nonlinear, or involves more than just a couple of uncertain parameters (Billinton and Allan, 1992).

For the estimation of domino accident frequencies by this algorithm, the frequency of primary events for each equipment and the escalation probabilities between equipment in case of an accident shall be specified by the user. This means that a detailed consequence assessment is necessary prior to the application of the method. Equipment failure frequencies may be generic data or may be calculated from the development of event trees.

In a nutshell, the idea behind this algorithm is to select an equipment randomly and then check its possibility to generate a primary event comparing its failure frequency with a random variable generated by the algorithm. If the primary event occurs, the probability of escalation involving target equipment items shall be checked. By giving the chance of experiencing failure or survival to every equipment, the actual lifetime of each item is imitated if a sufficient number of simulations are carried out. Finally, the number of failures for different items during this simulated lifetime is correlated to the domino frequencies. Details of the algorithm can be found elsewhere (Abdolhamidzadeh et al., 2010).

The nature of the MCS and consequently also the new algorithm is highly iterative and needs automation. A software package called FREEDOM (an acronym for FREquency Estimation of DOMino accidents) is developed on the basis of the proposed algorithm.

14.3.7.2 Technical Features

FREEDOM is coded in MATLAB software to benefit from the predefined functions available in this package. The other reason behind using MATLAB was the fact that input and output data in this algorithm are in the format of matrices which are easily defined and processed in MATLAB software.

The user has the opportunity to select the number of iterations or so-called experiments. Also, this option is provided for the user to select the level of domino accident study (considering only first level and/or higher levels). In this way, the user can optimize the run-time and also the level of the result's accuracy.

14.3.7.3 Advantages vs. Disadvantages

The very fundamental advantage of using FREEDOM in the estimation of domino accident frequencies is its independence from the complexity of the case under study. By the application of this method, limitations associated with analytical formulation are overcome. FREEDOM also enables the user to assess frequencies for the higher level domino accidents.

However, there are some disadvantages. The nature of the MCSs and consequently of the FREEDOM algorithm is highly iterative. One of the characteristics of MCS is that a higher number of runs yield a higher accuracy of the predictions. However, an extremely high number of iterations may lead to an extremely high run-time and to high requirements with respect to computational resources. There is no guideline or rule of thumb for determining the optimal number of iterations. This number shall be determined by the user based on the analysis of the stability of the results given by the method.

FREEDOM is not a complete DSS. This software package cannot be solely used in domino accident prevention and decision making, since no consequence analysis is provided. Moreover, the output results are not ready to use. The best application for this package is as an "add-on" to be used in conjunction with other complete domino DSS tools.

14.4 Overview

In comparison with the other fields of process safety and risk management, there are a limited number of software packages available that have the ability to deal with domino effects. Nonetheless, even these few tools can be categorized into different classes based on different perspectives. Two distinct approaches can be identified while reviewing the approaches which have been used by the existing software. The first approach is employing extensive data, time and computational resources to achieve quantitative domino risk assessment. Tools such as DOMIFFECT or the Domino version of Aripar-GIS software are of this kind. Obviously, the output result is presented in the format of widely used and well-recognized risk indexes such as individual or societal risk.

The alternative approach is ranking the installations in any specific area based on domino risk. In this approach, software developers tend to design the tools to generate ready-to-use decision support data. However, even among this class there are software tools such as DomPrevPlanning, which use relatively simple input data for obtaining a simplified assessment, and tools such as Domino XL 2.0, which employs extensive data and computational resources to generate the domino risk prioritization.

As proposed by Reniers and Dullaert (2007), in many cases the optimal approach to be used by the risk analysts is adopting both options simultaneously.

A different method of classification can be established based on the capabilities of each software to deal with complexities associated with domino risk assessment. These complexities, when compared to conventional risk assessment of stand-alone scenarios, lay both in consequence assessment and frequency estimation. A complete assessment of the consequences of such complicated scenarios is a difficult aim even using advanced tools such as CFD codes (Cozzani et al., 2005), as discussed in Chapter 11. The contemporary presence of multiple events may cause synergetic effects that none of the existing software packages are capable to take into account. The presence of contemporary events makes the frequency assessment of domino scenarios mathematically cumbersome. Among the available software, only the Domino version of the Aripar-GIS software is able to deal with multiple secondary events. Besides the contemporary secondary event, there is always the possibility of escalation to higher level domino rather than the so-called first level. None of the existing software considers more than second-level domino scenarios. Although conceptually some of the methodologies (such as those used in the development of DOMIFFECT and in the Domino version of the Aripar-GIS software) have the ability to be extended for considering the higher level dominos, this option is currently not exploited in order to keep the requirement of computational resources at a reasonable level.

There are a few software packages among the available DSS toolkits that take into account the effect of preventive and protective safety systems on domino risks. These tools such as Domino XL 2.0 or DomPrevPlanning give the opportunity to the risk analyst to examine the effect of different levels of preventive and protective systems and therefore come up with the best strategy to prevent domino effects.

The use of advanced technologies such as a GIS platform makes some of these tools more easy-to-use gadgets especially in communication. The domino version of the Aripar-GIS software and the STARS Domino tool are the only software of their kind in which GIS technology is employed. The output of a holistic review on the existing software features and capabilities is summarized in Table 14.2.

Table 14.2 Summary of Features and Capabilities of the Available Domino DSS

Toolkit	Domino Escalation Criteria		Frequency Estimation Approach			Consequence Modeling Approach			Level of Domino Accident		Output Results	Risk Based Result			Graphical Features		
	Threshold	Probit Equation	Analytical Formulation	Simulation Techniques	Consider Contemporary Events	Classical Models	CFD	Consider Contemporary Events	First	Second and Higher	Installation Ranking	Individual Risk	Societal Risk	Post-Processing Necessary	Input Site Map	Output Site Map	GIS Platform
DOMIFFECT		▨	▨			▨			▨			▨				▨	
DOMINO XL 2.0	▨		None	None	None	▨			▨		▨						
Aripar-GIS (Domino version)		▨	▨			▨			▨			▨	▨		▨	▨	▨
DomPrevPlanning	▨		▨			▨			▨		▨						
STARS Domino	▨		▨		▨	▨		▨	▨				▨	▨	▨		▨
FREEDOM	None	None	None	None	None	None	None	None		▨				▨	None	None	None

14.5   Perspectives in the Development and Application of DSS for Domino Assessment

The application of decision support tools in the area of process safety is growing fast. However, the use of DSS in domino risk management has not been well established. There is not yet a widely used or a commercially available software package with the ability to deal with domino effect. This shortage is rooted in the complex nature of domino accidents. The complexity leads to two different approaches in domino risk software development that are both not favorable for the industrial risk analysts. The first approach is oversimplification of the problems in order to make the software easy to use. This task is carried out limiting the number of required input data and also using simplified models. However, the results are not usually accurate and reliable enough for industrial decision making. On the contrary, there are some software packages in which the developers have tried to use the most state-of-the-art and meticulous models and techniques. This approach itself may lead to several disadvantages such as extensive requirement of input data, lack of user-friendliness or unreasonable analysis time. The user of such software also requires a deep knowledge and expertise in the field. It seems that there should be a balance between the level of complexity of the methodology proposed on the one hand and user-friendliness of these toolkits on the other hand.

In the previous section, a thorough review was carried out concerning the weak and strong points of the available software. An ideal DSS to deal with domino effects can benefit from the strong points of the existing software, while the weak points must be eliminated. One of the necessary characteristics of a perfect DSS for domino risk management is "completeness". An effective DSS for domino risk assessment should be an integrated set of computer tools that enables the user to perform the complete process of domino risk assessment. Software which only estimates domino effect consequences for instance, is not efficient enough. An effective DSS should enable its user to consider all possible levels of domino accidents. The software tools limited to "first level" escalation introduce some simplification in the analysis.

A barrier toward taking into account higher level domino effect is the limitation associated with the use of analytical methods for frequency estimation. Analytical methods, when it comes to deal with a large number of units playing a role in a domino chain, have inherent limitations. However, there are some alternative approaches such as the one proposed by Abdolhamidzadeh et al. (2010), which is based on MCS to estimate domino scenario frequencies. A software package called FREEDOM has been developed for this purpose. Using simulation techniques, current simplifications used in available toolkits are no more necessary, even if the tool is demanding from the point of view of computational resources required and currently does not consider the issue of consequence analysis.

Another promising approach in domino frequency estimation is the application of Bayesian probabilities (Khakzad et al., 2013), that was discussed in Chapter 10.

Consequence assessment of domino accidents is very complex and can be considered a further bottleneck toward an accurate risk estimation. There are several uncertainties associated with domino effect phenomena that the conventional

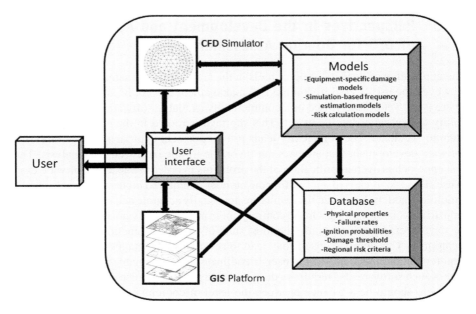

Figure 14.5 Proposed structure of an integrated DSS for domino risk assessment.

consequence and damage models cannot easily handle. To overcome these limitations, the application of CFD for modeling may be a viable route, although presently the resources required for the use of CFD in the consequence assessment of domino scenario are very high, as discussed in Chapter 11.

Another software technology that showed advantages when being applied to domino risk assessment is the use of GIS. Again, an ideal domino DSS shall be equipped with a GIS platform. This feature not only enormously facilitates the data input phase but also allows an easy effect visualization of results. As it was mentioned earlier one of the main characteristics of a perfect DSS is its ability for proper communication. Taking into account the common shortcomings and strong and innovative features of the available software and also looking at the future trends a framework for an integrated domino risk assessment DSS is outlined in Figure 14.5. The integration with CAD tools used in 3D design of chemical and process plants is a further option that may facilitate the calculation and representation of results.

The software tools conventionally used for process and plant design activities constitute another frontier for the assisted prevention of domino effect since the early phases of the project life cycle. Automatic routines for the evaluation of key performance indicators can be introduced in such software, following the approach suggested by Shariff et al. (2012) to support inherently safety design. Specific indicators for domino hazard (e.g. Tugnoli et al., 2008, 2012a,b) will provide a semi-quantitative feedback on the envisaged safety performance of the systems under design, orienting design activities toward the safer options well before the actual risk assessment is performed.

14.6   Conclusions

The application of DSS in domino effect prevention seems quite promising, since risk-based decisions involving domino effects are usually classified as semistructured or unstructured problems. However, there are only a few toolkits available for domino risk management. This limited number of existing software packages are hardly—if ever—used in industry. This is due to some common shortcomings and limitations associated with almost all of these software packages. However, there are still several advantages in using such DSS tools. There are a number of innovations and strong points embedded in the existing software packages. These features were thoroughly reviewed in this section for all the available toolkits. By overcoming the common shortcomings, using the strong points and also the application of advancements in technology, the next generation of domino DSS may be used more frequently and efficiently.

At the end, it should be mentioned that the technical judgment of experts cannot be replaced by the output of none of these available DSS toolkits. All the existing software are designed and developed only to facilitate the act of decision making which is very critical when it comes to manage the risk of domino effect.

References

Abdolhamidzadeh, B., Abbasi, T., Rashtchian, D., Abbasi, S.A., 2010. A new method for assessing domino effect in chemical process industry. Journal of Hazardous Materials 182 (1–3), 416–426.

Abdolhamidzadeh, B., Abbasi, T., Rashtchian, D., Abbasi, S.A., 2011. Domino effect in process industry accidents: an inventory of past events and identification of some patterns. Journal of Loss Prevention in the Process Industries 24 (5), 575–593.

AMINAL, 2004. Manual Failure Frequency Figures for drafting Safety Reports, Cel Veiligheidsrapportering. Ministry for the Flemish Community (in Dutch).

Antonioni, G., Spadoni, G., Cozzani, V., 2009. Application of domino effect quantitative risk assessment to an extended industrial area. Journal of Loss Prevention in the Process Industries 22, 614–624.

Ballocco, G., Carpignano, A., Di Figlia, G., Nordvik, J.P., Rizzuti, L., 2001. Development of new tools for the consequence assessment of major accidents. In: European Safety and Reliability Conference ESREL.

Billinton, R., Allan, R., 1992. Reliability Evaluation of Engineering Systems: Concepts and Techniques, second ed. Plenum Press, New York.

Bonvicini, S., Ganapini, S., Spadoni, G., Cozzani, V., 2012. The description of population vulnerability in quantitative risk analysis. Risk Analysis 32, 1576–1594.

CCPS, Center for Chemical Process Safety, 2000. Guidelines for Chemical Process Quantitative Risk Analysis. American Institute of Chemical Engineering, New York.

Contini, S., Bellezza, F., Christou, M.D., Kirchsteiger, C., 2000. The use of geographic information systems in major accident risk assessment and management. Journal of Hazardous Materials 78 (1–3), 223–245.

Cozzani, V., Salzano, E., 2004. The quantitative assessment of domino effects caused by overpressure, part I. probit models. Journal of Hazardous Materials 107 (3), 67–80.

Cozzani, V., Gubinelli, G., Antonioni, G., Spadoni, G., Zanelli, S., 2005. The assessment of risk caused by domino effect in quantitative area risk analysis. Journal of Hazardous Materials 127 (1–3), 14–30.

Cozzani, V., Gubinelli, G., Salzano, E., 2006a. Escalation thresholds in the assessment of domino accidental events. Journal of Hazardous Materials 129 (1–3), 1–21.

Cozzani, V., Antonioni, G., Spadoni, G., 2006b. Quantitative assessment of domino scenarios by a GIS-based software tool. Journal of Loss Prevention in the Process Industries 19 (5), 463–477.

Darbra, R.M., Palacios, A., Casal, J., 2010. Domino effect in chemical accidents: main features and accident sequences. Journal of Hazardous Materials 183 (1–3), 565–573.

Delvosalle, C., Fievez, C., Brohez, S., 2002. A methodology and a software (Domino XL) for studying domino effects. In: Chisa 2002 15th International Congress of Chemical and Process Engineering.

Directive 2012/18/EU, European Parliament and Council Directive 2012/18/EU of 4 July 2012 on Control of Major-Accident Hazards Involving Dangerous Substances, Amending and Subsequently Repealing Council Directive 96/82/EC. Official Journal of the European Communities, L 197/1, Brussels, 24.7.2012.

Directive 96/82/EC, Council Directive 96/82/EC of 9 December 1996 on the Control of Major-Accident Hazards Involving Dangerous Substances. Official Journal of the European Communities, L 10/13, Brussels, 14.1.97.

Druzdzel, M.J., Flynn, R.R., 1999. Decision Support Systems. Encyclopedia of Library and Information Science. A. Kent, Marcel Dekker, Inc.

Egidi, D., Foraboschi, F.P., Spadoni, G., Amendola, A., 1995. The ARIPAR project: analysis of the major accident risks connected with industrial and transportation activities in the Ravenna area. Reliability Engineering and System Safety 49, 75–89.

Forgionne, G.A., Kohli, R., 1996. HMSS: a management support system for concurrent hospital decision making. Decision Support Systems 16 (3), 209–229.

Gheorghe, A., Vamanu, D., 2004. Decision support systems for risk mapping: viewing the risk from the hazards perspective. Journal of Hazardous Materials 111 (1–3), 45–55.

Gorry, A., Scott-Morton, M.S., 1971. A framework for information systems. Sloan Management Review 13 (1), 56–79.

Keen, P.G.W., 1981. Value analysis: justifying decision support systems. Management Information System Quarterly 5 (1), 1–15.

Keen, P.G.W., Scott Morton, M.S., 1978. Decision Support Systems: An Organizational Perspective. Addison-Wesley Pub. Co, Reading, Mass.

Khakzad, N., Khan, F., Amyotte, P., Cozzani, V., 2013. Domino effect analysis using Bayesian networks. Risk Analysis 33 (2), 292–306.

Khan, F.I., Abbasi, S.A., 1997a. TOPHAZOP: a knowledge-based software tool for conducting HAZOP in a rapid, efficient yet inexpensive manner. Journal of Loss Prevention in the Process Industries 10 (5–6), 333–343.

Khan, F.I., Abbasi, S.A., 1997b. MAXCRED: a new software package for rapid risk assessment. Journal of Loss Prevention in the Process Industries 10 (2), 91–100.

Khan, F.I., Abbasi, S.A., 1998. DOMIFFECT (DOMIno eFFECT): user-friendly software for domino effect analysis. Environmental Modelling and Software 13 (2), 163–177.

Khan, F.I., Abbasi, S.A., 1999. HAZDIG: a new software package for assessing the risks of accidental release of toxic chemicals. Journal of Loss Prevention in the Process Industries 12 (2), 167–181.

Khan, F.I., Abbasi, S.A., 2001a. An assessment of the likelihood of occurrence, and the damage potential of domino effect (chain of accidents) in a typical cluster of industries. Journal of Loss Prevention in the Process Industries 14 (4), 283–306.

Khan, F.I., Abbasi, S.A., 2001b. Estimation of probabilities and likely consequences of a chain of accidents (domino effect) in Manali Industrial Complex. Journal of Cleaner Production 9 (6), 493–508.

Kourniotis, S.P., Kiranoudis, C.T., Markatos, N.C., 2000. Statistical analysis of domino chemical accidents. Journal of Hazardous Materials 71 (1–3), 239–252.

Leonelli, P., Bonvicini, S., Spadoni, G., 1999. New detailed numerical procedures for calculating risk measures in hazardous materials transportation. Journal of Loss Prevention in the Process Industries 12 (6), 507–515.

Mendonça, D., Beroggi, E.G., Van Gent, D., Wallace, W.A., 2006. Group decision support system (GDSS) for emergency response. Safety Science 44 (6), 523–535.

Papazoglou, I.A., Bonanos, G.S., Nivolianitou, Z.S., Duijm, N.J., Rasmussen, B., 2000. Supporting decision makers in land use planning around chemical sites. Case study: expansion of an oil refinery. Journal of Hazardous Materials 71 (1–3), 343–373.

Power, D.J., 2002. Decision Support Systems: Concepts and Resources for Managers. Quorum Books, Westport.

Power, D.J., 2007. A Brief History of Decision Support Systems. DSSResources.com, World Wide Web. http://DSSResources.COM/history/dsshistory.html.

Reniers, G.L.L., Dullaert, W., 2007. DomPrevPlanning: user-friendly software for planning domino effects prevention. Safety Science 45 (10), 1060–1081.

Reniers, G.L.L., Ale, B.J.M., Dullaert, W., Foubert, B., 2006. Decision support systems for major accident prevention in the chemical process industry: a developers' survey. Journal of Loss Prevention in the Process Industries 19 (6), 604–620.

RIVM, 2003. Instrument Domino Effecten. Rijksinstituut voor Volksgezondheid en Milieu, Bilthoven (NL).

Shariff, A.M., Leong, C.T., Zaini, D., 2012. Using process stream index (PSI) to assess inherent safety level during preliminary design stage. Safety Science 50 (4), 1098–1103.

Spadoni, G., Egidi, D., Contini, S., 2000. Through ARIPAR-GIS the quantified area risk analysis supports land-use planning activities. Journal of Hazardous Materials 71 (1–3), 423–437.

Sprague, R.H., Carlson, E.D., 1982. Building Effective Decision Support Systems. Prentice-Hall, Englewood Cliffs, NJ.

Sprague, R.H., Watson, H.J., 1993. Decision Support Systems: Putting Theory into Practice. Prentice Hall, Englewood Cliffs, N.J.

Stephanopoulos, G., Han, C., 1996. Intelligent systems in process engineering: a review. Computers and Chemical Engineering 20 (6–7), 743–791.

Tugnoli, A., Khan, F., Amyotte, P., Cozzani, V., 2008. Safety assessment in plant layout design using indexing approach: implementing inherent safety perspective; part 1 – guideword applicability and method description. Journal of Hazardous Materials 160 (1), 100–109.

Tugnoli, A., Landucci, G., Salzano, E., Cozzani, V., 2012a. Supporting the selection of process and plant design options by Inherent Safety KPIs. Journal of Loss Prevention in the Process Industries 25 (5), 830–842.

Tugnoli, A., Cozzani, V., Di Padova, A., Barbaresi, T., Tallone, F., 2012b. Mitigation of fire damage and escalation by fireproofing: a risk-based strategy. Reliability Engineering and System Safety 105, 25–35.

Uijt de Haag, P.A.M., Ale, B.J.M., 1999. Guidelines for Quantitative Risk Assessment (Purple Book). Committee for the Prevention of Disasters, The Hague (NL).

Van Den Bosh, C.J.H., Weterings, R.A.P.M., 1997. Methods for the Calculation of Physical Effects (Yellow Book). Committee for the Prevention of Disasters, The Hague (NL).

Zopounidis, C., Doumpos, M., Matsatsinis, N.F., 1997. On the use of knowledge-based decision support systems in financial management: a survey. Decision Support Systems 20 (3), 259–277.

Conclusions

Domino effect(s) may cause accident scenarios with devastating consequences in the chemical and process industry. This book explains and discusses the context and state-of-the-art of technical, scientific and managerial knowledge concerning such accident scenarios, caused by the escalation of a primary event to a complex accident scenario, often involving multiple plant units. Single chemical plants as well as chemical industrial clusters should be considered when dealing with domino scenarios.

In Part 1 ("Causes of domino effects"), historical surveys on past accidents were used in Chapter 2 to assess the significance of such accidents and their most important features. The importance of the diverse domino sequences was assessed by using the relative probability event tree, and a number of different primary scenarios emerged as possible domino triggers. The analysis of past accidents suggests that domino effects should always be considered in risk assessment related to major accident hazards.

Chapter 3 discusses the fundamental elements of escalation possibly resulting in a domino accident. A generic and unambiguous definition for "domino effect" is provided, taking into consideration a myriad of definitions that have been suggested in the past for the term.

Chapters 4–6 are dedicated to the analysis of the main direct causes of escalation in domino scenarios: blast waves, radiation and fragment projection. Models available for the calculation of escalation probability due to such scenarios are reported, and their potentialities and limitations are discussed.

Chapter 7 introduces less obvious and more controversial domino scenarios such as indirect escalation causes and external events. Indirect causes of escalation, even if not frequent, may be extremely important since they are likely to be overlooked, in particular in complex industrial clusters. Besides indirect effects, natural events or intentional malicious acts of interference can also escalate to cause severe domino accidents.

In Part 2 ("Prevention of domino effects from a technological perspective"), a framework and a classification of the available techniques that may be used for domino assessment are provided (Chapter 8). Three approaches having a different level of detail are outframed. Chapter 9 discusses the threshold-based approach for domino assessment. Escalation thresholds are widely used as a baseline approach to the assessment of the hazard posed by domino scenarios. These should be intended as conservative values of physical effects (thermal radiation, maximum overpressure, etc.) below which the escalation is deemed not credible.

In Chapter 10, the quantitative risk assessment (QRA) approach is discussed. A procedure for the QRA of domino accidents, suitable for the calculation of individual and societal risk indexes, is outlined. Alternative approaches proposed for the quantitative assessment of domino effects based on Bayesian techniques and Monte Carlo simulations are also discussed.

Chapter 11 provides a brief insight on the distributed parameters modeling approach for assessing domino scenarios. The potentialities of distributed parameter models for the specific assessment of domino accidents triggered by fire and over-pressure are discussed.

In Part 3 ("Prevention of domino effects from a managerial perspective"), the role of design in reducing domino hazards is explored. Risk-based design has an important part in the inherent prevention of domino effects, limiting the possibility of escalation by both physical distances between units and the introduction of robust safety barriers. Criteria for the assessment of appropriate safety distances were developed and implemented in layout design (Chapter 12). Since conceptual tools alone frequently fail in solving the tradeoffs related to the conflicting needs in the design improvement, an inherent safety metric was introduced to support design activities.

Chapter 13 argues that multidisciplinary knowledge and know-how, an adequate mindset, eye for detail and for the big picture, a short-term as well as a long-term vision, thorough collaboration efforts, a generalistic perspective with specialistic knowledge, are required to adequately manage domino effects prevention, protection and mitigation. Recommendations are provided on tackling at managerial level the risks related to possible domino effects within a chemical industrial company and in a chemical park.

Finally, in Chapter 14, the application of Decision Support Systems (DSS) for the prevention of domino effects is discussed. The definition, architecture, benefits and essential characteristics of DSS toolkits are described. The application, scope and approach of available computerized tools and software packages that allow managing domino effects are discussed and analyzed thoroughly. A framework for an integrated DSS to prevent domino effects is suggested.

As evident from the outline given above, the ambition of the present book is to document the state-of-the-art to date in the assessment and management of domino effect(s), as well as the knowledge gaps and the needs.

On the one hand, the main conclusion that may be drawn is that the relevant research work carried out in the last 15 years provided a framework to approach the quantitative assessment of domino effects and to support the management of risk due to domino scenarios. Simplified methods based on up-to-date thresholds, QRA tools and distributed parameter modeling approaches are now available to allow a sound assessment of domino hazards. Tools for safe design and decision support systems make the analysis and control of escalation hazard possible.

On the other hand, the relevant work carried out still needs to be consolidated and completed. The new tools and models now available need to enter in day-to-day safety assessments carried out in the chemical and process industry. Commercial software tools including the models and DSSs discussed in this book still need to be developed. It is hoped that this book will pave the way to such process, contributing to enhance safety and sustainability in the chemical and process industry.

Nomenclature

Chapter 4

C_c	Constant (Eq. *)[m s$^{0.5}$] (Eqn (4.5))
C_i	Constant (Eq. *) (Eqn (4.4))
D	Diameter of tank (Eqn (4.4))
DS	Damage state
E	Modulus of elasticity
$f_{DS,i}$	Overall expected frequency of the damage state
$f_{LOC I,i}$	Overall expected frequency of the LOC intensity
f_p	Expected frequency of the primary event
h	Average thickness of tank shell (Eqn (4.4))
H	Tank height (Eqn (4.4))
I	Impulse
k_1, k_2	Probit coefficients (Eqn (4.1))
LOC	Loss of containment
P	Static overpressure
P_d	Expected damage probability following blast wave impact
$P_{DS,i}$	Probability of the ith damage state following the damage due to blast wave impact
r	Tank radius (Eqns 4.4, 4.5)
T	Fundamental natural period of the structure
Y	Probit function (Eqn (4.1))
ΔP°	Peak static overpressure (Eqn (4.1))
ρ	Density
τ_d	Duration of pressure load

Subscript

imp	Impulsive
conv	Convective

Chapter 5

A	Surface extension of each thermal node in the lumped parameters model (Table 5.4)
BLEVE	Boiling liquid expanding vapor explosion
c	Heat capacity (Eqns 5.3, 5.4)
CFD	Computation fluid dynamics
D	Vessel external diameter (Table 5.4)

d	Wall thickness (Table 5.4)
FEM	Finite elements modeling
F_w	Geometrical view factor (Eqn (5.5))
$F_{w,l}$	Geometrical view factor of the wall in contact with the liquid phase
g	Gravity constant ($= 9.81$ m/s^2)
h	Convective heat transfer coefficient
H_V	Enthalpy of vaporization (Table 5.4)
I	Heat flux on the external surface of the insulating coating (Table 5.4)
k	Isotropic thermal conductivity (Eqn (5.4))
k_x	Thermal conductivity for anisotropic material in the x direction (Eqn (5.3))
k_y	Thermal conductivity for anisotropic material in the y direction (Eqn (5.3))
k_z	Thermal conductivity for anisotropic material in the z direction (Eqn (5.3))
L	Liquid level inside the tank (Table 5.4)
LPG	Liquefied petroleum gas
m	Bulk mass of the stored substance (Table 5.4)
M_w	Molecular weight (Table 5.4)
P	Pressure
P_{des}	Design pressure (Figure 5.5)
PRV	Pressure relief valve
q	Generic heat flux (Table 5.4)
Q_{conv}	Convective heat transferred from the fired equipment wall into the lading (Eqn (5.6))
Q_{flame}	Total heat flux (kW/m^2) generated by the flame and exchanged by radiation (Eqn (5.5))
Q_{HL}	Thermal flow received by the fire (kW/m^2) (Eqn (5.5))
Q_{rad_i}	Radiant heat flux received or emitted in a single portion (i) of the inner surface of the target equipment (Eqn (5.7))
R	Universal gas constant
T	Temperature
T_i	Generic portion of vessel inner surface receiving/emitting radiant heat flux with (j) (Eqn (5.7))
T_j	Generic portion of vessel inner surface receiving/emitting radiant heat flux with (i) (Eqn (5.7))
T_{max}	Maximum wall temperature recorded in the fire tests (Figure 5.1c)
t	Time
T^*	Saturation temperature at the pressure P (Table 5.4)
T_b	Bulk fluid temperature (Eqn (5.6))
tem	Time for effective mitigation
ttf	Time to failure
V	Nominal volume (m^3) (Table 5.5)
x	First three-dimensional coordinate axis (Eqns 5.3, 5.4)
y	Second three-dimensional coordinate axis (Eqns 5.3, 5.4)
z	Third three-dimensional coordinate axis (Eqns 5.3, 5.4)
Z	Height occupied by the liquid or vapor phase in contact with the heated wall (Table 5.4)

Subscript

ext	External side (Table 5.4)
f	Flame (Table 5.4)
i	Insulating coating (Table 5.4)

int	Internal side (Table 5.4)
j	Generic node for lumped parameters thermal model (Table 5.4)
L	Liquid (Table 5.4)
LV	Liquid/vapor (Table 5.4)
n	Generic vessel node (Table 5.4)
s	Steel (Table 5.4)
V	Vapor (Table 5.4)

Greek symbol

β	Volumetric coefficient of thermal expansion (Table 5.4)
ε	Emissivity of the material (Table 5.4)
ε_i	Internal emissivity of the material (Eqn (5.7))
μ	Viscosity (Table 5.4)
ρ	Density (Eqns 5.3, 5.4)
σ	Stephan–Boltzmann constant ($= 5.6703 \times 10^{-8}$ W m^{-2} K^{-4})
σ_{adm}	Maximum allowable stress (MPa)
σ_{adm_norm}	Normalized maximum allowable stress (Eqn (5.2))
σ_t	Surface tension between a liquid and its vapor (Table 5.4)
τ_a	Atmospheric transmissivity (Eqn (5.5))

Chapter 6

A	Impact area of missile (Table 6.11)
a, b	Dimensional constants (Eqn (6.5))
a_0	Speed of sound in the gas contained in the bursting vessel
A_D	Section of the fragment on a plane perpendicular to the trajectory
a_p	Sound velocity in the missile material (Table 6.11)
a_t	Sound velocity in the target material (Table 6.11)
BHN_p	Brinnel hardness of missile material (Table 6.11)
BLEVE	Boiling liquid expanding vapor explosion
BLEVE-F	Fired BLEVE (Table 6.2)
BLEVE-NF	Unfired BLEVE (Table 6.2)
C_D	Fluid dynamic drag coefficient
CE	Confined explosion (Table 6.2)
CE	Reference fragment shape: cylinder (Table 6.6)
CR#	Fragmentation pattern ID code: cone roof tanks (Table 6.5)
CR	Reference fragment shape: cone roof (Table 6.6)
CV#	Fragmentation pattern ID code: cylindrical vessels (Table 6.5)
d	Equivalent diameter of the missile (Table 6.11)
D	Pipe external diameter (Table 6.11)
DF	Drag factor of the fragment
DF_a	Average drag factor
DF_{max}	Maximum value of the drag factor considering possible orientations of the fragment
DF_{min}	Minimum value of the drag factor considering possible orientations of the fragment
E	Minimum kinetic energy of the missile for perforation of a target of a given thickness (Table 6.11)

E_{70}	Charpy V-test energy for the target material at ambient temperature (70 °F) (Table 6.11)
ENMAT	Energetic material decomposition (Table 6.2)
EOI	Effective orientation interval
E_T	Charpy V-test energy for the target material at operating temperature (Table 6.11)
ETI	Effective trajectory interval
E_v	Explosion (expansion) energy
f_F	Frequency of a domino event caused by the impact of a fragment
f_p	Frequency of the primary event
g	Gravitational acceleration
h	Geometrical feature of reference fragment shape: height (Table 6.6)
ID	Identifier code
k	Drag coefficient of the fragment
L	Distance between supports (Table 6.11)
l	Geometrical feature of reference fragment shape: length (Table 6.6)
l_1	Geometrical feature of reference fragment shape: width (Table 6.6)
M	Mass of the fragment
ME	Physical explosion (Table 6.2)
M_V	Mass of the bursting vessel
n	Trajectory factor: 1 in the descending part, 2 in the ascending part (Eqn (6.2))
N_f	Number of fragments
p	Perimeter of the missile projection on the target surface (Table 6.11)
$\wp(\theta, \varphi)$	Probability distribution function of the fragment initial direction
P_0	Atmospheric pressure (Table 6.4)
P_1	Pressure inside the vessel at failure (Table 6.4)
P_{cp}	Probability of fragment generation after initial crack propagation
$P_{dam,F}$	Probability of irreversible effects following impact of fragment F
P_F	Probability of the fragment projection chain (generation, projection and damage)
P_{fp}	Conditional probability of a fragmentation pattern
P_{fs}	Conditional probability of alternative configurations in fragment shape
$P_{gen,F}$	Probability of fragment F to be generated
$P_{imp,F}$	Probability of impact of fragment F on a given target
PL	Reference fragment shape: plate (Table 6.6)
P_s	Scaled pressure (Table 6.4)
PT	Reference fragment shape: tube section (Table 6.6)
PTE1	Reference fragment shape: tube end section (Reference Shape #1) (Table 6.6)
PTE2	Reference fragment shape: tube end section (Reference Shape #2) (Table 6.6)
QRA	Quantitative risk assessment
r	Radius of the "nose" of the missile (Table 6.11)
R	Recommended number of fragments in QRA (Table 6.5)
r	Geometrical feature of reference fragment shape: radius (Table 6.6)
RE#	Fragmentation pattern ID code: rotor (Table 6.7)
RR	Run-away reaction (Table 6.2)
s	Maximum plate thickness (Table 6.11)
SC	Reference fragment shape: spherical cap (Table 6.6)
s_t	Pipe wall thickness (Table 6.11)
SV#	Fragmentation pattern ID code: spherical vessels (Table 6.5)
t	Maximum thickness of penetration (Table 6.11)

t	Geometrical feature of reference fragment shape: wall thickness (Table 6.6)
t	Time (Eqns 6.1, 6.2)
U	Minimum missile impact velocity for perforation of a target (Table 6.11)
u	Velocity of a fragment
U_{50}	Missile impact velocity resulting in 50% probability for perforation of a target (Table 6.11)
u_s	Scaled velocity (Table 6.4)
V	Volume of the vessel [m^3] (Eqns 6.3, 6.4)
V	Volume of the vapor in the busting vessel (Table 6.4)
VA	Vulnerable area
W	Missile weight (Table 6.11)
x	Horizontal coordinate on the flight plane of a fragment
y	Vertical coordinate on the flight plane of a fragment

Greek symbols

α	Ratio of kinetic energy and explosion energy
γ, ξ, ψ	Geometrical features of reference fragment shape: characteristic angles (Table 6.6)
$\Delta\theta$	Interval of horizontal angles (azimuths) for which the missile collides with the target object
$\Delta\varphi$	Intervals identifying elevation angles for which the impact with target takes place
$\Delta\varphi_1$	Interval of elevation angles (altitudes) for which the missile lands within the vulnerable area of the target
$\Delta\varphi_2$	Interval of elevation angles (altitudes) for which the missile collides with the target object while in flight before reaching the final destination, which would have led it to land beyond the target
$\Delta\varphi_{\text{min d}}$	Interval of elevation angles (altitudes) for which the "minimum distance" assumption is satisfied
θ	Horizontal angle (azimuth) in a polar coordinate system centered on the fragment source
θ_0	Limit angle identifying a solid sector around a preferential direction of projection
ρ	Geometrical feature of reference fragment shape: density (Table 6.6)
ρ_p	Density of missile material (Table 6.11)
ρ_R	Reference density, 0.283lb/in^3 (Table 6.11)
ρ_t	Density of target material (Table 6.11)
σ_R	Maximum allowable stress of the target material (Table 6.11)
φ	Elevation angle (altitude) in a polar coordinate system centered on the fragment source
Φ	Angle of missile "nose" (Table 6.11)

Chapter 7

DDU	Domino danger unit
DF	Distance factor
s_i^{out}	Outstrength of node i
ν_i	Installation corresponding to node i

Chapter 9

DS	Damage state
F	Explosion strength factor in the Multi-Energy method
I	Radiation intensity (kW/m^2)
LI	Category of loss of containment intensity
M_f	Mach number of flame
P	Maximum static overpressure (kPa)

Chapter 10

$D_{d,i}$	Doses considered in vulnerability models due to the secondary scenarios
D_{pe}	Dose considered in vulnerability models due to the physical effects caused by the primary event that triggers the domino scenario
f_{de}	Expected frequency of a domino event (events/year)
f_{pe}	Expected frequency (events/year) of the primary event (PE) of the domino sequence
$f_{pe,n}$	Expected frequency of the primary event in the absence of escalation
N	Number of secondary events considered in a domino quantitative assessment
$P^{(k,m)}$	The probability of a secondary scenario given by a generic combination m of k secondary events ($k \leq N$)
P_d	Conditional probability of escalation given the primary event
P_e	Total probability that an escalation will take place
$R_{i,de}$	Individual risk due to a domino scenario
V_{de}	Vulnerability (death probability) due to a domino event
ν	The total number of different domino scenarios that may be generated by the primary event considered
ν_k	The total number of domino scenarios in which the primary event triggers k contemporary secondary events

Chapter 11

CFD	Computational fluid dynamics
FEM	Finite elements modeling
LES	Large eddy simulation
LPG	Liquefied petroleum gas
RANS	Reynolds average Navier–Stokes equations
VCE	Vapor Cloud Explosion

Chapter 12

BLEVE	Boiling liquid expanding vapor explosion
BS	Baker–Sthrelow method (Table 12.1)
c	Constant (Eqn (12.1))

C	Cost (Eqn (12.17))
$C_{\text{Add-on}}$	Costs of add-on safety measures
C_{AL}	Cost of direct asset loss
C_{Control}	Cost of process control measures
$C_{\text{ConvSafety}}$	Cost of conventional safety
C_{DEC}	Domino escalation cost
CE	Critical event
C_{ECC}	Environmental cleanup cost
$cf_{i,k}$	Credit factor
C_{HHL}	Cost of human health loss
C_{Inherent}	Cost of inherent safety implementation
$C_{\text{InhSafety}}$	Cost of inherent safety
C_{Loss}	Value of expected losses
C_{PL}	Cost of production loss
CSCI	Conventional safety cost index
DCA	Domino chain actual hazard index
DI	Damage index
$d_{i,f,k}$	Escalation damage distance
$D_{k,j}$	Distance between equipment units
DS	Structural damage state
E	Extent of applicability of inherent safety guideword (Eqn (12.17))
$e_{i,k}$	Element of the unit escalation vector
F	Strength factor for the Multi-Energy method
HAZOP	Hazard and operability assessment
HCI	Hazard control index
HI	Hazard index
I	Heat radiation intensity (Table 12.1)
I2SI	Integrated inherent safety index
I2SI-LA	Integrated inherent safety index for layout assessment
ISCI	Inherent safety cost index
ISI	Inherent safety index
IS-KPI	Inherent safety key performance indicators
ISPI	Inherent safety potential index
LI	Loss intensity
LOC	Loss of containment
LSI	Loss saving index
ME	Multi-Energy method (Table 12.1)
M_f	Mach number
m_k	Total number of release modes
N	Number of units constituting a plant (Eqn (12.9))
$n_{i,f,k}$	Element of the escalation matrix
P	Maximum peak static overpressure (Table 12.1)
PHCI	Process and hazard control index
q_j	total number of source units considered
QRA	Quantitative risk analysis
R	Energy scaled distance (Table 12.1)
R-RM	Reference release mode
s_j	Escalation credibility factor (Eqn (12.15))
TDI	Target domino actual hazard index

UDI	Unit domino actual hazard index
UHD	Unit hazard domino index
u_k	Total number of target units considered
UPD	Unit potential domino index
VCE	Vapor cloud explosion

Subscript

a	Attenuation guideword
base option	Reference base layout option (Eqn (12.18))
f	Dangerous phenomenon
i	Reference release mode
j	Target unit
k	Source unit
l	Limitation of effects guideword
la	Limitation of the affected area
lb	Limitation of the damage potential to target buildings
le	Limitation of the effects of domino escalation
option n	*n*th layout option considered (Eqn (12.18))
si	Simplification guideword
system	Entire system

Greek symbol

$\alpha_{i,k}$	Inventory parameter (Eqn (12.6))

Tutorial for Tools for the Quantitative Assessment of Domino Effect Presented in Part II

Gabriele Landucci*, Giacomo Antonioni†, Gigliola Spadoni†

* Dipartimento di Ingegneria Civile e Industriale, Università di Pisa, Pisa, Italy,
† LISES—Dipartimento di Ingegneria Civile, Chimica, Ambientale e dei Materiali, Alma Mater Studiorum—Università di Bologna, Bologna, Italy

1   Introduction

This annex focuses on the application of the methodology for the quantitative assessment of risk due to domino scenarios described in Part 2 of this book and of available tools for the assessment of escalation probability presented in Part 1. An overall procedure for the quantitative assessment of domino effect is exemplified step-by-step through the application to case studies. Figure 1 shows the flowchart of the procedure applied for the quantitative assessment of risk due to domino scenarios, as discussed in Chapter 10.

A sample layout of a plant section of an industrial facility is hereby considered for the application and exemplification of the methodology. The layout is reported in Figure 2, and represents a storage tank farm including both atmospheric and pressurized storage tanks. The main relevant data of the equipment items present in the layout are reported in Table 1, which exemplifies the level of detail concerning equipment data required to apply the methodology outlined.

Several case studies were based on the layout reported in Figure 2 in order to illustrate the different possibilities and alternatives of the proposed methodology for quantitative domino risk assessment. In the application of the procedure to the case studies, reference will be made to the single steps reported in Figure 1.

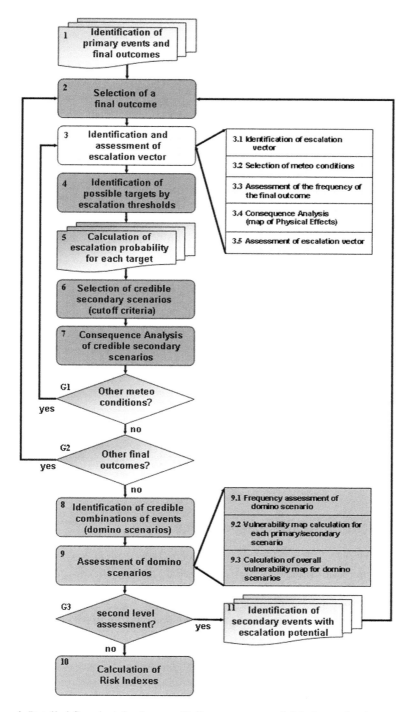

Figure 1 Detailed flowchart for the quantitative assessment of risk due to domino scenarios (G: gates).

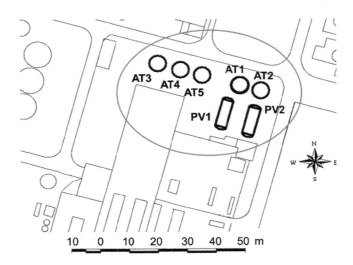

Figure 2 Layout considered in the case studies.
Source: Adapted from Cozzani et al. (2006a).

Table 1 Characteristics and Inventory of the Primary Risk Sources in the Considered Layout

Vessel ID	Type	Substance	Nominal Capacity (m^3)	Inventory (t)
PV1	Pressurized tank	Propane	120	54
PV2	Pressurized tank	Propane	120	54
AT1	Atmospheric tank	Ethanol	200	120
AT2	Atmospheric tank	Ethanol	200	120
AT3	Atmospheric tank	Hydrofluoric acid (40%)	200	181
AT4	Atmospheric tank	Hydrofluoric acid (40%)	200	181
AT5	Atmospheric tank	Hydrofluoric acid (40%)	200	181

2   Application of the Methodology

Step 1—Identification of primary events

The starting point of the methodology is focused on the identification and assessment of "primary events" associated to each equipment item in the layout of concern (Figure 2). As explained in Chapter 3, "primary events" are the accidental scenarios (e.g.: fires, explosions, and fragment projection) that are able to trigger an escalation by the damage of nearby equipment.

The quantitative assessment of domino scenarios is an activity that should be undertaken at the end of the assessment of conventional scenarios. It may then reasonably be assumed that hazard identification, frequency assessment and consequence analysis of primary scenarios are available when starting the assessment of domino hazard. Thus, in order to identify the primary events, it is sufficient to screen the results of the analysis of conventional scenarios and to identify those that have an escalation potential. This procedure is exemplified in the following.

Table 2 reports the list of loss of containment (LOC) events identified for each equipment item in the layout by hazard identification techniques (Center for Chemical Process Safety (CCPS), 2008) usually applied in conventional risk assessment activities (Lees, 1996; CCPS, 2000; CCPS, 2008; Uijt de Haag and Ale, 1999).

Step 2—Selection of final outcomes

In order to identify the primary events that may cause escalation, an event tree analysis (ETA) (Lees, 1996; CCPS, 2000; CCPS, 2008) was carried out for each LOC, determining the relevant final outcomes. Figure 3 reports the ETA results for pressurized vessels PV1 and PV2 (Figure 3(a)) and for atmospheric vessels AT1 and AT2 (Figure 3(b)). For the other equipment items (e.g. AT3, AT4 and AT5), only toxic vapor dispersion from the pool of released liquid was considered as a final outcome.

In order to have a limited number of overall domino scenarios, a single final outcome was considered for each equipment item included in the present analysis. This is not a necessary assumption but it allows a relevant simplification of the case studies, thus it is introduced in the present analysis to allow a more easy understanding of the procedure and of the results. It is worth to mention that the possibility of a catastrophic failure of the vessel containing pressurized liquefied gas followed by a fireball and not caused by an external fire was neglected.

The selected primary scenarios are marked with the red box in Figure 3, and are summarized in Table 3 with the correspondent frequency (namely, f_{pe}). For pressurized vessels, the vapor cloud explosion (VCE) was considered, while for atmospheric tanks containing flammable liquid (ethanol), the pool fire was selected. Since toxic dispersion was the only final outcome associated to the other atmospheric vessels, this was selected as primary event.

These final outcomes will be considered in Section 3, reporting the results of the case studies, both for the assessment of escalation probability and for the assessment of the consequences of primary scenarios in the conventional risk analysis carried out without considering domino effect.

Step 3—Identification and assessment of escalation vector

As shown in Figure 1, this step is divided into several substeps. First, an escalation vector needs to be associated to each primary scenario selected. Hence, overpressure was considered for the VCE scenario associated to vessels PV1 and PV2, while heat radiation was considered for the pool fire associated to vessels AT1 and AT2. As discussed in Chapters 3 and 7, toxic releases do not generate significant direct escalation vectors, hence the final outcomes associated to vessels AT3, AT4 and AT5 were not taken into account for this step.

The consequences of the primary events having an escalation potential then need to be evaluated for the quantification of the escalation vector. Conventional literature models for consequence analysis (Lees, 1996; CCPS, 2000; Van den Bosh & Weterings, 1997) were applied. For the sake of simplicity, a single set of

Table 2 Characterization of Loss of Containment Events (LOCs) Associated to the Vessels Considered in this Study

Vessel	Loss of Containment (LOC) Event	Release Type	LOC
PV1	Release from 10 mm hole in the vessel	Continuous	LOC_PV1_1
	Release of entire inventory in 10 min	Instantaneous	LOC_PV1_2
	Catastrophic rupture of the vessel and release of the entire inventory	Instantaneous	LOC_PV1_3
PV2	Release from 10 mm hole in the vessel	Continuous	LOC_PV2_1
	Release of entire inventory in 10 min	Instantaneous	LOC_PV2_2
	Catastrophic rupture of the vessel and release of the entire inventory	Instantaneous	LOC_PV2_3
AT1	Release from 10 mm hole in the vessel	Continuous	LOC_AT1_1
	Release of entire inventory in 10 min	Instantaneous	LOC_AT1_2
	Catastrophic rupture of the vessel and release of the entire inventory	Instantaneous	LOC_AT1_3
AT2	Release from 10 mm hole in the vessel	Continuous	LOC_AT2_1
	Release of entire inventory in 10 min	Instantaneous	LOC_AT2_2
	Catastrophic rupture of the vessel and release of the entire inventory	Instantaneous	LOC_AT2_3
AT3	Release from 10 mm hole in the vessel	Continuous	LOC_AT3_1
	Release of entire inventory in 10 min	Instantaneous	LOC_AT3_2
	Catastrophic rupture of the vessel and release of the entire inventory	Instantaneous	LOC_AT3_3
AT4	Release from 10 mm hole in the vessel	Continuous	LOC_AT4_1
	Release of entire inventory in 10 min	Instantaneous	LOC_AT4_2
	Catastrophic rupture of the vessel and release of the entire inventory	Instantaneous	LOC_AT4_3
AT5	Release from 10 mm hole in the vessel	Continuous	LOC_AT5_1
	Release of entire inventory in 10 min	Instantaneous	LOC_AT5_2
	Catastrophic rupture of the vessel and release of the entire inventory	Instantaneous	LOC_AT5_3

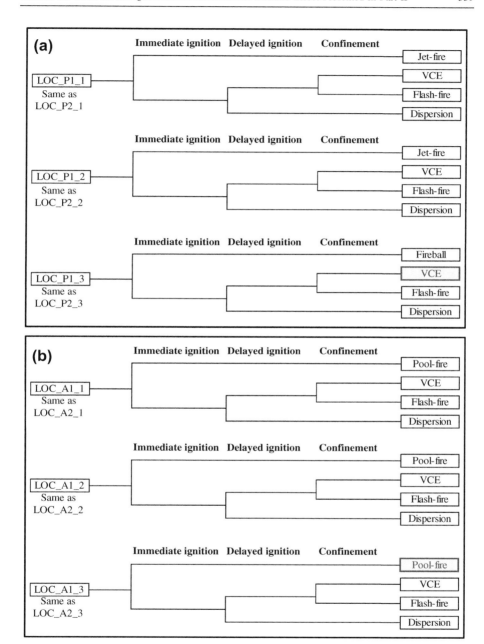

Figure 3 Results of event tree analysis for (a) pressurized vessels (PV1 and PV2) and (b) atmospheric vessels (AT1 and AT2).

Table 3 Selection of Primary Scenarios and Associated Frequency for this Study

Vessel	LOC	Selected Primary Event	Frequency of Primary Event f_{pe} (1/y)
PV1	LOC_P1_3	VCE	5×10^{-8}
PV2	LOC_P2_3	VCE	5×10^{-8}
AT1	LOC_A1_3	Pool fire	1×10^{-5}
AT2	LOC_A2_3	Pool fire	1×10^{-5}
AT3	LOC_A3_3	Toxic dispersion	1×10^{-4}
AT4	LOC_A4_3	Toxic dispersion	1×10^{-4}
AT5	LOC_A5_3	Toxic dispersion	1×10^{-4}

meteorological parameters was chosen to calculate the consequences of the releases (wind velocity of 5 m/s, stability class D), and a uniform distribution of wind directions was assumed.

Table 4 reports the main parameters used for consequence calculations of the primary events, and a summary of the physical effects was calculated for the scenarios.

Although it is evident that all the final outcomes considered in Table 4 have an escalation potential, in the following, only two primary scenarios will be considered: the pool fire due to a loss of containment from storage tank AT2 and the VCE due to the instantaneous release of all the inventory of vessel PV1. This assumption is only introduced to simplify the discussion of the following steps of the case studies. A complete assessment would require considering escalation assessment for all the pool fires and VCEs reported in Table 4.

Figure 4 reports the map of physical effects calculated for the final outcomes of the two primary scenarios further considered for escalation assessment: the pool fire from vessel AT2 (Figure 4(a)) and the VCE from vessel PV1 (Figure 4(b)).

Step 4—Identification of possible targets by escalation thresholds

After the consequence assessment of the final outcomes considered for the primary events, the identification of the credible domino targets was carried out by the use of escalation thresholds. In particular, as discussed in Chapter 9, several detailed escalation criteria were applied to support the identification of potential targets. Given the map of physical effects of each final outcome considered for the primary scenarios (Figure 4), all the equipment items positioned inside the areas where the physical effect exceed the specific threshold values were considered as secondary targets.

Table 5 summarizes the escalation vectors (the physical effects of the primary scenarios) and the detailed escalation criteria selected to identify the credible domino scenarios (Cozzani et al., 2004; Cozzani and Salzano, 2004; Cozzani et al., 2006b), respectively, for fires and explosions.

Table 6 shows the value of the physical effects calculated in the case of pool fire as primary scenario (Figure 4(a)). The fire strongly affects the pressurized liquefied petroleum gas vessels (PV1 and PV2) resulting in a partial flame engulfment with a heat load, which exceeds the considered threshold for pool fire radiation damage to

Threshold Damages were Derived from the "Green Book"

Input Parameters

Weather Conditions

Wind speed: 5 m/s
Stability class: D
Atmospheric conditions: 20 °C; 1.01 bar
Relative humidity: 50%
Surface roughness length: 951 mm*
Solar radiation flux: 400 kW/m^2

Consequence Assessment

Explosion modelling: TNT equivalency
TNT efficiency: 5%
Pool fire radiation fraction: 0.3
Pool fire duration: 30 min
Dispersion model: Gaussian
Source term for toxic dispersion: 0.13 kg/s[†]

Consequence Analysis Results for the Primary Scenarios

Scenario	Minor Damage			Moderate Damage			Severe Damage		
	Definition	Threshold Value	Damage Distance (m)	Definition	Threshold Value	Damage Distance (m)	Definition	Threshold Value	Damage Distance (m)
Pool fire in the catch basin of tanks AT1 or AT2	Reversible/ nonsevere effects	3 kW/m^2	63	Pain to unsheltered people, 2nd degree burn unlikely	5 kW/m^2	50	High lethality level	12.5 kW/m^2	35
VCE following the rupture of vessels PV1 or PV2	Reversible/ nonsevere effects	0.035 bar	286	Irreversible effects	0.07 bar	143	High lethality level	0.35 bar	29
Toxic dispersion following the rupture of tanks AT3–AT5	Negligible effects/odour	ERPG-1[‡] (2 ppm)	1080	Irreversible effects	IDLH[§] (30 ppm)	240	High lethality level	LC$_{50}$ 30 min[¶] (450 ppm)	51

*Associated to industrial area (Van den Bosh & Weterings, 1997).
[†]HF gas from the pool formed in the catch basin.
[‡]Emergency Response Planning Guidelines, first level concentration (Uijt de Haag and Ale, 1999).
[§]Immediately dangerous to life or health concentration.
[¶]Lethal concentration corresponding to the 50% death probability for 30 min of exposure.
Source: "Green Book" (Committee for the Prevention of Disasters, 1992).

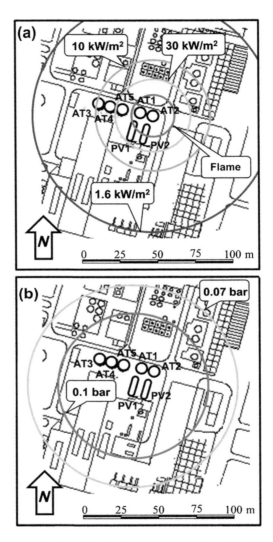

Figure 4 Consequence analysis of the final outcomes considered for the primary scenarios: (a) radiation intensities caused by the primary pool fire in the catch basin of tank AT2; (b) overpressure caused by the VCE following the release from vessel PV1.

pressurized equipment (Table 5). Thus, both vessels were identified as potential domino target and included in the analysis. Also tank AT1 is engulfed in the flames and may cause an escalation. Tanks AT4 and AT5, although only exposed to distant source radiation, receive radiation intensity higher than the considered threshold value, and thus should be considered in the analysis. Only tank AT3 is far enough from the pool fire to receive a radiation value under the threshold considered. Thus, this piece of equipment was not further considered for escalation assessment.

Table 5 Threshold Values Considered in this Study for Escalation
Due to Radiation and Overpressure

Selected Primary Scenario	Target Equipment	Threshold Values
Pool fire	Atmospheric	15 kW/m^2 for more than 10 min
	Pressurized	50 kW/m^2 for more than 10 min
VCE	Atmospheric	22 kPa
	Pressurized	17 kPa
	Elongated (toxic)	16 kPa
	Elongated (flammable)	31 kPa
	Auxiliary (toxic)	37 kPa
	Auxiliary (flammable)	Unlikely

Table 6 Evaluation of the Domino Targets Affected by the Primary Scenarios Considered
on the Basis of the Threshold Approach

Vessel	Vessel Type	Heat Radiation Damages (kW/m^2) Due to the Pool Fire from AT2 Tank		Overpressure Damages (kPa) Due to VCE from PV1 Vessel	
		Threshold Value	Calculated Value	Threshold Value	Calculated Value
PV1	Pressurized	50	55	17	*
PV2	Pressurized	50	Full engulfment[†]	17	151
AT1	Atmospheric	15	55	22	70
AT2	Atmospheric	15	*	22	144
AT3	Atmospheric	15	10	22	47
AT4	Atmospheric	15	16	22	63
AT5	Atmospheric	15	34	22	94

[*]Vessel considered as primary source.
[†]A maximum radiation of 80 kW/m^2 is considered.

Radiation values on equipment items outside the plant section considered are far lower than the escalation thresholds, thus indicating that the possibility of escalation events may reasonably be neglected.

In the case of the VCE from vessel PV1, the maximum peak overpressure exceeds the threshold value for all the vessels in the area considered (Figure 4(b)). Hence all the equipments were considered possible secondary targets with respect to this scenario.

Step 5—Calculation of the escalation probability for each target

The more crucial part of the quantitative domino assessment is the determination of escalation probability (namely, P_d) given the impact vector caused by the primary

Table 7 Models for Damage Probability Considered in this Study for Escalation Due to Radiation and Overpressure

Selected Primary Scenario	Target Equipment	Damage Probability Models
Pool fire	Atmospheric	$Y = 9.25 - 1.847 \cdot \ln(\text{ttf}/60)$ $\ln(\text{ttf}) = -1.13 \cdot \ln(\text{I}) - 2.667 \times 10^{-5}\,\text{V} + 9.877$
	Pressurized	$Y = 9.25 - 1.847 \cdot \ln(\text{ttf}/60)$ $\ln(\text{ttf}) = -1.29 \cdot \ln(\text{I}) + 10.97\,\text{V}^{0.026}$
VCE	Atmospheric	$Y = -18.96 + 2.44 \cdot \ln(\text{P}_\text{s})$
	Pressurized	$Y = -42.44 + 4.33 \cdot \ln(\text{P}_\text{s})$
	Elongated (toxic)	$Y = -28.07 + 3.16 \cdot \ln(\text{P}_\text{s})$
	Elongated (flammable)	$Y = -28.07 + 3.16 \cdot \ln(\text{P}_\text{s})$
	Auxiliary (toxic)	$Y = -17.79 + 2.18 \cdot \ln(\text{P}_\text{s})$
	Auxiliary (flammable)	Damage probability below cutoff

Y, probit value for escalation given the primary scenario; ttf, time to failure, seconds; I, radiation intensity on the target equipment, kW/m^2; V, equipment volume, m^3; P_s, peak static overpressure on the target equipment, kPa.

event. In Part 1 of this book, the use of specific vulnerability models to assess the escalation probability as a function of the impact vector intensity was discussed. Table 7 shows the models applied to assess escalation probability as a consequence of thermal radiation and overpressure.

In the case of radiation, the probit function is based on the estimation of the vessel time to failure (ttf, in seconds) (Chapter 5). The ttf was evaluated as a function of the vessel volume (V, in cubic meters) and of the radiation intensity (I, in kilowatt per square meter) to which each target is exposed, using the simplified correlations shown in Table 7. Table 8 reports the calculated ttf values for each vessel. Therefore, the radiation level determined in STEP 4 (Figure 4(a); Table 6) was hereby used to calculate the escalation probability of each piece of equipment.

Damage and escalation probability for the VCE scenario were calculated by the probit functions reported in Table 7 and discussed in detail in Chapter 4. The probit models correlate damage and escalation probability to maximum peak overpressure (P_s, in Pascal). The consequence map calculated for the VCE (Figure 4(b)) was used to obtain the value of P_s on each secondary target considered. Conversion of probit values to probability can be easily obtained by the specific function or by conversion tables (Lees, 1996).

Table 8 reports the results of the equipment vulnerability assessment. As shown in the table, the overpressure caused by the VCE generated by vessel PV1 results in almost unitary values of equipment vulnerability for all the other vessels in the layout considered. As a matter of fact, this event is able to cause damages to the entire facility, but for the sake of simplicity only the equipment listed in Table 8 were considered in the risk assessment. Thus, tank AT3 and the other nonlabeled equipment are excluded from the analysis in order to reduce the number of possible domino combinations, thus allowing a more easy understanding of the results.

Table 8 Evaluation of Escalation Probability and Associated Reference Secondary Scenarios for the Domino Targets Affected by the Considered Primary Scenarios

Selected Primary Scenario	Target	Evaluated Time to Failure (s)	Evaluated Probit	Escalation Probability P_d	Reference Secondary Scenario	Single Scenario Frequency f_{de} (1/y)
Pool fire in the catch basin of tank AT2	PV1	1415	3.41	0.056	Fireball	5.62×10^{-7}
	PV2	875	4.3	0.242	Fireball	2.42×10^{-6}
	AT1	214	6.90	0.971	Pool fire	9.71×10^{-6}
	AT4	844	4.37	0.263	Toxic dispersion	2.63×10^{-6}
	AT5	360	5.94	0.826	Toxic dispersion	8.26×10^{-6}
VCE following the rupture of vessel PV1	PV2	–	9.20	1.000	Fireball	5.00×10^{-8}
	AT1	–	8.26	0.999	Pool fire	5.00×10^{-8}
	AT2	–	10.02	1.000	Pool fire	5.00×10^{-8}
	AT4	–	8.04	0.9988	Toxic dispersion	5.00×10^{-8}
	AT5	–	8.98	1.0000	Toxic dispersion	4.99×10^{-8}

It is clear from Table 8 that the VCE considered is extremely severe if compared to the pool fire associated to AT2 that results in lower damage probabilities, at least for the equipment located far from the flame.

Step 6—Selection of credible secondary scenarios and frequency evaluation

After the determination of escalation probability, the ETA was again performed to identify the possible secondary final outcomes. Since this procedure was already applied in the description of STEP 2, only the final results are shown in Table 8. In particular, in order to reduce the complexity of the analysis, a single final outcome was considered for each secondary scenario. Therefore, the overall frequency of the secondary final outcomes associated to each vessel (f_{de}) was calculated as follows:

$$f_{de} = f_{pe} \cdot P_d \tag{1}$$

where f_{pe} is the frequency of the primary event $(1/y)$ and P_d is the escalation probability. The overall frequency values for each secondary scenario are reported in Table 8.

Step 7—Consequence analysis of credible secondary scenarios

After the identification of the final outcomes related to each secondary scenario caused by escalation, the associated physical effects were evaluated. The same models for consequence analysis used for the assessment of primary events consequences were applied (Van den Bosh and Weterings, 1997), and the same assumptions were introduced for meteorological conditions (e.g. stability class D; wind velocity 5 m/s). Table 9 reports the main parameters used for consequence calculations of the identified secondary events and a summary of the physical effects calculated for the scenarios.

Step 8—Identification of credible combinations of events and combinations frequency

After the identification and quantitative assessment of the single escalation scenarios, the actual overall domino scenarios should be identified and assessed. Actually, as discussed in Chapter 3, a primary scenario may trigger more than one secondary scenario. Thus, the possibility of having the simultaneous damage of several equipment items resulting in several simultaneous secondary scenarios should be considered. Therefore, the procedure discussed in Chapters 3 and 10 was applied to the identification of each possible domino combination.

It is quite evident that the procedure results in the identification (and in the assessment in Step 9) of a very high number of scenarios. Therefore, a specific software tool was needed in order to identify and assess, in the following step of the procedure, all the possible combinations of secondary events that may take place in the overall domino scenario. The integration of the present approach with the potentialities of a geographical information system (GIS) platform allowed a straightforward application of the methodology. The Aripar-GIS software was used in the present assessment. An extended description of the software is reported elsewhere (Spadoni et al., 2000, 2003; Cozzani et al., 2006a).

Table 9 Input Parameters and Results of the Consequence Analysis for the Secondary Scenarios Reported in Table 8. Threshold Damages for Humans were Derived from the "Green Book".

Input Parameters

Weather Conditions

Wind speed: 5 m/s
Stability class: D
Atmospheric temperature: 20 °C
Relative humidity: 50%
Surface roughness length: 951 mm*
Solar radiation flux: 400 kW/m²

Consequence Assessment

Pool fire radiation fraction: 0.3
Pool fire duration: 30 min
Fireball radiation fraction: 0.4
Dispersion model: Gaussian
Source term for toxic dispersion: 0.13 kg/s[†]

Consequence Analysis Results for the Primary Scenarios

Scenario	Minor Damage			Moderate Damage			Severe Damage		
	Definition	Threshold Value	Damage Distance (m)	Definition	Threshold Value	Damage Distance (m)	Definition	Threshold Value	Damage Distance (m)
Pool fire in the catch basin of tanks AT1 or AT2	Reversible/ nonsevere effects	3 kW/m²	63	Pain to unsheltered people, 2nd degree burn unlikely	5 kW/m²	50	High lethality level	12.5 kW/m²	35
Fireball following the rupture of vessels PV1 or PV2	Reversible/ nonsevere effects	3 kW/m²	700	Pain to unsheltered people, 2nd degree burn unlikely	5 kW/m²	550	High lethality level	12.5 kW/m²	330
Toxic dispersion following the rupture of tanks AT3–AT5	Negligible effects/ odor	ERPG-1[‡] (2 ppm)	1080	Irreversible effects	IDLH[§] (30 ppm)	240	High lethality level	LC_{50} 30 min[¶] (450 ppm)	51

* Associated to industrial area (Van den Bosh & Weterings, 1997).
[†] HF gas from the pool formed in the catch basin.
[‡] Emergency Response Planning Guidelines, first level concentration (Uijt de Haag and Ale, 1999).
[§] Immediately dangerous to life or health concentration.
[¶] Lethal concentration corresponding to the 50% death probability for 30 min of exposure.
Source: "Green Book" (Committee for the Prevention of Disasters, 1992).

Step 9—Assessment of domino scenarios

The overall probability and the frequency of each domino scenario identified in Step 8 are, respectively, calculated as follows:

$$P_{\text{d}}^{(k,m)} = \prod_{i=1}^{N} \left[1 - P_{\text{d},i} + \delta(i, \mathbf{J}_m^k)\left(2 \cdot P_{\text{d},i} - 1\right) \right] \tag{2}$$

$$f_{\text{de}}^{(k,m)} = f_{\text{pe}} \cdot P_{\text{d}}^{(k,m)} \tag{3}$$

The application of Eqns (2) and (3) is discussed in detail in Section 10.2.4, as well as the symbols used. Table 10 shows the results obtained for the conditional probability of overall domino sequences considering the two primary scenarios analyzed, e.g. a pool fire from tank AT2 and a VCE following the rupture of vessel PV1. As shown in the table, a clear ranking of domino sequences emerges from a probabilistic point of view, with probabilities ranging on more than two orders of magnitude. It is worth to remark that in the case of domino triggered by VCE, the combination in which all the possible secondary events take place simultaneously has a probability very near to 1. This causes the probabilities of all the other combinations to be very low, falling below 10^{-4}. The opposite situation takes place for the domino triggered by fire, in which more combinations need to be evaluated due to the lower values of escalation probability. A cutoff value was used to reduce the number of combinations reported in the table and further considered in the analysis. The value of the probability cutoff was set at 10^{-4}.

The severity of the overall domino scenarios is then assessed. First (Step 9.2), the vulnerability (death probability) map (Lees, 1996; Uijt de Haag and Ale, 1999) is calculated for the primary and all the possible secondary scenarios. Probit models were used to convert the dose of physical effect into damage probability to humans. Probit models for lethal effects on humans used in this study are summarized in Table 11.

A uniform population distribution of 10^{-2} persons/m^2 was assumed to calculate the impact of the accidental scenarios. Although this is a rather high value (typical of an urban area and not of an industrial zone), this value was assumed to better evidence the contribution of escalation and secondary scenarios to risk indexes. Table 12 reports the number of fatalities calculated by the Aripar-GIS software for all the primary and the secondary events considered in this study. As evident from the table, the highest potential impact is related to the fireball caused by the collapse of pressurized vessels (e.g. PV1 and PV2) induced by fire. This highlights that secondary scenarios may well be more severe with respect to the primary events.

In order to obtain the vulnerability maps of the overall domino scenarios, where a primary and one or more than one secondary scenario are supposed to take place (almost) simultaneously, the death probability maps were combined as follows:

$$V_{\text{d}}^{(k,m)} = \min\left[\left(V_{\text{p}} + \sum_{i=1}^{n} \delta(i, \mathbf{J}_m^k) V_{\text{d},i} \right), 1 \right] \tag{4}$$

Table 10 Calculated Probabilities of Escalation Events and of Domino Scenarios Caused by Pool Fire and VCE Following the Rupture of AT2 and PV1 Vessels Respectively

Combinations Triggered by Pool Fire

Targets PV1	PV2	AT1	AT4	AT5	Combination Probability
	X				8.50×10^{-4}
	X			X	4.03×10^{-3}
				X	1.26×10^{-2}
	X	X			2.84×10^{-2}
		X			8.91×10^{-2}
	X	X		X	1.35×10^{-1}
		X		X	4.23×10^{-1}
	X		X		3.03×10^{-4}
			X		9.50×10^{-4}
X					1.58×10^{-4}
	X		X	X	1.44×10^{-3}
X	X			X	2.39×10^{-4}
			X	X	4.51×10^{-3}
X				X	7.49×10^{-4}
	X	X	X		1.02×10^{-2}
X	X	X			1.69×10^{-3}
		X	X		3.18×10^{-2}
X		X			5.29×10^{-3}
	X	X	X	X	4.82×10^{-2}
X	X	X		X	8.01×10^{-3}
		X	X	X	1.51×10^{-1}
X		X		X	2.51×10^{-2}
X			X	X	2.67×10^{-4}
X	X	X	X		6.02×10^{-4}
X		X	X		1.89×10^{-3}
X	X	X	X	X	2.86×10^{-3}
X		X	X	X	8.95×10^{-3}

Combinations Triggered by VCE

Targets PV2	AT1	AT2	AT4	AT5	Combination Probability
X	X	X	X	X	9.79×10^{-1}
X	X	X		X	1.18×10^{-3}

In the case of escalation triggered by pool fire, the tank AT3 was excluded since the predicted radiation was lower the threshold for escalation; in the case of escalation triggered by VCE, the same vessel was excluded in order to reduce the number of possible combinations.

Table 11 Models for Human Vulnerability (Lees, 1996; Committee for the Prevention of Disasters, 1992) Used in the Case Studies

Vulnerability Vector	Probit Equation	Dose
Radiation	$Y = -14.9 + 2.56 \ln(D)$	$D = I^{1.33} t_e$
Overpressure	$Y = 1.47 + 1.37 \ln(D)$	$D = P_s$
Toxic release	$Y = -9.82 + 0.71 \ln(D)$	$D = C^2 t_e$

Y, probit value for fatality; I, radiation intensity, kW/m²; P_s, peak static overpressure, psi; C, toxic concentration, ppm; t_e, exposure time, min.

Table 12 Expected Number of Fatalities Calculated for the Accidental Scenarios Considered in this Study

Scenario	Type of Scenario	Substance Involved	Risk Source	Expected Fatalities
VCE	Primary event	Propane	PV1, PV2	20
Pool fire	Primary and secondary event	Ethanol	AT1, AT2	40
Toxic dispersion	Primary and secondary event	Hydrogen fluoride	AT3, AT4, AT5	2
Fireball	Secondary event	Propane	PV1, PV2	150

The derivation and application of Eqn (4) is discussed in detail in Section 10.2.5. The vulnerability maps obtained by this procedure were used in Step 10 to calculate the individual and societal risk indexes.

Step 10—Calculation of risk indexes

The Aripar-GIS software was used to perform the risk indexes calculation, and the results were visualized using the GIS interface of the software.

3 Results of the Case Studies

This section presents the outcomes of the application of Steps 9 and 10 of the methodology to the case study selected. In order to present effectively the results of risk assessment, the domino scenarios triggered by the two primary events considered in the case studies will be discussed separately. Hence, domino effects triggered by pool fire and by VCE will be discussed respectively in Sections 3.1 and 3.2.

3.1 Domino Effected Triggered by Pool Fire Radiation

Figure 5 reports the results of the risk assessment procedure for the escalation scenarios triggered by the pool fire considered as a primary event (Section 2). The map

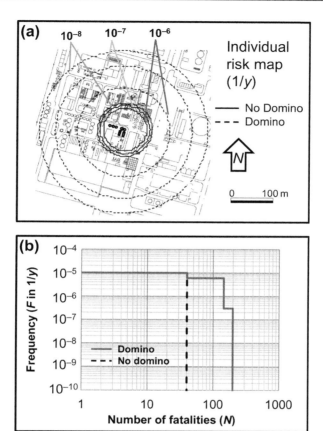

Figure 5 (a) Individual risk and (b) societal risk calculated for case study 1 with and without the contribution of domino scenarios.
Source: Adapted from Cozzani et al. (2006a).

obtained for local-specific individual risk is reported in Figure 5(a), while Figure 5(b) shows the *F–N* curve obtained from societal risk calculation. More details on the procedures for the definition and calculation of individual and societal risk indexes are reported elsewhere (CCPS, 2000; Lees, 1996; Uijt de Haag and Ale, 1999).

Figure 5(a) shows the changes in the individual risk due to the contribution of domino scenarios. The figure evidences that the high severity of the fireballs that may take place as secondary events (due to the fire-induced rupture of PV1 and PV2, as shown in Table 8) cause an important increase in the individual risk values. The diameter of the area where individual risk is higher than 10^{-6} events/year is almost doubled, increasing from about 100 to 200 m.

Figure 5(b) shows the corresponding increase in the societal risk. The figure points out that domino effect resulted in additional steps of the *F–N* curve, having a higher number of fatalities but a lower frequency. In spite of the high number of possible domino scenarios, the *F–N* curve only shows two additional steps. These are related to the domino scenarios in which fireballs were present (involving PV1 or PV2 vessels), which have a number of expected fatalities equal to that of the fireball. A severity higher than that due to the (almost) simultaneous fireballs caused by the failure of vessels PV1 and PV2 due to the primary fire is not possible, since the impact areas of all the other scenarios is within those of the two fireballs. This is confirmed by the analysis of the 50% lethality map (Figure 6), in which the extension of the damage area of the fireball is compared with one of the primary events.

The results obtained evidenced the importance of considering the possibility of domino events, in which multiple secondary are triggered by a single primary scenario. As a matter of fact, the overall severity of the domino scenarios, represented by the total number of expected fatalities *N*, is greatly influenced by the number of contemporary secondary scenarios triggered by the escalation. Moreover, the results obtained evidenced that domino effect was responsible of a relevant increase in the individual and societal risk indexes caused by the primary event considered.

3.2 Domino Effected Triggered by VCE

This section is concerned with the analysis of the secondary events triggered by the VCE due to the instantaneous release of the entire inventory of vessel PV1. The risk indexes calculated are shown in Figure 7.

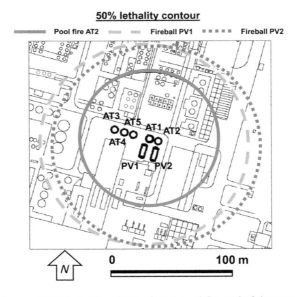

Figure 6 Impact area at 50% lethality of the primary pool fire and of the secondary fireballs in case study 1.
Source: Adapted from Cozzani et al. (2006a).

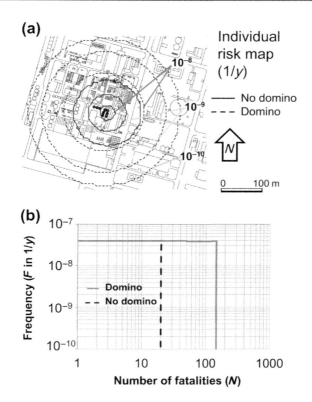

Figure 7 (a) Individual risk and (b) societal risk calculated for case study 2 with and without the contribution of domino scenarios.
Source: Adapted from Cozzani et al. (2006a).

Figure 7(a) evidences the increases in the individual risk caused by the possible domino scenarios. The severity of the secondary events and the very high propagation probability resulted in a relevant increase of the risk indexes. This is confirmed by the higher N values in the $F–N$ curve (Figure 7(b)), to which, however, an almost constant value of frequency is associated. Thus, it is evident from the results that in the case of severe primary events, domino effect may cause a relevant increase in the risk indexes. However, it must be remarked that, in general, severe primary events have rather low expected frequencies. Thus, if the absolute values of risk indexes are compared for the two primary scenarios analyzed, the contribution of domino effect results by far more important for the high-frequency low-severity pool fire considered in the previous section, even in the presence of lower propagation probabilities.

This is confirmed by the analysis of potential life loss (PLL) index (Lees, 1996; CCPS, 2000; Uijt de Haag and Ale, 1999) reported in Table 13 for the two case studies and by the analysis of the correspondent PLL escalation factor (EF) defined as follows:

$$EF = PLL_D/PLL_P \tag{5}$$

Table 13 PLL of Primary Scenarios (PLL_P), PLL Including Domino effect (PLL_D) and PLL Escalation Factor (EF) in Case Studies Considered. PLL is Expressed in Fatalities/year

Case Study	PLL_P	PLL_D	EF
Case study 1: Domino triggered by pool fire radiation	3.90×10^{-4}	1.04×10^{-3}	2.66
Case study 2: Domino triggered by VCE	7.60×10^{-7}	5.69×10^{-6}	7.49

where PLL_P is the PLL evaluated only considering the primary event, while PLL_D is evaluated including the domino escalation contribution.

As it can be seen, the primary pool file (case study 1) leads to a PLL value three orders of magnitude higher than the one connected to the primary VCE (case study 2). If escalation is considered, the PLL increment in case study 2 is more severe with respect to case study 1 (the escalation factor EF is almost tripled) but, due to the higher occurrence frequency, the escalation induced by pool fire leads in this case to a higher risk contribution.

4   Conclusions

This annex presented an exemplification of the methodology for the quantitative assessment of domino effect in quantitative risk assessment. The extended description of the approach applied is reported in Chapter 10, while the models used for escalation assessment are discussed in Chapters 4 and 5.

A simplified layout of an industrial tank farm was considered for the analysis of domino scenarios. Primary events associated to the equipment items considered were identified and a threshold-based approach was used to select the credible secondary target. Models for equipment damage probability were then used to estimate the escalation probability for each possible secondary scenario. The simultaneous escalation to more than one secondary scenario was then considered, identifying the more credible overall domino scenarios.

Finally, individual and societal risk indexes were calculated considering two possible primary scenarios identified for the facility of interest. The results allowed the analysis of the contribution of domino scenarios to industrial risk.

The contribution of domino effect resulted in both cases in the increase of risk, due to the secondary scenarios triggered by escalation. This confirms that a quantitative risk assessment of domino scenarios provides important information to address decision making in the management of industrial risk and in land-use planning.

References

Cozzani, V., Salzano, E., 2004. Threshold values for domino effects caused by blast wave interaction with process equipment. Journal of Loss Prevention in the Process Industries 17 (6), 437–447.

Cozzani, V., Gubinelli, G., Russo, G., Salzano, E., Zanelli, S., 2004. An assessment of the escalation potential in domino scenarios. In Proc. 11th Int. Symp. on Loss Prevention and Safety Promotion in the Process Industries, PCHE, Prague, p. 1153.

Cozzani, V., Antonioni, G., Spadoni, G., 2006a. Quantitative assessment of domino scenarios by a GIS-based software tool. Journal of Loss Prevention in the Process Industries 19 (5), 463–477.

Cozzani, V., Gubinelli, G., Salzano, E., 2006b. Escalation thresholds in the assessment of domino accidental events. Journal of Hazardous Material 129 (1–3), 1–21.

CCPS – Center for Chemical Process Safety, 2000. Guidelines for Chemical Process Quantitative Risk Analysis, second ed. American Institute of Chemical Engineers, New York.

CCPS – Center for Chemical Process Safety, 2008. Guidelines for Hazard Evaluation Procedures, third ed. American Institute of Chemical Engineers, New York.

Committee for the Prevention of Disasters, 1992. Methods for the Determination of Possible Damage to People and Objects Resulting from Releases of Hazardous Materials (CPR 16E). TNO "Green Book", The Hague. (Committee for the Prevention of Disasters).

Lees, F.P., 1996. Loss Prevention in the Process Industries, second ed. Butterworth-Heinemann, Oxford.

Spadoni, G., Contini, S., Uguccioni, G., 2003. The new version of ARIPAR and the benefits given in assessing and managing major risks in industrialised areas. Process Safety and Environmental Protection 81 (1), 19–30.

Spadoni, G., Egidi, D., Contini, S., 2000. Through ARIPAR-GIS the quantified area risk analysis supports land-use planning activities. Journal of Hazardous Materials 71 (1–3), 423–437.

Uijt de Haag, P.A.M., Ale, B.J.M., 1999. Guidelines for Quantitative Risk Assessment (Purple Book). Committee for the Prevention of Disasters, The Hague.

Van Den Bosh, C.J.H., Weterings, R.A.P.M., 1997. Methods for the Calculation of Physical Effects (Yellow Book), third ed. Committee for the Prevention of Disasters, The Hague.

Tutorial for the Main Methodologies in Part III

Genserik Reniers, Robby Faes†*

* Centre for Economics and Corporate Sustainability (CEDON), HUB, KULeuven, Brussels, Belgium,
† 3M Environmental Health & Safety, Zwijndrecht, Belgium

Chapter Outline

This tutorial serves to indicate to the reader how domino effects can be *managed*. Besides the obvious technological aspects of dealing with escalating events (discussed in the previous tutorial), which are a matter for technicians, software specialists and engineers, the managerial aspects are equally important and provide an answer to the question how prevention managers can start and/or optimize *managing* the prevention, protection and mitigation of these possibly devastating events.

In this tutorial, two types of domino effects are discerned: internal domino effects (involving only one single company), and external domino effects (involving more than one company). We make the distinction because both types of knock-on effects require a specific managerial approach. It is evident that companies should first be concerned with managing internal domino effects, since if these possible events are prevented, a lot of external domino effects are also prevented. However, since zero-accident does not exist, and one can never completely discard the possibility of external domino effects, such possible cross-plant events are discussed as well in this tutorial. Moreover, the cross-organizational managerial requirements to tackle such external events are quite different from those installed in single companies to deal with internal events.

Some of the primary scenarios leading to a major accident of an installation within a chemical company will have the potential to have an effect on the other installations situated within the plant, and possibly cause an internal domino effect. So before working on external domino effects, one can gain experience working on internal domino effects.

1 Managing Internal Domino Effects

Chemical and process companies working with hazardous goods are obliged by law to have an understanding of the different major accidents scenarios that are possible for the installations operated within their premises. Usually, taking precaution measures for dealing with the primary "easy-to-read" major risks is not a big problem for plant management. However, taking adequate precautions against rare escalating events are much more difficult to explain to such top management. A motivation based on (highly uncertain) facts will most probably be necessary, but uncertainties are difficult to understand by managers. In any case, the motivation of plant management is essential to deal with internal domino effects to free the necessary budgets.

When top management is motivated, the prevention managerial actions to be taken are usually well known, and are similar to those required to deal with any ("easier to read") major hazards. Sound process safety management that includes managing internal escalating events implements the topics and measures listed in Table 13.1 (Chapter 13). After all, a free-standing plant (thus a company not being in a multiplant configuration) should take the same kind of measures to avoid a primary scenario from taking place. Making every company installation that can be part of a major accident scenario safe is already a big step forward in managing internal domino effects. As an example, a chemical reactor that can handle runaways and external fire will do so also if the heat radiation comes from another installation within the same plant, or will do so even if the heat radiation has a source external to the company. As a counterexample, one of the events that one has to look at separately is the effect of pressure waves.

Let us now have a look at the motivational factors to convince higher management for dealing with internal domino effects. Ten factors should convince management such that dealing with internal domino effects should be considered a must within any chemical or process plant:

1. Being an innovating company, not only on Research & Development but also on Process Safety Management, leads to a sustainable and profitable company.
2. Business Continuity Management is already influenced by an initial runaway or other possible scenario. Preventing this initial event can actually be considered the same as a start in dealing with domino effects. Hence, management is already concerned with internal domino effects by managing the "easy-to-read" scenarios, but it should be expanded.
3. Dealing with internal domino effects helps to avoid a reduced return on investment.
4. Dealing with internal domino effects helps to avoid problems with operating permits or a license to operate from the public.
5. Preventing dangerous situations is a corporate expectation and part of Responsible Care.
6. The company is liable for its employees, contractors, visitors, neighboring companies and external population. If a company fails to do so, it is risking damaging its image/reputation, and even destroying it. Building a reputation is much more difficult than destroying one.
7. Expectations of the authorities. Failing to comply will result in an official report and in some cases in a considerable fine. Noncompliance is also not good for the company image.
8. Dealing with internal domino effects helps to avoid short- and long-term environmental impacts.
9. Safety, in general, and investments in domino effects prevention, in particular, should be regarded as investments in the survival of a company, and not as unwanted expenses.
10. The financial impacts of domino effect precaution measures can be reduced by working together at the cluster level. Financial agreements on prevention, protection and mitigation, and also on mutual assistance can keep the cost within the acceptable range.

2   Managing External Domino Effects

Assessing probabilities and/or consequences of chemical risks may lead to the conclusion that it is acceptable to tolerate a risk. In the case of so-called external domino effect risks, decisions are likely to be based on the perception that the occurrence probability is so low that it can be ignored. Basically, this means that the decision maker will do whatever is necessary to recover from an external domino accident when it occurs, but he will not do all that is necessary to prepare for it in advance. However, evidence suggests that escalation accidents (which have actually happened) indicate the devastating effects to be so extreme that it is often not possible to restore production and/or storage. The nature of cross-company domino effects further suggests that multiple plant cooperation and multiple plant management may be the only approaches to truly prevent such devastating events.

Most importantly, strong collaboration links between companies need to be established to efficiently and effectively deal with external domino effects. The emergence of some type of cross-plant safety culture would be a logical consequence. However, the change from individual safety cultures toward a multiple plant safety culture would appear to be extremely difficult to overcome mainly due to lack-of-trust factors. Several cross-plant domains need to be elaborated and integrated into a cluster environment: people (collaboration), procedures (management systems), and technology (software, etc.). However, all three domains already exist and are already elaborated, be it in different ways, within the single plants. Therefore, plants are not inclined to change current practices and/or habits, although they could ameliorate thanks to benchmarking and cross-fertilization. One of the hurdles is that when drafting multiplant procedures or when planning cross-plant risk analyses to avoid external domino effects, companies' confidential information has to be used. Therefore, a configuration needs to be established on a supraplant level that takes these confidentiality issues into account.

Joint safety procedures—external domino risk identification and evaluation and continuous safety optimization—should be based on individual plant information and on collaboration between personnel at different levels of the companies. When conceptualizing a workable framework to enhance cross-company collaboration between safety experts, these professionals need to be able to communicate on the same level and with the same knowledge. Using commonly known risk analysis techniques serves this purpose. Software used to examine domino risks diverges significantly and there is a lack of user-friendly software to plan domino effects precautions in a cluster of chemical companies. Nonetheless, current software which is available to deal with domino effect risks does exist and can be used by companies on a cluster level. Future research will continue to advance the external domino effect software.

Glossary

Accident An event characterized by an unwanted outcome for human health, environment or/and business.

Accident scenario The sequence of events and end-point consequences that may follow a loss of containment or, more generally, a critical event in a process plant.

Active fire protection Systems or barriers which require external automatic activation for shielding and protecting a target from fire attack or for mitigating the fire strength. These systems are composed of three subsystems in chain: a fire or gas detection system, a logic solver and an actuation system.

Ambient Reference to conditions of the atmosphere surrounding the system of interest.

Atmospheric tank Cylindrical vertical vessel with flat bottom and fixed or floating roof operating at atmospheric pressure, typically employed for the storage of large inventories of liquid.

Atmospheric transmissivity Fraction of incident thermal radiation passing unabsorbed through a path of unit length in the atmosphere. It is a function of the concentration in air of infrared absorptive gases (mainly water and carbon dioxide).

Basic Process Control System A system responding to input signals from the process and/or from an operator and generating output signals, causing the process to operate in the desired manner. The Basic Process Control System consists of a combination of sensors, logic solvers, process controllers, and final control elements which automatically regulate the process within normal production limits.

Bayesian network A directed acyclic graph for reasoning under uncertainty in which the nodes represent variables and are connected by means of directed (causal) arcs.

Blast wave The rapid change in air pressure that propagates away from the region of an explosion. Weak pressure waves propagate with the speed of sound.

Boiling Liquid Expanding Vapor Explosion (BLEVE) Instantaneous vaporization and expansion of a pressurized liquefied gas above its normal (atmospheric) boiling point. This phenomenon is usually associated with the catastrophic rupture of a vessel with blast and projectiles are the immediate hazards. In the case of flammable substances, a BLEVE is associated with a fireball in case of immediate ignition or a flash fire or vapor cloud explosion in case of delayed ignition. If the release is toxic then toxic cloud dispersion is a hazard.

Computational fluid dynamics (CFD) Distributed parameter models (see glossary) aimed at simulating the physic of fluid flows by solving Navier–Stokes equations for the conservation of mass, momentum and scalar quantities, typically in a three dimensions.

Confined explosion see Explosion (confined).

Consequence The direct, undesirable result of an accident, usually measured by the physical effects of the accident or by effects on health or/and environment, loss of property, or business costs.

Convective heat transfer Transfer of heat from one point to another within a moving fluid, gas or liquid, by the mixing of one portion of the fluid with another. It also includes heat transfer from a moving fluid to a surface.

Damage state Reference category of equipment damage following the impact of an escalation vector. A scaling value for the target equipment damaged by the pressure wave or heat radiation produced by the primary scenario.

Dangerous substance A substance in which toxicity, flammability, instability or explosivity might induce hazard(s) for people or/and equipment.

Deflagration The word "deflagration" refers to a manner of propagation of the flame through gas or vapor cloud. The flame speed in a deflagration will be subsonic compared to the speed of sound in the unburnt mixture.

Detonation The detonation phenomenon is much more powerful and more destructive than the deflagration phenomenon. It involves a supersonic front accelerating through a medium that eventually drives a shock front propagating directly in front of it.

Dispersion Mixing and spreading of gases in air, causing the formation of a gas cloud that moves with the wind and is progressively diluted by air.

Dispersion model A mathematical model describing how released material is transported and dispersed in the atmosphere from a release.

Distributed Parameters Modeling Modeling of a system based on detailed models developed considering distributed parameters (as opposed to lumped or integral models).

Domino accident An accident in which a primary unwanted event propagates within an equipment ("temporally"), or/and to nearby equipment ("spatially"), sequentially or simultaneously, triggering one or more secondary unwanted events, in turn possibly triggering further (higher order) unwanted events, resulting in overall consequences more severe than those of the primary event.

Domino cardinality The number of the domino event links in a chain or sequence of domino events, starting from the initiating event having domino cardinality "1".

Domino chain Specific domino sequence in which a simple propagation pattern of the primary event is assumed: the primary event causes a single secondary event, that in turn may further cause a single tertiary event, etc.

Domino effect Propagation of a primary event to cause further secondary events.

Domino event An accident in which a domino effect takes place.

Domino scenario The overall accident scenario that takes place due to the escalation of a primary event resulting in a domino accident.

Domino sequence The sequence of events resulting in a domino scenario (primary event, propagation to generate secondary scenarios, etc.).

Elastic deformation Reversible change in the shape or size of an object due to an applied force. Once the forces are no longer applied, the object returns to its original shape.

Emergency depressurization/blowdown Controlled emergency venting of hazardous material inventory from units to the atmospheric vent system or to the flare in case of emergency situations (flammable gas releases, fires, etc.).

Emergency response External fire brigade or site emergency team intervention for suppressing/mitigating fires and to cool the eventual nearby target equipment.

Emergency shutdown (ESD) Automatic system which provides isolation of process equipment by intercepting process lines with shutdown valves activated by the operator or by a specific logic triggered by fire or gas detector signal in emergency situations.

Equipment item Elementary constitutive part of a technical section of a chemical plant (for example, storage tank, reactor, distillation column, boiler, heat exchanger, rail tank car, etc.).

Establishment The entire area under the control of an operator and where dangerous substances are present in one or more installations and equipment items, including common or related infrastructures or activities.

Escalation (direct) Propagation and intensification of overall consequences of a primary event due to direct damage of secondary equipment caused by the effect of radiation, blast wave or fragment impact.

Escalation (indirect) The escalation of a primary event in which the secondary scenarios are not caused by a direct damage of equipment items. Operator errors, loss of control due to loss of communication, and structural damage of civil structures (warehouses, support structures, etc.) can be identified as the most likely mechanisms of indirect escalation.

Escalation vector A vector of physical effects generated by the final outcome of the primary accident scenario.

Escalation Propagation of a primary event resulting in the intensification of the overall consequences.

Explosion The sudden and violent release of energy in a given system (confined, partially confined, in the open atmosphere) with the release of noise, pressure and/or heat. The rate at which the energy is released determines the violence of the explosion.

Explosion (Confined) The sudden and violent release of energy in a confined or partially confined system, which may fail structurally, with the production of fragments.

Explosion (Unconfined) A Violent expansion in which the energy produced by the explosion is transmitted outward as a blast wave.

Failure A system or component failure occurs when the delivered service deviates from the intended service, leading to an unwanted event.

Fault tree analysis (FTA) Technique able to quantitatively evaluate an undesired event likelihood based on the use of a logic model that graphically portrays the combinations of causes leading to the undesired event.

FEM (Finite elements modeling) Numerical technique for finding approximate solutions to partial differential equations and their systems. In stress analysis, fluid flow, heat transfer, and other areas, it is employed by subdividing continuous domains into discrete (finite) elements connected by nodes, which allow for the interpolation of the field quantity over the entire system. In each node, a set of simultaneous algebraic equations is generated and solved.

Film boiling Boiling regime characterized by the formation of a thin layer of vapor close to the heating surface due to the extremely severe temperature difference respect to the liquid. Due to the lower thermal conductivity the vapor layer insulates the heating surface. Heat is transferred by radiation.

Fireball Flame caused by the immediate ignition of a flammable vapor or gas cloud generated after a catastrophic loss of containment, e.g. caused by a relevant storage or process vessel failure. Fire able to burn sufficiently rapidly for the burning mass to rise into the air as a cloud or ball.

Fired domino effect Domino effect scenario triggered by an external fire affecting nearby process or storage equipment leading to their failure by heat up.

Flash fire Combustion of a flammable gas/vapor and air mixture in which flame passes through the mixture at subsonic velocity without the production of a blast wave due to the low confinement and/or congestion of the cloud, or to the low reactivity of the flammable mixture.

Fragment Section physically detached from the original structure originated in the rupture of a piece of equipment. Formation of fragments usually involves the formation and propagation of cracks through the structural material of the original unit. A fragment can be projected as a missile.

Fragmentation pattern A reference fragmentation mode for an equipment item, describing the expected position of the cracks and the number of the fragments generated by the rupture of the item.

Frequency In the framework of process hazard identification and quantitative risk assessment, should be intended as the probability of the reference event to take place over a given time interval.

Hazard A potential source of harm or the inherent potential of a material or activity to harm people, property, or the environment.

Heat of combustion Ideal amount of energy that can be released by burning a unit amount of fuel.

Heat load Thermal load imposed by a fire to a target object combination of convection from the hot combustion products passing over the object surface, in case of partial or full flame engulfment, and radiation emitted by the flame itself to the object surface.

Industrial cluster A set of industrial activities sited in a single narrow area, run in general by different companies.

Inherent safety Risk reduction strategy which aims at the elimination of the hazard rather than at reducing the likelihood of negative consequences of an accident scenario. Whenever complete elimination of hazard is not practicable, the inherent safety strategy promotes the reduction of the severity of the potential accidents by following a set of reference guide words (intensification, moderation, substitution, simplification, limitation of the effects).

Installation A technical unit within an establishment in which dangerous substances are produced, used, handled or stored. It shall include all the equipment, structures, pipework, machinery, tools, private railway sidings, docks, unloading quays serving the installation, jetties, warehouses or similar structures, floating or otherwise, necessary for the operation of the installation.

Isorisk curve Represents the levels of individual risk around the installation analyzed. An isorisk curve connects all the geographical locations around a hazardous activity with an equal individual risk, i.e. all the locations with the same overall probability of lethality [14].

Jet/torch fire Flame resulting from the combustion of flammable gas/vapor and/or liquid released from an orifice with significant momentum.

Liquid stratification Development of a temperature profile among the liquid phase of a fire-heated vessel, in which the liquid near the top of the tank will be at a higher temperature than the liquid lower down. A critical phenomenon affecting the pressure rise in the tank, which is dictated by the warmest liquid.

Likelihood Measure of the expected probability (or frequency) of the occurrence of an event.

Line of defense Safety function controlling a chemical process.

Loss of containment Event resulting in the release of material from an equipment item to the atmosphere.

Loss intensity Intensity of loss of containment (e.g. release rate) after equipment damage.

Major accident An occurrence such as a major fire, explosion or toxic release resulting from uncontrolled developments in the course of the operation of any establishment where hazardous substances are handled or stored, and leading to serious danger to human health and/or the environment, immediate or delayed, inside or outside the establishment, and involving one or more dangerous substances.

Maximum allowable stress Parameter indicating the effective resistance of the construction material of a vessel to mechanical and thermal stress, function of the material mechanical properties and of the temperature.

Missile Fragment projected as a result of the rupture of a piece of equipment. Upon rupture, a sufficient amount of kinetic energy is transferred to the fragment to make it a flying projectile. Missiles are a possible cause of damage to people or equipment.

NaTech Natural hazard triggering a technological disaster.

Nucleate boiling Boiling regime characterized by the growth of bubbles on a heated surface, which rise from discrete points on a surface, whose temperature is only slightly above that of the liquid.

Overpressure (peak) The maximum pressure of the pressure wave, related to the atmospheric pressure. It is the main blast parameter which characterizes the pressure wave propagating in the atmosphere.

Passive fire protection Systems or barriers which do not require either power or external activation to shield a target from fire attack. These technologies can be classified on the basis of the protective action: (1) coatings: directly applied on the protected steelwork and (2) thermal shields: barriers that separate the protected steelwork from the flame sources.

Physical effects Effects of fires (usually expressed as intensity of flame radiation), of blast waves (usually expressed by static peak maximum overpressure and duration of positive impulse), or toxic dispersions (usually expressed as the concentration of the toxic substance in air as a function of time).

Plastic deformation Change in the shape or size of an object due to an applied force with irreversible effects usually characterized by a strain-hardening region, a necking region and finally, fracture (also called rupture) increasing the force action.

Pool fire Flame resulting from the combustion of material evaporating from a layer of liquid at the base of the fire.

Potential life loss (PLL) The number of fatalities per year expected in the area of interest due to all the risk sources considered in the analysis.

Preliminary hazard analysis (PHA) Procedure for the identification and analysis of process hazards in early design phases.

Pressure relief valve (PRV) Valve used to control or limit the pressure buildup in a system or vessel.

Pressure wave See Blast wave.

Pressurized liquefied gas Gas that has been liquefied by compression to a pressure equal or higher than the saturated vapor pressure at the desired storage temperature (typically the ambient temperature).

Pressurized tank Vessel operating a pressure higher than the atmospheric one, containing vapor, gas and/or liquid.

Probability Measure of the likelihood of an event. The measure may be based on observations or expert opinions. Hence, the probability is the expression for the likelihood of occurrence of an event or an event sequence during an interval of time or the likelihood of the success or failure of an event on test or on demand. Probability is expressed as a dimensionless number ranging from 0 to 1.

Probability of failure on demand (PFD) Probability that a system or component being not in service is not available or fails when requested to operate on demand.

Process equipment This category includes:

- equipment designed for the processing or the physical or chemical separation of substances (reactors, distillation columns, absorption columns, liquid–liquid extractions, centrifuges, etc.);
- intermediate storage equipment integrated into the process;
- utilities (pumps, compressors, gas expansion facility, etc.);
- equipment designed for energy production and supply (furnaces, boilers, etc.);
- all the pipes which are part of the above-quoted equipment.

Propagation In case of a spatial domino effect, the propagation indicates the involvement of other units or equipment items, present at different positions with respect to that of the primary accident. In case of a temporally-propagating domino effect, there is propagation in time within the same unit or equipment item.

Quantitative risk assessment (QRA) Consists of a set of methodologies for estimating the risk posed by a given system in terms of human loss or, in some cases, economic loss. Hence, a QRA involves the development of a quantitative assessment of risk combining estimates of initiating event frequency, of independent protection layers and of potential consequences.

Radiation In the present framework, it should be intended as the emission of energy from a flame in the form of electromagnetic radiations.

Radiative heat transfer Transfer of heat from an emitting body to another (receiver), not in contact with it, by means of wave motion through space.

Risk (Individual) The expected frequency of the reference damage occurring as a consequence of any accident scenario considered in the analysis, to a person who is permanently present (24 h a day per year) in a given point of the area considered, with no protection and no possibility of being sheltered or evacuated.

Risk (Societal) The overall risk for the local community due to the presence of the risk sources of concern, usually expressed using frequency–fatality (F–N) plots.

Risk Uncertainty about, and severity of, the consequences (or outcomes) of an activity with respect to something that humans value.

Risk state A scaling value for the intensity of the loss of containment of hazardous materials from target equipment damaged by the pressure wave or heat radiation produced by the primary scenario.

Runaway reaction Fugitive reaction deriving from the loss of control of a chemical process.

Safety distance The minimum segregation distance at which the physical effects from possible accident scenarios do not result in relevant damage of a given target. In the case of domino escalation, the target typically considered is a unit of equipment and the reference damage is a loss of containment large enough to result in severe secondary consequences.

Scenario (Primary) An accident scenario that starts a domino effect propagating and escalating to other process or storage units, triggering one or several secondary accident scenarios.

Scenario (Secondary) An accident scenario caused by the impact of an escalation vector generated by a primary accident scenario.

Shock wave See blast wave.

Stress intensity Distribution of the equivalent intensity of combined stress, which allows representing multiaxial loading conditions by a simple scalar function.

Structure instability/buckling Following a little deformation, the structure loses its initial configuration assuming another configuration of equilibrium and losing the integrity.

Tank fire A pool fire formed on the top of a liquid contained in a vessel (frequently a floating roof tank as a consequence of the ignition of vapors leaking from rim seal and/or of the failure of the floating roof).

Thermal nodes modeling Thermal nodes or zones models allow for dividing a complex domain into homogenous lumped regions in which the physical quantities are averaged. Heat and mass balances are then evaluated by the interaction of the considered regions.

Threshold (Damage) Value of physical effects above which the damage of the secondary equipment of interest is expected.

Threshold (Escalation) Value of physical effects above which the escalation of a primary scenario is expected due to the damage of a secondary unit of interest and to the escalation potential of the secondary scenarios that may follow the damage.

Time for effective mitigation (tem) Time to effectively perform mitigation actions by active safety systems or by emergency team operations. In the case of fired domino effect, these actions include time to detect fire, give alarm and to start the emergency operations; time required to start the mitigation action (e.g. start of the emergency water deluges, water cooling by emergency teams, etc.); and finally safe fire suppression.

Time to failure (ttf) The time that occurs between the fire starting and the eventual equipment failure bringing a loss of containment.

Vapor cloud explosion (VCE) The explosion of a large cloud of flammable gas or vapor in air. After ignition, the flame accelerates through semiconfined or congested structures or obstructions. Near- and far-field pressure effects are significantly high.

View factor Synthetic parameter which quantifies the geometric relationship between the emitting and receiving surfaces, indicating the fraction of heat received by a target due to its geometry, orientation, distance from the emitting surface and emitting body geometry.

Vulnerability model Model for the calculation of vulnerability as a consequence of an adverse effect (in the present context, a dose calculated on the basis of radiation intensity, blast wave maximum static overpressure, toxic concentration, etc.).

Vulnerability Death probability as obtained by a model for human vulnerability, usually indicated as "probit" model. More in general, vulnerability may indicate the probability of a reference damage (e.g. probability of injury, probability of equipment damage, etc.).

Worst-case accident The worst (more severe) accident that may be identified for the system of interest, independently from its likelihood.

Worst credible accident The worst (more severe) accident that may be considered credible for the system of interest from

Index

Printed and bound by CPI Group (UK) Ltd, Croydon, CR0 4YY

08/05/2025

01864815-0001